CONTEMPORARY MATHEMATICS

718

Topology and Quantum Theory in Interaction

NSF-CBMS Regional Conference
in the Mathematical Sciences
Topological and Geometric Methods in QFT
July 31–August 4, 2017
Montana State University, Bozeman, Montana

David Ayala
Daniel S. Freed
Ryan E. Grady
Editors

AMS
AMERICAN MATHEMATICAL SOCIETY

2010 *Mathematics Subject Classification.* Primary 18A05, 53C44, 55Q10, 55T15, 81S40, 81Txx.

Library of Congress Cataloging-in-Publication Data

Cataloging-in-Publication Data has been applied for by the AMS. See http://www.loc.gov/publish/cip/.

∞ The paper used in this book is acid-free and falls within the guidelines
established to ensure permanence and durability.
Visit the AMS home page at `https://www.ams.org/`

10 9 8 7 6 5 4 3 2 1   23 22 21 20 19 18

Contents

Preface

In early August 2017, two of us (Ayala and Grady) organized a CBMS conference on *Topological and Geometric Methods in Quantum Field Theory* at Montana State University (MSU) in Bozeman. The third editor of this volume (Freed) gave a series of ten lectures. Supplementary lectures by several other mathematicians and physicists were also given. Many students and postdocs attended, as did several more senior mathematicians. The beautiful natural setting and relaxed atmosphere were perfect catalysts for interactions among the participants.

This volume brings together texts based on some of the supplementary lectures and related contributed manuscripts. They give a window into some aspects of the vigorous interaction between topology and physics.

We thank Beth Burroughs, Jane Crawford, and Katie Sutich in the Department of Mathematical Sciences at MSU for massive assistance during the planning and execution of the CBMS conference. We also thank the College of Letters and Science and the VP of Research at Montana State University for support. Finally, we are grateful to Christine M. Thivierge, Sergei Gelfand, and the AMS Editorial Committee for encouraging us to publish this volume.

David Ayala
Dan Freed
Ryan Grady

May 2018

Introduction

Quantum theory was developed at the beginning of the 20th century to explain microscopic phenomena in physics, which defied explanation using classical theories. The quantum mechanics of particles inspired the invention of new mathematics, in particular, the rapid development of operator theory by von Neumann in the late 1920s. This provided a solid foundation not only for quantum mechanics but also for many topics in mathematics unrelated to physics. At about that time physicists also initiated the quantum mechanics of fields, in particular, the electromagnetic field, whose classical behavior is governed by Maxwell's equations. From the beginning this subject was plagued by foundational problems. Computational progress came a few decades later in the hands of Feynman, Schwinger, Tomonaga, and others: infinities were tamed and physical predictions—some to spectacular accuracy—were theoretically derived and later experimentally verified. In the 1960s quantum field theory was extended from the electromagnetic field to fields that carry the weak and strong forces. The resulting Standard Model of Glashow, Salam, and Weinberg is the gold standard for fundamental theories of nature with experimental confirmation. Despite these spectacular successes in physics, and despite substantial mathematical work, there is no solid mathematical foundation for quantum field theory analogous to von Neumann's operator theory for quantum mechanics. Over the past several decades mathematical engagement with quantum field theory has turned to the geometric side, which has led to a fruitful interaction among several branches of mathematics and theoretical physics. This volume is part of that enterprise.

In mechanics there is a duality between *states* and *operators*: they pair to give a probability measure on the real line. Approaches to quantum theory can be crudely divided into *Schrödinger* and *Heisenberg* types accordingly as to whether they emphasize states or operators. Thus, in the Schrödinger picture of quantum mechanics states evolve in time, whereas in the Heisenberg picture the operators evolve. There are two mathematical axiom systems for quantum field theory in flat Minkowski spacetime, both developed in the 1960s. We might categorize the Wightman axiom system as Schrödinger type and the Haag–Kastler axiom system as Heisenberg type. Both emphasize analytic aspects of quantum fields and observables, and they were the basis for many further developments and constructions. While this mathematics was advancing, in the 1970s and 1980s physicists pushed in new directions, exploring more geometric theories (σ-models and Yang–Mills theories, among others). Then topological features with important physical ramifications were identified and studied. This work attracted the attention of a few geometers, in particular, Isadore Singer, who foresaw the broad impact that quantum field theory would have on mathematics. In the 1980s string theory, and the closely related two-dimensional conformal field theories, came to the fore. Mathematicians from other fields, such as representation theory, began to engage with this physics as well. In that context Graeme Segal proposed a Schrödinger-type axiom system for quantum field theory, which emphasizes its geometric aspects. Many variations and extensions have been explored by mathematicians since, especially for topological field theories. Michael Atiyah pointed out the structural connection in the topological case with classical bordism theory, an observation with profound consequences. More recently, a mathematical Heisenberg-type axiom system has been developed by Kevin Costello and Owen Gwilliam. It too originates in two-dimensional conformal field theory—the chiral algebras of Beilinson–Drinfeld—and the whole concept is a geometric version of the Haag–Kastler approach to quantum field theory. Examples of factorization algebras are based on Costello's approach to perturbative quantum field theory.

These modern geometric formulations of basic structures in quantum field theory have impacted several parts of mathematics: algebraic and symplectic geometry, low-dimensional topology, geometric representation theory, category theory, etc. They are not definitive definitions; there are constant revisions, variations, refinements, etc. Much remains to be done, and especially the dynamical aspects of quantum field theory remain out of reach. Still, these numerous mathematical ramifications make clear that quantum field theory, once properly formalized, will be an important mathematical structure.

In a different direction one can ask to test these axiom systems against *physics*. Can they be used to rederive known facts about quantum theories or, better yet, be the framework in which to solve open problems? The main lecture series at the CBMS conference in Bozeman[1] recounts a solution to a classification problem in condensed matter theory, which is ultimately based on Segal's field theory axioms. The first three papers in this volume amplify various aspects of these lectures. Another paper in this volume develops some category theory, which lies behind the cobordism hypothesis, the major structure theorem for topological field theories. The final three papers are related to Costello's approach to perturbative quantum field theory and the Costello–Gwilliam factorization algebras. Two of these papers use this framework to recover fundamental results about some physical theories: two-dimensional σ-models and the bosonic string. Perhaps it is surprising that such sparse axiom systems encode enough structure to prove important results in physics. These successes can be taken as encouragement that the axiom systems are at least on the right track toward articulating what a quantum field theory is.

With the broad context understood we now introduce each of the papers with more specificity. David Morrison starts us off by recounting crucial episodes in the interaction among geometry, topology, and gauge theory. As in his lecture at the CBMS conference, Morrison uses these episodes to frame general observations on the relationship between mathematics and physics over a 65-year period. On the journey from Dirac quantization to Seiberg–Witten equations a geometer meets many old friends: fiber bundles, Chern classes, algebraic bundles, index theory, nonlinear elliptic equations, and Chern–Simons invariants. The applications to physics trigger a backreaction in which physical theories profoundly influence mathematics. The surprises generated by this fertile mix continue unabated.

[1]These lectures will be published in a separate volume in the American Mathematical Society CBMS Regional Conference Series in Mathematics.

The next three papers in the volume are motivated by work of Freed–Hopkins, which we now summarize briefly. Just as in geometry one constructs moduli spaces of geometric objects with fixed discrete parameters, so one can contemplate moduli spaces of quantum systems of a fixed type. After removing a locus of "singular" systems (phase transitions, gapless theories) one defines path components to be *phases*: quantum systems connected by a path should exhibit the same qualitative features at low energies. This is a very general picture applicable to quantum field theories, string theories, as well as to quantum systems defined discretely on a lattice. It is this latter type that has recently been the subject of intensive research in condensed matter physics. Physicists move fluidly between discrete lattice models and continuous field theories, and for the particular case of "invertible" systems the mathematical understanding of the field theory side is effective in producing a complete classification. The invertibility moves the problem to stable homotopy theory, and the final result is expressed as the dual to a Thom bordism group, which depends on the parameters of dimension and symmetry type. This formula tests perfectly against classifications derived using physics arguments, often based on lattice models.

Ingmar Saberi's manuscript, loosely based on a lecture of Max Metlitski at the CBMS conference, leads us through basic insights and examples relevant to the classification of phases. From the beginning he emphasizes that lattice systems and field theories are two approximate models of an underlying physical reality, neither expected to be exact. Saberi covers basics of quantum mechanics and field theory before turning to phases of matter. Throughout he illustrates with detailed examples, such as the Heisenberg spin chain and Kitaev's toric code. He includes a thorough discussion of the central problem—classification of phases—which gives perspective and grounding to the more abstract discussions elsewhere.

Arun Debray and Sam Gunningham develop one example in detail, using Gunningham's lecture at the CBMS conference as a starting point. The spacetime dimension in question is 2 and the Wick-rotated symmetry group is Pin_2^-, which is to say the theories of interest include fermionic states and exhibit time-reversal symmetry. The paper begins with field theory, which as stated above leads to bordism theory, and so there is a review of basics of spectra and the Pontrjagin–Thom link to bordism. The relevant invertible field theories have partition functions, which are bordism invariants; for the generating theory the partition function is the Arf–Brown invariant of a closed Pin^- surface for which the authors recount several constructions. The invertible *Arf–Brown topological field theory* with that partition function is realized with codomain the Picard 2-groupoid of complex central simple superalgebras. The last section of the paper switches gears to the lattice perspective. There is a geometric account of the Majorana chain—a lattice model—which in a sense made precise and proved has a precise relationship to the Arf–Brown theory at low energies.

Computations of Thom bordism groups for general symmetry types use well-developed techniques in stable homotopy theory. This is the basis for computations from field theory, which agree with classification results based on lattice models. Agnès Beaudry and Jonathan Campbell give a thorough introduction to these homotopy theoretic techniques, and they use them to compute pertinent examples. This account, based on Beaudry's lecture at the CBMS conference, will be valuable to any student of the Adams spectral sequence, whether for this application to physics or for other purposes. They too begin with a review of spectra in stable homotopy theory. The mod 2 Steenrod algebra plays a central role since the mod 2 cohomology over a space is an $\mathcal{A}$-module. There is a distinguished subalgebra $\mathcal{A}_1$ that suffices for the computations done here. The $\mathcal{A}_1$-module structure is encoded concisely in pictorial form, as is illustrated in relevant examples. The next step is some homological algebra over $\mathcal{A}_1$, which is used as input into the Adams spectral sequence. The entire process is meticulously explained, illustrated, and applied.

The next paper in the volume, by David Ayala and John Francis, is a contribution to higher category, part of their ambitious program centered around topological field theories and the cobordism hypothesis. This technical article is a first step in an anticipated fantastic duality between certain topological quantum field theories: between certain σ-models (with algebraic target) and state-sum theories (factorization homology). A perturbative instance of this was established in the authors' earlier work, which intertwined Koszul duality for $\mathcal{E}_n$-algebras and Poincaré duality for n-manifolds. The anticipated nonperturbative situation then concerns Koszul duality for higher categories. The input for this deformation theory of higher categories, instantiated as Koszul duality, is a *flagged n-category*, which is introduced and characterized in this work. A notable example of such is the bordism n-category.

The remaining three papers pivot from Schrödinger to Heisenberg, that is, from Segal's axioms based on bordisms to the Costello–Gwilliam axioms for factorization algebras and, more prominently, the Costello approach to perturbation theory. The first, by Owen Gwilliam and Theo Johnson-Freyd, is a novel take on the origin of Feynman diagrams, which are a fundamental tool in traditional perturbative quantum field theory to compute asymptotic expansions of oscillatory integrals. In Costello's approach it is rather Batalin–Vilkovisky (BV) algebras that are the more fundamental starting point. Gwilliam and Johnson-Freyd bridge the gap by first defining a particular example of a BV algebra in the form of an explicit chain complex. They demonstrate that the computation of its lowest degree homology can be done using Feynman diagrams.

The final two papers resume the theme of applying mathematical axiom systems for quantum theory back to physics. The first, by Ryan Grady and Brian Williams, proves the formula for the β-function of the two-dimensional σ-model, first derived in Friedan's thesis. Up to a constant it is the Ricci curvature, which means that the renormalization group flow is closely related to Ricci flow. (Indeed, that link is one inspiration for Perelman's work.) To begin, Grady and Williams formulate an effective BV theory that describes the low-energy regime of the σ-model. Via an explicit cohomological computation, they show that the obstruction to quantization vanishes at one-loop, i.e., a solution to the quantum master equation is constructed modulo $\hbar^2$. The authors then briefly recall the essentials of the β-function in Costello's formalism. Several illustrative computations in scalar field theory are given before the article finishes with the identification of the β-function and the Ricci tensor in the two-dimensional σ-model.

The last paper in this volume, by Owen Gwilliam and Brian Williams, takes up the bosonic string, the simplest string theory. A distinguishing—and disqualifying—feature is that it is only consistent in spacetime dimension 26. This is one of the results they recover in this paper using Costello's perturbation theory. Their first task is to formulate the bosonic string in the BV formalism. The natural home for the construction is derived algebraic geometry, as is explained. The authors also check that it is consistent with the Polyakov action in the traditional approach. After investigating deformations, they turn to quantization on a disk. The quantum master equation leads to the dimension restriction: a BV quantization only exists in that dimension. With local quantization in hand, there are local algebras of operators in the form of factorization algebras. The cosheaf of algebras of observables is proved to be locally constant, so essentially topological. This recovers results of Getzler, Lian–Zuckerman, and others on Gerstenhaber algebras in bosonic string theory. The paper concludes with global aspects. In particular, the usual anomaly of the bosonic string arises in the Costello–Gwilliam framework.

In aggregate these papers provide an entrée to modern geometric formulations of quantum field theory and their applications. As befits a mathematical theory in its early stages, there are many heuristic explanations and even more examples treated in detail. We invite the active reader to use them as a springboard to further explorations.

Contemporary Mathematics
Volume **718**, 2018
https://doi.org/10.1090/conm/718/14490

Geometry and physics: An overview

David R. Morrison

ABSTRACT. We present some episodes from the history of interactions between geometry and physics over the past century.

I was asked by the conference organizers to survey the modern interactions between geometry, topology, and physics in a one-hour lecture. Rather than attempting to be comprehensive, I have chosen to provide some vignettes drawn from recent history which emphasize the impact each field has had upon the other. As part of the story, I will trace a particular circle of ideas from physics to math, back to physics, and on to math once again, and show the significant impacts at each stage along the way. My general theme is gauge theory, geometry, and topology.

1. Dirac quantization

> The most powerful method of advance that can be suggested at present is to employ all the resources of pure mathematics in attempts to perfect and generalize the mathematical formalism that forms the existing basis of theoretical physics, and after each success in this direction, to try to interpret the new mathematical features in terms of physical entities.
>
> Paul Dirac (1931)

Paul Dirac laid out the manifesto quoted above for the proper interaction between mathematics and physics in the course of a beautiful 1931 paper [**Dir31**] devoted to the theoretical investigation of magnetic monopoles.

It had long been observed that Maxwell's equations for electromagnetism can be made symmetric between electricity and magnetism by introducing particles carrying a net magnetic charge: the so-called magnetic monopoles.[1] The quantum version of Maxwell's theory, however, relies on the existence of an electromagnetic potential and prior to Dirac's work it was believed that monopoles would prevent

2010 *Mathematics Subject Classification.* Primary 57R22, 81T13, 14D21, 81T50, 57R57, 81T45.

This research was partially supported by NSF grant PHY-1620842.

[1]I hasten to point out that magnetic monopoles have not (yet) been observed in nature.

such a potential from being defined. Dirac showed that by using a careful interpretation of the role of the potential, quantum physics could be formulated using it (in spite of the definitional problem). Morever, there was a surprise: the existence of a magnetic monopole forces the "charge quantization" of electrically charged particles such as the electron.

There was some interesting topology in Dirac's original argument, a variant of which I will now review. Consider the wave function of an electron in the presence of a magnetic monopole of magnetic charge g, located at the origin in $\mathbb{R}^3$, and consider a vector potential A for the magnetic field. One might suppose that the vector potential could be defined throughout $\mathbb{R}^3 - \{\vec{0}\}$, but that is not possible for the following reason. If we consider a sphere of radius r and a circle γ_θ at fixed spherical polar coordinate angle θ which bounds a spherical cap Σ_θ containing the north pole, then the line integral

$$\int_{\gamma_\theta} A\, d\varphi = \int_{\Sigma_\theta} \nabla \times A\, d\sigma = \int_{\Sigma_\theta} B\, d\sigma \tag{1}$$

calculates the magnetic flux through Σ_θ (by Stokes' theorem). That flux, which can be calculated to be $2\pi g(1 - \cos\theta)$ increases continuously as θ ranges from 0 to π. But at $\theta = \pi$ the circle has shrunk to zero size so that the integral must vanish, which is a contradiction!

Dirac's interpretation was that the vector potential must become singular along some semi-infinite string anchored at the monopole. If we direct such a string from the origin through the south pole of the sphere, the contradiction is removed. The integral around a zero-size circle at the south pole cannot be computed directly due to the singularity in the vector potential, but takes the value $4\pi g$ thanks to the integral over the sphere.

The magnetic field causes a change in the phase of the wave function of an electric particle around a circle. For an electric particle of charge e, if the change in phase of the wave function around a circle is zero at the north pole, and it must take the value

$$\frac{e}{\hbar c}\int_{S^2} B\, d\sigma = \frac{4\pi g e}{\hbar c} \tag{2}$$

at the south pole. In a quantum theory, a change of phase in the wave function is physically indetectable if it is a multiple of 2π. The conclusion is that in order for the quantum theory to be well-defined, we must have

$$\frac{4\pi g e}{\hbar c} = 2\pi N \tag{3}$$

for some integer N. In other words, if there is a single monopole of magnetic charge g, then all electric charges must be integer multiples of $\hbar c/2g$. This is Dirac's famous quantization condition.

Conversely, since we know experimentally the smallest possible electric charge e_0, all magnetic charges must be integer multiples of $\hbar c/2e_0$.

In modern topological terms,[2] the vector potential A should only be locally defined, with a change between northern hemisphere and southern hemisphere specified by a change of potential in a neighborhood of the equator, which amounts to a change by a variable element of $U(1)$ (known as an *abelian gauge transformation*).The vector potential can be regarded as a globally defined object if it is

[2]This mathematical argument first appeared in the physics literature [**WY75**].

treated as a section of a bundle over $\mathbb{R}^3 - \{\vec{0}\}$ (or over S^2) whose transition functions are given by that variable element.

The abelian gauge transformation is specified by a map $S^1 \to U(1)$. Since $\pi_1(U(1)) \cong \mathbb{Z}$, the topological type of the bundle is determined by an integer (the first Chern class of the bundle). Physically, this integer is identified with the multiple of $\hbar c/2e_0$ which gives the magnetic charge g of the monopole in question.

Dirac did not formulate his result in terms of a bundle, but his treatment is surprisingly modern, given the relative novelty of topological concepts at that time. This result turns out to be only the first step in an important series of interactions between topology, geometry, and physics.

2. Missed opportunities

> As a working physicist, I am acutely aware of the fact that the marriage between mathematics and physics, which was so fruitful in past centuries, has recently ended in divorce.
>
> Freeman Dyson (1972)

In his 1972 Gibbs lecture [**Dys72**], Freeman Dyson lamented the divide between mathematics and physics which existed at that time, drawing a number of examples from history. One of his principal examples centered around Feynman's approach to the study of relativistic quantum field theory.

The years immediately following World War II saw rapid advances in the study of relativistic quantum field theory, led by Dyson, Feynman, Schwinger, and Tomonaga. Feynman's approach was based on his "sum over histories" idea. In the Lagrangian formulation of classical physics, the evolution of a physical system from a starting time to an ending time is the one which minimizes the physical *action* of the system. Such a minimum can be determined using the techniques of the calculus of variations to find the path through the configuration space which extremizes the action. In Feynman's "sum over histories" approach to quantum physics, *all* paths must be considered, and the probability that a particular path is followed is proportional to the exponential of the negative of the action. Thus, the probability will be highest along the classical path, but the quantum theory requires consideration of contributions from other paths.

To determine physically measurable quantities, it is necessary to integrate over the space of all paths, and this is where the mathematical trouble arises: it is unknown how to carry out such integrals. One of the difficulties is a choice of measure on the space of paths which determines how different paths are to be weighted.

Nevertheless, by considering theories which are perturbations of "free" theories, Feynman was able to give a prescription for an asymptotic series describing the (purported) answer to the path integral, as an infinite sum indexed by the famous *Feynman diagrams*. The mathematical consistency of the path integral values as determined by Feynman diagrams has never been established, but in the hands of skilled practitioners unique answers are produced. Remarkably, when this approach is used to study quantum electrodynamics, it is incredibly precise: for example, the best theoretical and experimental values of the anomalous magnetic dipole moment of the electron agree to at least eight significant figures (see [**PS95**, Chapter 6]).

As Dyson remarks, Feynman himself was not concerned with mathematical rigor. The fact that physicists were obtaining such spectacular agreement between theory and experiment likely contributed to a similar attitude among many physicists from that era: they did not need modern mathematics to accomplish their goals. And so Dirac's manifesto languished.

3. Yang–Mills theory and connections on fiber bundles

> ... we are concerned with the necessary *concepts to describe the physics of gauge theories.* It is remarkable that these concepts have already been studied as mathematical constructs.
>
> Tai Tsun Wu and Chen Ning Yang (1975)

In 1954, during the era of minimal communication between mathematics and theoretical physics, C. N. Yang and R. L. Mills [**YM54**] introduced *gauge transformations* consisting of locally varying symmetries taking values in a compact Lie group[3] G, and studied physical theories which are invariant under such gauge transformations. These generalized the already-familiar *abelian gauge transformations* from electromagnetism – the same ones we encountered in Section 1 – for which $G = U(1)$. These gauge theories (or "Yang–Mills theories") eventually became the basis of the Standard Model of particle physics, the formulation of which was finalized in the mid 1970s using the group[4] $G = (SU(3) \times SU(2) \times U(1))/\mathbb{Z}_6$.

In the late 1960s and early 1970s, Yang got acquainted with James Simons, then the mathematics department chair at SUNY Stony Brook where Yang was a professor of physics. In the course of their conversations,[5] Yang and Simons came to recognize that there were important similarities between formulas which were showing up in Yang's work, and formulas which appeared in parts of mathematics which Simons was familiar with. Simons identified the relevant mathematics as the mathematical theory of *connections on fiber bundles*, and recommended that Yang consult Steenrod's foundational book on the subject [**Ste51**] (which coincidentally was published just a few years prior to the work of Yang and Mills). Yang found the book difficult to read, but through further discussions with Simons and other mathematicians (including S.-S. Chern) he came to appreciate the power of the mathematical tools which fiber bundle theory offered. By 1975, Yang had co-authored a paper with T. T. Wu [**WY75**] (quoted at the head of this section) which applied those methods to problems in physics. Within their paper, Wu and Yang provided a dictionary between the parallel concepts in physics and mathematics, allowing the application of topological and geometric techniques to the study of Yang–Mills theory.

---

[3]To be precise, [**YM54**] treats the case $G = SU(2)$ but the notion was soon generalized to an arbitrary compact group.

[4]For a mathematical account of the standard model gauge group and the connection between its representations and the elementary particles, see [**BH10**].

[5]Yang and Simons shared memories of this period in a joint interview in 2008 [**Zim08**].

Simons communicated these newly uncovered connections with physics to Isadore Singer at MIT who in turn discussed them with Michael Atiyah of Cambridge University. Similar observations were made independently by others. For example, in Moscow Ludwig Faddeev was inspired to give a geometric interpretation of Yang–Mills theory [**TAA+17**] by the book [**Lin57**].

A new chapter in the interaction between mathematics and physics was about to open.

4. Unreasonable effectiveness

> ... mathematical concepts turn up in entirely unexpected connections. Moreover, they often permit an unexpectedly close and accurate description of the phenomena in these connections.
>
> Eugene Wigner (1960)

A key example of Wigner's "unreasonable effectiveness of mathematics" principle [**Wig60**] occurred in the mid 1970s, not long after Dyson's Gibbs lecture. Thanks to the opening of communication between mathematicians such as Atiyah and Singer on the one hand and the gauge theory community in physics on the other hand, when Polyakov proposed in [**Pol75**] the importance of studying instantons in Yang–Mills theory, mathematicians were ready to assist in finding such instantons.

A *Yang–Mills instanton* is a solution on $\mathbb{R}^4$ to the Euclidean version[6] of the Yang–Mills equations for a compact Lie group G, which are the variational equations for the norm-squared

$$\|F\|^2 = \int_{\mathbb{R}^4} \operatorname{tr}(F \wedge \star F) \tag{4}$$

of the curvature F of a connection A on a principal G-bundle, where the Hodge star operator is used to define the norm. Such a solution is only interesting to physicists if it has a suitably controlled behavior far from the origin, and the initial assumption made by mathematicians in studying the problem is that[7] the solution extends to S^4.

After some progress had been made on the problem in the physics community, it was given a purely mathematical formulation by Atiyah, Hitchin, and Singer [**AHS77**], who established the dimension of the space of solutions. Not long thereafter the problem was solved in general by Atiyah, Drinfeld, Hitchin and Manin [**AHDM78**] via what came to be known as the "ADHM construction." These papers used techniques – the Penrose twistor transform, and the algebraic geometry of vector bundles – which were then unknown to physicists.

Let me briefly explain the topological setting of the Yang–Mills instanton problem (already described in one of the earliest papers [**BPST75**]), which can be viewed in two ways. From the perspective of a solution on $\mathbb{R}^4$, all G-bundles are trivial but there is an asymptotic behavior of an instanton which is determined by

[6] One of the important techniques for studying quantum field theory considers these Euclidean theories as a kind of analytic continuation from ordinary time t to "imaginary time" it, via a procedure known as *Wick rotation*. It is beyond the scope of this lecture to explain why this is relevant to physics.

[7] It was later shown by Uhlenbeck [**Uhl82a**] that this is automatically true if $\|F\|^2$ is finite.

the behavior of the connection on the S^3 at infinity. There must be a gauge transformation on S^3, i.e., a map $S^3 \to G$, which trivializes the connection there. Thus, the topology of an instanton is measured by $\pi_3(G)$, which is isomorphic to $\mathbb{Z}$ for any compact semisimple Lie group. The resulting integer k is called the *instanton number*.[8]

From the perspective of a solution on S^4, the map $S^3 \to G$ specifies the bundle by giving gluing data along the equator. The topological classication of principal G-bundles on S^4 is via $\pi_3(G)$ and determines the instanton number as before. To see how this is related to the curvature of a connection, it is convenient to remember that the Hodge star operator $\star$ on a Riemannian four-manifold squares to the identity on 2-forms. Thus, F can be decomposed into its *self-dual* and *anti-self-dual* parts:

$$F = F_{\text{sd}} + F_{\text{asd}} \tag{5}$$

where $\star F_{\text{sd}} = F_{\text{sd}}$ and $\star F_{\text{asd}} = -F_{\text{asd}}$. We then have

$$\begin{aligned} \|F\|^2 &= \int_{S^4} \operatorname{tr}(F \wedge \star F) = \|F_{\text{sd}}\|^2 + \|F_{\text{asd}}\|^2 \\ 8\pi^2 k &= \int_{S^4} \operatorname{tr}(F \wedge F) = \|F_{\text{sd}}\|^2 - \|F_{\text{asd}}\|^2, \end{aligned} \tag{6}$$

reflecting the topology of the solution. We also see from this that the minimal action solutions must be either self-dual or anti-self-dual depending on the sign of k.

I was fortunate enough to attend the *Loeb lectures* delivered at Harvard University by Michael Atiyah in the spring of 1978 in which he explained the ADHM construction.[9] The lectures were held in the physics department, and probably constituted the largest meeting that had been held up until that time at Harvard between mathematicians and physicists.[10] My memory is that the audience was roughly half and half: there were large numbers of mathematicians as well as large numbers of physicists. To an algebraic geometry graduate student such as myself, it was an amazing experience to see my quite abstract corner of mathematics applied to the "real world" of theoretical physics. The lectures were frustrating in one sense: Atiyah made the pedagogical choice of treating algebraic geometry as a "black box" for the purpose of the lectures, so we didn't get to hear about the details of the algebraic geometry! (I believe that some of the physicists were frustrated by this as well, since they missed the opportunity to learn about the algebraic geometry.) However, we did learn about the twistor transform which was new material to many of us.

[8] Changing the orientation of the four-manifold changes the sign of the instanton number, and there is some ambiguity in the literature about how this is defined. We have attempted to be self-consistent in this paper.

[9] Atiyah subsequently published notes based on the Loeb lectures as well as similar lectures delivered at two other places [**Ati79**].

[10] Harvard was and is strong in the more traditional areas of mathematical physics, but the ADHM construction was something new and different and attracted a big audience from outside that community.

5. Anomalies and index theory

> From a more mathematical standpoint, the study of anomalies has elicited very interesting applications of index theory in quantum field theory. [... A] powerful form of the Atiyah–Singer index theorem (the index theorem for families of elliptic operators) has been used to provide a global understanding of the non-Abelian anomaly as well as the gravitational anomalies.
>
> Luis Alvarez-Gaumé (1986)

Let me now turn to another chapter in the math–physics dialogue of the 1970s and 1980s: the calculation of anomalies in quantum field theories and quantum theories of gravity. Although not directly related to our main story it is another signification interaction of the era.

Classical physical theories are determined by the equations of motion of the theory, but quantum theories require an understanding of the physical "action" on a broad configuration space of possible physical fields. Moreover, Feynman's formulation requires an understanding of a measure on the space of paths through that configuration space as well.

In the 1960s, while attempting to understand how symmetries of classical theories of particle physics act on the associated quantum theory, it was discovered that they might not: there could be an "anomaly" in the quantum theory which prevented the action from being well-defined. This phenomenon was originally expressed in terms of Feynman diagrams, and came to be regarded as the statement that, although the symmetry group preserves the Lagrangian, it fails to preserve the measure on the space of paths.

This formulation is unsatisfying to mathematicians, who know that the measure hasn't been properly defined in mathematics and so who rightfully wonder how an ill-defined thing can fail to be preserved by a group action?

There is another interpretation of the anomaly, however, in terms of the Dirac operators of the quantum theory. The equations of motion for the non-self-dual bosonic fields in a (Wick-rotated) physical theory involve the Laplacian $\Delta : \mathcal{V} \to \mathcal{V}$ acting on a space of functions or differential forms on spacetimes (sometimes bundle-valued). However, for fermionic fields in the theory, the equations of motion involve a Dirac operator $\nabla : \mathcal{V} \to \mathcal{W}$ which is a "square root" of the Laplacian and typically does not map the classical space of spinors to itself. Quillen [**Qui85**] introduced a "determinant line bundle" associated to the Dirac operator ∇, and Bismut and Freed [**BF86**] equipped it with a connection. The theory is anomaly-free if the line bundle is trivial and the connection is flat. Parallel developments also occured in Moscow, where Faddeev and Shatashvili [**FS84**] gave a formulation in terms of group cohomology.

The beauty of this approach is the connection to the Atiyah–Singer index theorem, as mentioned in the quote from [**AG86**] at the head of this section. The original Atiyah–Singer theorem can be used to interpret anomalies of abelian group actions, but the anomalies of non-abelian group actions (and gravitational anomalies) generally require the families index theorem. In all cases one is calculating topological obstructions to the triviality of the determinant bundle and the connection on it.

6. Donaldson invariants

> The surprise produced by Donaldson's result was accentuated by the fact that his methods were completely new and were borrowed from theoretical physics, in the form of the Yang-Mills equations.
>
> Michael Atiyah (1986)

As discussed in Section 4, the idea of Yang–Mills instantons was able to move from physics to mathematics thanks to renewed communication between mathematicians and physicists beginning in the mid 1970s. The next step, however, was truly remarkable. In the early 1980s, Simon Donaldson studied the Yang–Mills instanton equations on arbitrary compact four-manifolds, and using them, was able to make very unexpected progress in the study of differentiable four-manifolds [**Don83**].

For a fixed principal G-bundle P with instanton number k over a four-manifold X, where G is a compact Lie group with Lie algebra $\mathfrak{g}$, Donaldson considered the set $\mathcal{A}$ of connections on P (which is an affine space acted upon by the space $\Omega^1(X,\mathfrak{g})$ of $\mathfrak{g}$-valued 1-forms on X and can be regarded as a torsor for $\Omega^1(X, \mathrm{ad}(P))$), modulo the automorphism group $\mathcal{G}$ of P. On the orbit space $\mathcal{A}/\mathcal{G}$ the self-dual connections can be identified as the kernel of the operator $d_A^- : \Omega^1(X,\mathfrak{g}) \to \Omega^2_-(X,\mathfrak{g})$ which is the composition of covariant differentiation with projection to the anti-self-dual part. In his early work, Donaldson studied the kernel of d_A^- on $\mathcal{A}/\mathcal{G}$ in the case $G = SU(2)$ and $k = 1$, obtaining a moduli space $\mathcal{M}_1 \subset \mathcal{A}/\mathcal{G}$ of self-dual connections.[11] He found that if the intersection form on X is negative definite, then the moduli space $\mathcal{M}_1$ is a 5-manifold away from a finite collection of singularities corresponding to reducible connections (i.e., connections compatible with a decomposition of the associated vector bundle $E = P \otimes_{SU(2)} \mathbb{C}^2$ into a sum of two line bundles $L \oplus L^{-1}$). Analyzing these singularities carefully led to restrictions on the intersection form on second cohomology, and in particular showed that if the intersection form is positive definite, then it is diagonalizable. This had many remarkable consequences including the failure of the smooth version of the h-cobordism conjecture in dimension four as well as the existence of an "exotic" differentiable structure on (topological) $\mathbb{R}^4$. In his proofs, Donaldson relied on earlier work of Taubes [**Tau82**] and Uhlenbeck [**Uhl82a, Uhl82b**]. The significance of these results are dramatically strengthened when they are combined with Freedman's near-simultaneous solution of the topological Poincaré conjecture in dimension 4 [**Fre82**].

This work led to the award of a Fields medal to Donaldson at the 1986 ICM, during which Donaldson's work was presented by Atiyah [**Ati87**]. As Atiyah emphasized in the quote at the head of this section, the input from physics was one of the most remarkable aspects of the work. The conversation which had begun with the ADHM solution of a problem in physics had now become a two-way conversation!

[11] More precisely, the relevant space is given by the kernel of a small perturbation of the operator d_A^-, which makes the space more regular.

Donaldson's work on four-manifolds did not end with the awarding of the Fields medal. He extended the work in a number of directions, including a definition of polynomial invariants on the cohomology of X of arbitrary degree, based on the moduli space $\mathcal{M}_k$ of instantons with second Chern class k [**Don90**].

7. Topological quantum field theory

> [Witten's paper [**Wit88**]], which introduces TQFT in the context of Donaldson's theory of 4-manifolds and Floer's theory of 3-manifolds, could well emerge as one of the most significant works in late 20th century topology.
>
> Daniel Freed (1988)

The problem which Donaldson solved was not really a problem from physics: it was a problem from mathematics whose techniques were inspired by physics. There were many developments stemming from Donaldson's original work, including a related theory in three dimensions developed by Andreas Floer.

A good starting point for Floer's theory is the Chern–Simons functional, which on a three-manifold Y equipped with a connection A on a G-bundle over Y is the quantity

$$\mathrm{CS}(A) = \frac{1}{4\pi}\int_Y \mathrm{tr}(A \wedge F + \frac{2}{3}A \wedge A \wedge A) \in \mathbb{R}/\mathbb{Z}. \tag{7}$$

Equivalently, if A_0 is the trivial connection on the trivial G-bundle over Y and $A_t = (1-t)A + tA_0$ then

$$\mathrm{CS}(A) = \frac{1}{8\pi^2}\int_{Y\times[0,1]} \mathrm{tr}(F \wedge F). \tag{8}$$

In its interpolation between A and A_0 at the two ends of $Y \times [0,1]$, the connection A is an *instanton* in the sense the word is used in physics (see [**Col85**]), albeit an instanton of Euclidean signature.

Floer used a small perturbation of the function CS as a Morse function on the space of connections $\mathcal{A}$. Although $\mathcal{A}$ is infinite dimensional and the formal Hessian of CS at a critical point has infinite sets of both positive and negative eigenvalues, it is of *Dirac type* and essentially coincides with the operator $\star d$ acting on $\Omega^1/d\Omega^0$, suitably extended to Lie algebra valued forms. What is well-defined is the *index*, the difference between the positive and negative eigenvalues. Moreover, as in Witten's earlier interpretation of Morse theory in terms of quantum tunneling [**Wit82**], given two critical points P and Q the solutions to $dA/dt = -\operatorname{grad}\mathrm{CS}$ on $Y \times \mathbb{R}$ with connection A_P as $t \to -\infty$ and connection A_Q as $t \to \infty$ are identified with boundary operators in a chain complex. The resulting homology theory (which is only defined mod 8 for $G = SU(2)$ due to the index theorem in 4 dimensions) is Floer homology [**Flo88**].

The critical points of (slightly perturbed) CS are identified with irreducible representations $\pi_1(Y) \to G$ and account for the connection between Floer homology and the Casson invariant (which counts such representations, with appropriate signs). Moreover, the Floer theory provides a natural setting for Donaldson theory on four-manifolds with boundary. including Donaldson's polynomial invariants.

Atiyah [**Ati88**] put Donaldson's and Floer's work together (also combining them with some ideas about Heegard splitting to extend the theory to dimension two), obtaining a non-relativistic quantum field theory. Witten [**Wit88**] went one step further, and found the proper physical setting for the work of Donaldson and Floer.

The relativistic quantum field theory which describes Donaldson's and Floer's results is a topological twist of the usual[12] supersymmetric quantum field theory associated to the $SU(2)$ gauge group, a new notion which Witten introduced in order to provide the physical setting. Starting from a supersymmetric theory with certain supercharges, the action of those supercharges on the physical fields was modified in a way whch made them independent of the choice of metric on space-time. The correlation functions in the corresponding field theory turned out to precisely be the Donaldson polynomial invariants!

Witten used path integrals to motivate his construction, and the theory itself clearly belongs to physics, not mathematics. But as Dan Freed remarks in the review of [**Wit88**] quoted at the head of this section [**Fre89**], the implications for topology itself were profound. In fact, the intrinsic study of topological field theories (independent of the precise details of Donaldson theory) has become an important aspect of twenty-first century mathematics.

8. Seiberg–Witten theory

> In the last three months of 1994 a remarkable thing happened: this research area was turned on its head by the introduction of a new kind of differential-geometric equation by Seiberg and Witten: in the space of a few weeks long-standing problems were solved, new and unexpected results were found, along with simpler new proofs of existing ones, and new vistas for research opened up.
>
> Simon Donaldson (1995)

In 1994, Seiberg and Witten made some of the first progress in understanding quantum field theories from a non-perturbative perspective (in particular, studying properties which are not necessarily tied to the path-integral formalism [**SW94a, SW94b, SW94c**]). The particular theory which they studied first – the $\mathcal{N} = 2$ supersymmetric gauge theory in four dimensions – was the same one which Witten had earlier shown could be topologically twisted to yield the Donaldson–Witten theory. Now, Seiberg and Witten were able to find a new description of the infrared behavior of the $SU(2)$ gauge theory which took the form of a $U(1)$ gauge theory coupled to a magnetic monopole.

This work in physics had an immediate consequence in mathematics (after twisting) in the form of new topological invariants analogous to the Donaldson invariants [**Wit94**]. These Seiberg–Witten invariants, as they came to be called, were substantially easier to compute than the Donaldson invariants and progress

[12]More precisely, the supersymmetric quantum field theory with twice the minimal amount of supersymmetry.

was quickly made on many difficult conjectures which had been left open by the original Donaldson theory. Donaldson himself expressed great astonishment at the speed of progress [**Don96**], as quoted at the head of this section.

As John Morgan said during a lecture at the 1995 Cornell Topology Festival [**Mor**]: "The physicists keep coming up with amazing equations for us to solve. Once we know the equations, we can get lots of mathematics out of them, but why can't we find the equations ourselves?"

The exchange of ideas between physics and mathematics had now proceeded through at least four stages: from progress in solving the instanton equation in physics, to Donaldson's application of those ideas to the understanding of four-manifolds, to Witten's construction of topological quantum field theory, to the Seiberg–Witten study of the infrared properties of that theory and the spectacular mathematics which resulted!

9. Conclusions

> I am sure that [the interaction of math and physics] is going to continue and I believe the reason it will continue is that quantum field theory and string theory ... have rich mathematical secrets.
>
> Edward Witten (2014)

The story I have presented about interactions between mathematics and physics did not end in 1994, but has continued to develop fruitfully in many directions. For lack of time, I did not mention string theory at all in this lecture, but many of the important interactions since the mid 1980s have involved string theory as well as quantum field theory. As Edward Witten predicted in a 2014 interview conducted by Hirosi Ooguri [**Oog15**] and quoted at the head of this section, this interaction is likely to continue for a long time to come! Witten went on to say: "When some of these secrets come to the surface, they often come as surprises to physicists because we do not really understand string theory properly as physics – we do not understand the core ideas behind it. At an even more basic level, the mathematicians are still not able to fully come to grips with quantum field theory and therefore things coming from it are surprises. So for both of those reasons, I think that the physics and math ideas generated are going to be surprising for a long time."

Acknowledgments

I am grateful to Arun Debray and Sean Pohorence for sharing the notes they took during the lecture, and to Andy Neitzke for a very useful remark. I am also grateful to Greg Moore and Samson Shatashvili for their comments on an earlier draft of this article.

References

[AG86] Luis Alvarez-Gaumé, *An introduction to anomalies*, Fundamental problems of gauge field theory (Erice, 1985), NATO Adv. Sci. Inst. Ser. B Phys., vol. 141, Plenum, New York, 1986, pp. 93–206. MR878768

[AHDM78] M. F. Atiyah, N. J. Hitchin, V. G. Drinfel'd, and Yu. I. Manin, *Construction of instantons*, Phys. Lett. A **65** (1978), no. 3, 185–187, DOI 10.1016/0375-9601(78)90141-X. MR598562

[AHS77] M. F. Atiyah, N. J. Hitchin, and I. M. Singer, *Deformations of instantons*, Proc. Nat. Acad. Sci. U.S.A. **74** (1977), no. 7, 2662–2663. MR0458424

[Ati79] M. F. Atiyah, *Geometry on Yang-Mills fields*, Scuola Normale Superiore Pisa, Pisa, 1979. MR554924

[Ati87] Michael Atiyah, *On the work of Simon Donaldson*, Proceedings of the International Congress of Mathematicians, Vol. 1, 2 (Berkeley, Calif., 1986), Amer. Math. Soc., Providence, RI, 1987, pp. 3–6. MR934209

[Ati88] Michael Atiyah, *New invariants of* 3- *and* 4-*dimensional manifolds*, The mathematical heritage of Hermann Weyl (Durham, NC, 1987), Proc. Sympos. Pure Math., vol. 48, Amer. Math. Soc., Providence, RI, 1988, pp. 285–299, DOI 10.1090/pspum/048/974342. MR974342

[BF86] Jean-Michel Bismut and Daniel S. Freed, *The analysis of elliptic families. I. Metrics and connections on determinant bundles*, Comm. Math. Phys. **106** (1986), no. 1, 159–176. MR853982

[BH10] John Baez and John Huerta, *The algebra of grand unified theories*, Bull. Amer. Math. Soc. (N.S.) **47** (2010), no. 3, 483–552, DOI 10.1090/S0273-0979-10-01294-2. MR2651086

[BPST75] A. A. Belavin, A. M. Polyakov, A. S. Schwartz, and Yu. S. Tyupkin, *Pseudoparticle solutions of the Yang-Mills equations*, Phys. Lett. B **59** (1975), no. 1, 85–87, DOI 10.1016/0370-2693(75)90163-X. MR0434183

[Col85] Sidney R. Coleman, *The uses of instantons*, Aspects of Symmetry, Cambridge University Press, 1985, pp. 265–350.

[Dir31] P. A. M. Dirac, *Quantised singularities in the electromagnetic field*, Proceedings of the Royal Society of London A: Mathematical, Physical and Engineering Sciences **133** (1931), no. 821, 60–72.

[Don83] S. K. Donaldson, *An application of gauge theory to four-dimensional topology*, J. Differential Geom. **18** (1983), no. 2, 279–315. MR710056

[Don90] S. K. Donaldson, *Polynomial invariants for smooth four-manifolds*, Topology **29** (1990), no. 3, 257–315, DOI 10.1016/0040-9383(90)90001-Z. MR1066174

[Don96] S. K. Donaldson, *The Seiberg-Witten equations and* 4-*manifold topology*, Bull. Amer. Math. Soc. (N.S.) **33** (1996), no. 1, 45–70. MR1339810

[Dys72] Freeman J. Dyson, *Missed opportunities*, Bull. Amer. Math. Soc. **78** (1972), 635–652, DOI 10.1090/S0002-9904-1972-12971-9. MR0522147

[Flo88] Andreas Floer, *An instanton-invariant for 3-manifolds*, Comm. Math. Phys. **118** (1988), no. 2, 215–240. MR956166

[Fre89] Daniel S. Freed, *Review of* [**Wit88**], MR953828, 1989.

[Fre82] Michael H. Freedman, *The topology of four-dimensional manifolds*, J. Differential Geom. **17** (1982), no. 3, 357–453. MR679066

[FS84] L. D. Faddeev and S. L. Shatashvili, *Algebraic and Hamiltonian methods in the theory of non-Abelian anomalies*, Theoret. and Math. Phys. **60** (1984), no. 2, 770–778, DOI 10.1007/BF01018976.

[Mor] John Morgan, *Applications of Seiberg–Witten invariants to the topology of 4-manifolds*, Lecture at the 1995 Cornell Topology Festival, unpublished.

[Lin57] André Lichnerowicz, *Théorie globale des connexions et des groupes d'holonomie* (French), Edizioni Cremonese, Roma, 1957. MR0088015

[Oog15] Hirosi Ooguri, *Interview with Edward Witten*, Notices Amer. Math. Soc. **62** (2015), no. 5, 491–506, DOI 10.1090/noti1248. MR3328919

[Pol75] A. M. Polyakov, *Compact gauge fields and the infrared catastrophe*, Phys. Lett. **B59** (1975), 82–84.

[PS95] Michael E. Peskin and Daniel V. Schroeder, *An introduction to quantum field theory*, Addison-Wesley Publishing Company, Advanced Book Program, Reading, MA, 1995. Edited and with a foreword by David Pines. MR1402248

[Qui85] D. Quillen, *Determinants of Cauchy-Riemann operators on Riemann surfaces*, Funktsional. Anal. i Prilozhen. **19** (1985), no. 1, 37–41, 96, English translation: Functional Anal. Appl. **19** (1985), no. 1, 31–34. MR783704

[Ste51] Norman Steenrod, *The Topology of Fibre Bundles*, Princeton Mathematical Series, vol. 14, Princeton University Press, Princeton, N. J., 1951. MR0039258

[SW94a] N. Seiberg and E. Witten, *Electric-magnetic duality, monopole condensation, and confinement in $N = 2$ supersymmetric Yang-Mills theory*, Nuclear Phys. B **426** (1994), no. 1, 19–52, DOI 10.1016/0550-3213(94)90124-4. MR1293681

[SW94b] N. Seiberg and E. Witten, *Erratum: "Electric-magnetic duality, monopole condensation, and confinement in $N = 2$ supersymmetric Yang-Mills theory"*, Nuclear Phys. B **430** (1994), no. 2, 485–486, DOI 10.1016/0550-3213(94)00449-8. MR1303306

[SW94c] N. Seiberg and E. Witten, *Monopoles, duality and chiral symmetry breaking in $N = 2$ supersymmetric QCD*, Nuclear Phys. B **431** (1994), no. 3, 484–550, DOI 10.1016/0550-3213(94)90214-3. MR1306869

[TAA+17] L. A. Takhtadzhyan, A. Yu. Alekseev, I. Ya. Aref′eva, M. A. Semenov-Tyan-Shanskiĭ, E. K. Sklyanin, F. A. Smirnov, and S. L. Shatashvili, *L. D. Faddeev's scientific legacy. A survey of his work* (Russian, with Russian summary), Uspekhi Mat. Nauk **72** (2017), no. 6(438), 3–112, DOI 10.4213/rm9799; English transl., Russian Math. Surveys **72** (2017), no. 6, 977–1081. MR3748689

[Tau82] Clifford Henry Taubes, *Self-dual Yang-Mills connections on non-self-dual 4-manifolds*, J. Differential Geom. **17** (1982), no. 1, 139–170. MR658473

[Uhl82a] Karen K. Uhlenbeck, *Removable singularities in Yang-Mills fields*, Comm. Math. Phys. **83** (1982), no. 1, 11–29. MR648355

[Uhl82b] Karen K. Uhlenbeck, *Connections with L^p bounds on curvature*, Comm. Math. Phys. **83** (1982), no. 1, 31–42. MR648356

[Wig60] Eugene P. Wigner, *The unreasonable effectiveness of mathematics in the natural sciences*, Comm. Pure Appl. Math. **13** (1960), 1–14.

[Wit82] Edward Witten, *Supersymmetry and Morse theory*, J. Differential Geom. **17** (1982), no. 4, 661–692 (1983). MR683171

[Wit88] Edward Witten, *Topological quantum field theory*, Comm. Math. Phys. **117** (1988), no. 3, 353–386. MR953828

[Wit94] Edward Witten, *Monopoles and four-manifolds*, Math. Res. Lett. **1** (1994), no. 6, 769–796, DOI 10.4310/MRL.1994.v1.n6.a13. MR1306021

[WY75] Tai Tsun Wu and Chen Ning Yang, *Concept of nonintegrable phase factors and global formulation of gauge fields*, Phys. Rev. D (3) **12** (1975), no. 12, 3845–3857, DOI 10.1103/PhysRevD.12.3845. MR0426712

[YM54] C. N. Yang and R. L. Mills, *Conservation of isotopic spin and isotopic gauge invariance*, Physical Rev. (2) **96** (1954), 191–195. MR0065437

[Zim08] Bill Zimmerman, *Interview with Chen Ning Yang and James Harris Simons*, Stony Brook Masters Series, March 28, 2008, Available at https://www.youtube.com/watch?v=zVWlapujbfo.

Departments of Mathematics and Physics, University of California, Santa Barbara, California 93106

Email address: drm@math.ucsb.edu

Contemporary Mathematics
Volume **718**, 2018
https://doi.org/10.1090/conm/718/14480

An introduction to spin systems for mathematicians

Ingmar Saberi

ABSTRACT. We give a leisurely, albeit woefully incomplete, overview of quantum field theory, its relevance to condensed matter systems, and spin systems, which proceeds through a series of illustrative examples. The goal is to provide readers from the mathematics community a swift route into recent condensed matter literature that makes use of topological quantum field theory and ideas from stable homotopy theory to attack the problem of classification of topological (or SPT) phases of matter. The toric code and Heisenberg spin chain are briefly discussed; important conceptual ideas in physics, that may have somehow evaded discussion for those with purely mathematical training, are also reviewed. Emphasis is placed on the connection between algebras of nonlocal operators and the appearance of nontrivial TQFTs in the infrared.

Introduction

It seems to have become a fashionable truism in some mathematical circles that physics is impossible to understand. This is despite a deep and growing relationship between the two fields; physical reasoning has come to be viewed as some sort of mysterious oracle, which from time to time emits an interesting and profound idea such as mirror symmetry, but whose inner workings are beyond comprehension.

This is, of course, not true; physicists, after all, are made rather than born, and the theorists among them learn their trade in a way that is (to any of the enormous number of people without interest in either discipline) indistinguishable from a mathematician's training. The real stumbling block in establishing mutual understanding between these two camps seems to lie in the radically different ways we are taught to position ourselves in an epistemological relationship to the world. For a mathematician, every system is closed: one does not care in what kind of system one is working, as long as the axioms are clear and things are well-defined. Once the assumptions are fixed, the borders of the system cannot change.

For a physicist, on the other hand, every system is open, and (more to the point) approximate. One never really expects that the mathematical problem one formulates and then solves will provide an *exact* or *complete* description of a physical system. Rather, it is some kind of picture of a real situation, which represents some features and perhaps leaves some others out. Many such pictures of the same situation may exist; one expects a relationship between them because they are pictures of the same situation, rather than because of anything intrinsic to either picture in itself.

2010 *Mathematics Subject Classification.* Primary 57R56, 81T10, 81T27.

Speaking narrowly, the purpose of this contribution is to give an introductory discussion of spin systems, emphasizing why such systems are of physical interest and what sorts of questions about them physicists would like to have answers to. This motivates some discussion of the notion of a *phase*, and so serves to introduce the central theme of the conference, Dan Freed's lectures on recent work using ideas from homotopy theory and topological field theory to compute groups of invertible phases. More broadly, though, we would like to try and cast at least a bit of light on how spin systems and field theories can (and often do) appear interchangeably. The central reason for this is that they are two different pictures of the same situation.

One way of expressing what that situation is would be to say that physicists care about field theory because it is a way of talking about systems with *many* degrees of freedom, labeled by points in a geometric space, and interacting in a way that respects the notions of homogeneity and locality associated to the geometry of that space. Examples of such systems, like Maxwell's electromagnetism, are all around us, but spin systems are also models of many degrees of freedom that are designed to respect the same notions. This idea is perhaps unfamiliar to those whose road into field theory starts with TQFT (in which there are no local degrees of freedom present at all), but it underlies the way physicists switch back and forth between various classes of models, secure in the belief that they represent similar physics. This constant change of perspectives can be a bit nauseating for those who are not used to it, and so we try to offer some discussion as to why one might believe it's reasonable. In a style typical of physics, this discussion will proceed through a series of examples.

We've also tried to highlight the importance of symmetry to the study of phases, and show why it is so crucial that one's concept of symmetry be expanded to include algebras of operators that are nonlocal. We feel that this is the most straightforward way to understand the crucial role of anyon systems in $2+1$-dimensional physics, as well as gesture to the importance of line and surface operators in the study of field theory in general. As such, we have tried to emphasize the role of anyons, and show their relationship to algebras of nonlocal symmetries.

Essentially none of the material in this article is new. The particular choice of topics and their presentation, however, is novel, and we feel it fills a gap in the existing literature. Most existing treatments of the material we discuss are either too elementary or too dependent on physical language and intuition to be of great use to a mathematically literate audience who are nonetheless physical novitiates. With this in mind, we have chosen to include a brief exegesis of topics (including basic quantum mechanics) that are found in countless standard physics textbooks. Emphasis has been placed on key ideas ("system, state, measurement") and pieces of physical or epistemological reasoning that have precise meaning, but belong a bit more to the realm of natural philosophy than to mathematics *per se*.

A couple of points on notation: we will follow conventions that are more typical of the physics literature than of mathematics, in the hope that this will prove to be instructive exposure rather than merely a source of confusion. As such, letters in bold type will denote spatial vectors. Indices from the middle of the Roman alphabet (i, j, and so on) will denote the components of such a vector, and so can be thought of as belonging to the set $\{1, 2, \ldots, d\}$, where d is the spatial dimension of the system at hand; we will also sometimes denote the index set by $\{x, y, z\}$. i will therefore appear both as an index and as the imaginary unit; it should always be clear from context which use is intended. Indices from the middle of the Greek alphabet (μ, ν, ...) will denote the components of "four-vectors," i.e., sections of the tangent bundle to the spacetime. One will often use the same letter in both contexts, for both a four-vector and its spatial part; the index value zero refers to time, so that expressions like

$$x^\mu = (x^0, \mathbf{x})$$

are common. Greek indices from the beginning of the alphabet (α, β) will label lattice sites; we will reserve Latin indices from the beginning of the alphabet (a, b, ...), as well as other Greek letters such as γ, for other index sets, to be defined specific to a given context.

Index placement is used to distinguish vectors from dual vectors; thus, an object like x^μ is associated to the tangent bundle, and p_μ to the cotangent bundle. The evaluation map is denoted by an implicit summation, of the form $p_\mu x^\mu$, which is to be understood whenever repeated indices appear. Often, indices are "raised and lowered" using a metric, whose components are denoted $g_{\mu\nu}$; thus, for example, the expression x_μ is understood to mean $g_{\mu\nu}x^\nu$. Again, both raised indices and exponents will appear without further comment, but context should be a sufficient guide. We will carefully obey the index placement rule for four-vectors, but will be sloppy about it otherwise (for example, with quantities like the angular momentum, which are not, strictly speaking, spatial vectors, although often represented as such).

A similar convention differentiates between vectors and dual vectors in a quantum-mechanical context: we frequently write elements of the Hilbert space $\mathscr{H}$ with symbols of the form $|a\rangle$, and denote elements of $\mathscr{H}^\vee$ by $\langle b|$. The evaluation pairing then looks like $\langle b|a\rangle$, while a symbol of the form $|a\rangle\langle b|$ denotes the obvious element of $\mathrm{End}(\mathscr{H}) \cong \mathscr{H} \otimes \mathscr{H}^\vee$. When not clear from context, "operators"—elements of $\mathrm{End}(\mathscr{H})$—may wear a circumflex. The labels we have written here as a, b may be any labels or indices characterizing the intended vector, but will often be eigenvalues of a particular operator labeling its eigenvectors. Again, letters may be used in more than one of these contexts simultaneously, so that an eigenvalue equation may appear something like

$$\hat{a}\,|a\rangle = a\,|a\rangle\,.$$

A similar convention extends to the action of operators on arbitrary states, so that one also sees equations like

$$\hat{a}\,|\psi\rangle = |\hat{a}\psi\rangle$$

(which should be understood as defining the meaning of the expression at right).

These conventions, or slight variants, are very commonly (though not universally) obeyed, and are well-known enough as to be used in many papers without explanation. For that reason, we hope that the brief leitfaden and sample usage in this article will render further encounters with the physics literature less alienating.

1. General principles of quantum mechanics: a reminder

We begin our discussion with a small crash course in the principles and axioms of quantum mechanics. In keeping with the idea that the relation between quantum and classical physics is crucial for understanding the former, we include some discussion of classical systems, hoping to emphasize physical ways of thinking that may be less familiar to our intended audience, as well as the epistemological concepts and ideas or intuitions about nature that motivate the use of various mathematical structures.

The concept of *state* in physics connotes a set of information about a system, which is sufficient to completely determine the future time-evolution of that system, at least to the greatest extent possible within the theory being considered. The theory then consists of a definition of the state of a system, together with the set of rules that govern its time-evolution. Since measurement is the only possible epistemic interaction one can have with the system, the state is something that can be determined in principle by repeated measurements, in any fully satisfactory theory of physical phenomena. In quantum mechanics, the requirement continues to hold, up to an important subtlety about how one should interpret the word "determined;" we will return to this subtlety shortly.

The space of states, therefore, admits a one-parameter group of automorphisms, representing evolution in time. (If information is lost under time-evolution, so that only a semigroup of endomorphisms corresponding to *forward* time-evolution is well-defined, one might say that the system "lacks *unitarity;*" we will ignore such systems in what follows.) Points in the state space (at a particular moment in time) are therefore in one-to-one correspondence with trajectories of the time evolution.

Two other concepts that are important are *reductionism* and *locality*. Reductionism means that it should be possible to separate any system into constituent pieces or subsystems, such that the total dynamics can be described in terms of the independent dynamics of the subsystems, together with interactions between them. Locality is the principle that the interactions between subsystems should become negligible when the subsystems are sufficiently far apart, so that it is possible to model a physics experiment taking place in a room without writing the dynamics of other nearby buildings or of the Andromeda galaxy: the subsystem can be studied on its own. As such, these notions are critical for any workable and predictive theory: they ensure that we do not need all information about the whole universe in order to make predictions about some small part of it. So we should have a notion of forming a composite system out of two subsystems, and a notion of a total dynamics in which the subsystems do not interact.

To give an example, let's review how these notions are manifested in classical physics. The reader is referred to the literature for more detail, and in particular to [**2**, **17**].

The classical state of (for example) a point particle consists of its position and velocity, and the way in which these are measured is taken to be obvious. The rule for the evolution of the state is Newton's equation, $F = m\ddot{q}$. Knowledge of the position and velocity is sufficient initial data to specify, by solving Newton's equation, the position $q(t)$ at all later times. Of course, one then also knows the velocity $\dot{q}(t)$; in other words, the evolution of the state in time is completely determined.

There are numerous ways of reformulating classical mechanics that demonstrate more closely its connections to quantum mechanics; while these reformulations are equivalent in principle, they highlight different important features of and perspectives on the theory. One of these is the *principle of least action*, which states that the time-evolution takes place along a trajectory that extremizes the *action functional*

$$S = \int dt\, L(q, \dot{q}). \tag{1}$$

Here the function L is called the *Lagrangian*. For a particle moving in a potential, it is the difference of the particle's kinetic and potential energies:

$$L = \frac{1}{2} m \dot{q}^2 - U(q), \tag{2}$$

where U denotes potential energy. It is an instructive exercise to check that the condition for S to be extremal (the Euler-Lagrange equation) is then nothing other than Newton's equation for the conservative force $F = -\,\partial U/\partial q$.

Another formulation, which is especially instructive for our purposes, requires taking the Legendre transform of the Lagrangian to obtain another functional called the *Hamiltonian*. That is, one defines a new coordinate on the state space by the relation

$$p = \frac{\partial L}{\partial \dot{q}}, \tag{3}$$

and then sets

$$H(q, p) = p\dot{q} - L(q, \dot{q}), \tag{4}$$

where it is implicit that $\dot{q}$ on the right-hand side is eliminated in favor of p, using (3). After doing this, the second-order equations of motion appearing above are replaced by a set of two first-order equations:

$$\dot{q} = \frac{\partial H}{\partial p}, \quad \dot{p} = -\frac{\partial H}{\partial q}. \tag{5}$$

The first of these is immediate from (4), and merely re-expresses the constraint (3) defining the canonical momentum p. The second is Newton's equation, rewritten in the first-order form $F = dp/dt$.

Hamilton's formulation has several advantages. First off, it shows that the space of states, or phase space $\mathscr{S}$, is symplectic: q and p are Darboux coordinates, and the time evolution is Hamiltonian flow. Secondly, it gives a clear notion of observables, which are simply functions on the phase space. The space of observables therefore carries a Poisson bracket, which in Darboux coordinates is simply expressed as

$$\{f, g\} = \left(\frac{\partial f}{\partial q}\frac{\partial g}{\partial p} - \frac{\partial f}{\partial p}\frac{\partial g}{\partial q} \right). \tag{6}$$

(In cases where more than one pair of variables is present, a sum over these pairs is understood.) Here f and g are any elements of the space of observables, $C^\infty(\mathscr{S})$. It is important to note that $\{q, p\} = 1$. What is more, the dynamics of an arbitrary observable are simple to state in this language: they are easily written as

$$\frac{df}{dt} = \{f, H\}. \tag{7}$$

Again, it is an instructive exercise to work out these statements in detail. Constants of the motion, or conserved quantities (which are by definition observables whose value does not change under time evolution) are therefore observables that Poisson-commute with the Hamiltonian. A system is said to be *completely integrable* when n mutually commuting such quantities can be found ($2n$ being the dimension of the phase space).

Furthermore, it is clear what it means to concatenate two subsystems in a noninteracting fashion: since the state is characterized by a pair of states, one for each subsystem, the phase space is the Cartesian product, and the noninteracting dynamics are governed by the Hamiltonian $H_1 + H_2$.

We can now turn to how the same epistemological ideas are represented in a quantum-mechanical theory. Luckily, much of the Hamiltonian formalism we've reviewed can be carried over to quantum mechanics in a way that makes the parallelism or correspondence clear. We'll start at the beginning, with a reminder of how the concept of the state is represented in quantum theories. It is helpful to introduce the notion of state and the notion of measurement (which, of course, are closely related) side by side.

Let us suppose that some quantity a can be measured, and the result of this measurement is an element of a set $\Sigma = \{a_\gamma\}$ of possible outcomes. (Σ is a subset of $\mathbb{R}$, which may be continuous or discrete; for clarity, we will treat the discrete case, although the principles remain the same.) Suppose, furthermore, that a measurement of a completely characterizes the system: there are no other quantities available to measure. Such a system might be, for example, an electron, where what is measured is its spin along a fixed axis; Σ then consists of only two values, "up" and "down" (or ± 1).[1]

The source of essentially all of the strangeness in quantum mechanics is the following postulate: Given a complete set Σ of possible outcomes for measurements, the quantum-mechanical state space is not Σ itself; rather, it is the formal linear span of Σ over $\mathbb{C}$, subject to the constraint of L^2-normalizability when Σ is infinite. To speak more properly, it is the space $\mathscr{H} = L^2(\Sigma)$ of normalizable complex-valued functions on that set. This space is equipped with an orthogonal decomposition as a direct sum of one-dimensional subspaces, spanned by the elements of Σ, and labeled by the corresponding real numbers a_γ which are the outcomes of the measurement. A decomposition of a complex vector space into an orthogonal direct sum labeled by real numbers is equivalent to the data of a Hermitian operator acting in that space; we are thus led to the idea that observables in quantum mechanics are represented by Hermitian operators.

This is, on the one hand, a familiar kind of statement to mathematical ears. It is a very standard gambit to linearize a problem by thinking about a space of functions: the functions on, for example, a set or topological space form a vector space, often equipped with additional algebraic structure, and a group action on the set gives a linear G-representation on the function space. So one might be tempted to think of the passage to quantum mechanics as a way of replacing a symplectic geometry problem by a (potentially easier) linear algebra problem, and this is in some loose sense true.

On the other hand, this postulate is an overwhelmingly peculiar and counterintuitive thing to say about the physical world. For instance, it immediately follows from this that even a system which is completely determined by a single measurement admitting only two possible outcomes has *uncountably* many distinct possible states.

[1] It should be carefully noted that, in cases where more than one measurement must be made, Σ can consist of a *smaller* set of measurements than those that define $\mathscr{S}$; the source of this subtlety should become clear momentarily.

The attentive reader will immediately point out a problem here. We've said that the state is determined in principle by measurements; how can a system in which any measurement has at most two outcomes have more than two possible states?

The key lies in the identification of measurable quantities with Hermitian operators, or, what is the same, with a choice of basis in the Hilbert space and a marking thereof by outcomes of the measurement. While the dimension of the Hilbert space indicates the number of possible outcomes for any given measurement, different choices of basis are always possible and correspond to physically distinct measurements. Furthermore, these different measurements are *mutually incompatible:* being in an eigenstate of one such operator implies that one is not in an eigenstate of the other, and therefore that the other takes on no particular definite value.

Of course, a measurement can still be made in such a state; how is one to understand what its outcome will be? The answer is provided by perhaps the most surprising postulate of quantum mechanics, the *Born rule*. Its content is as follows: Suppose an observable $\hat{a}$ is to be measured, with possible eigenvalues a_γ. Imagine that a system is prepared in the state

$$|\psi\rangle = \sum_\gamma c_\gamma \, |a_\gamma\rangle \, ,$$

for some constants c_γ. Following this, $\hat{a}$ is measured. The outcome will be one of the a_γ; if the measurement is repeated, the probability of obtaining a_γ will be $|c_\gamma|^2$. Furthermore, *after* the measurement is performed, the state of the system is $|a_\gamma\rangle$, as evidenced by the measured value of $\hat{a}$.

Thus, measuring a state in quantum mechanics doesn't tell you what the state was previously; rather, it *prepares* the state corresponding to the outcome of the measurement. Nonetheless, a general state ψ can be determined, assuming many identical copies of it can be obtained or similarly prepared: repeated measurements obviously identify the values of $|c_\gamma|$, while other experiments are possible in principle to access the relative phase of c_γ and $c_{\gamma'}$. For the sake of space, this is all we can say about this subject here.

As a simple example of this framework, if the Hilbert space is two-dimensional, any two-by-two Hermitian matrix is an observable. The space of such matrices is spanned (along with the identity, which is a trivial measurement) by the *Pauli matrices:* these are a set of standard Hermitian matrices σ_i, obeying the commutation relations

$$[\sigma_x, \sigma_y] = 2i\sigma_z$$

(together with all cyclic rearrangements). The $i\sigma_i$ thus form a basis of $\mathfrak{su}(2)$. By convention, they are written in an eigenbasis of σ_z, so that

$$\sigma_x = \begin{bmatrix} 0 & 1 \\ 1 & 0 \end{bmatrix}, \quad \sigma_y = \begin{bmatrix} 0 & -i \\ i & 0 \end{bmatrix}, \quad \sigma_z = \begin{bmatrix} 1 & 0 \\ 0 & -1 \end{bmatrix}.$$

We will also sometimes use the matrices $s_i = \sigma_i/2$, which differ by a normalization from the standard Pauli matrices. The algebra of all observables in such a system is thus equivalent to $\mathfrak{su}(2)$, which is sometimes called the *angular momentum algebra*, and the Hilbert space is its defining representation, often termed the *spin-*$1/2$ *representation* in physics. One thus often thinks of such a two-dimensional Hilbert space as representing the spin of an electron, or any other spin-$1/2$ particle in three-dimensional space. The conclusion is that the spin of such a particle may be measured along any axis in three-dimensional space—but with the understanding that any two such measurements along non-collinear axes are *mutually incompatible*, and states of definite (for example) x-spin do not correspond to states with any definite z-spin. For further discussion of the physical behavior this implies, and the experiments that confirmed it, we recommend the discussion of the Stern–Gerlach experiment in [**26**]. The existence of non-basis states implies even stranger phenomena when a system can be decomposed into subsystems; these go by the name of *entanglement*, and we will return to them shortly.

What does the algebra of the Pauli matrices have to do with angular momentum? Well, it's obviously the Lie algebra of the three-dimensional rotation group, but the same algebra can also be derived by remembering that angular momentum in three dimensions is defined by the equation

$$\mathbf{L} = \mathbf{q} \times \mathbf{p},$$

and computing the Poisson brackets of the components of this pseudovector using (6). One finds that

$$\{L_x, L_y\} = L_z. \tag{8}$$

So the failure of the components of angular momentum to commute was *already present* in classical mechanics, reflecting the fact that rotations around different axes don't commute, but encoded in subtle fashion in the Poisson bracket. (The normalization of the s_i was chosen to correspond precisely to the algebra (8), in fact.) Even the famous canonical commutation relation, implying the uncertainty relation between position and momentum, was already written down above, in its classical form $\{q, p\} = 1$! Quantization thus asks (among other things) for a deformation of the algebra of classical observables, which was just $C^\infty(\mathscr{S})$, to a noncommutative algebra with commutation relations defined by the rule

$$[\hat{f}, \hat{g}] = i\hbar\{f, g\} \tag{9}$$

(at least at leading order in $\hbar$).

Even the rules of time evolution can be obtained by this method: from (7), we obtain

$$i\hbar \frac{d\hat{f}}{dt} = [\hat{f}, H], \tag{10}$$

which is equivalent to the Schrödinger equation. One usually thinks of time-evolution as acting on the states, but of course it is equally natural to think of a one-parameter subgroup of $\mathrm{Aut}(\mathscr{H})$ acting on $\mathscr{H}$ directly, or by conjugation on $\mathrm{End}(\mathscr{H})$—i.e., on operators instead of states. Since one can only measure things of the form $\langle\psi|\, \hat{a}\, |\phi\rangle$, the difference is immaterial, like that between active and passive coordinate transformations; in physics, one refers to these two formulations as the *Schrödinger* and *Heisenberg pictures*. And, of course, the content of (10) is the same in both pictures; it merely says that H is the generator of time-evolution, i.e. that $i\hbar\, \partial/\partial t = H$, where the action of H is interpreted either in the fundamental representation or in the adjoint action on $\mathrm{End}(\mathscr{H})$.

A couple of further subtleties need to be remarked on immediately. First of all, a state is not an element of $\mathscr{H}$; rather, it is a one-dimensional subspace. So the space of physical states is really the space of lines in $\mathscr{H}$, i.e., its projective space $P\mathscr{H}$. But a theorem of Wigner ensures that symmetries of $P\mathscr{H}$ can always be lifted to $\mathscr{H}$, so that they are linear and unitary (or possibly antilinear and antiunitary). So it is typical to always think in terms of $\mathscr{H}$, while remembering that scalar multiplication is an equivalence relation on physical states.

Secondly, we assumed for the sake of exposition that an allowed value a of the measurement we called $\hat{a}$ characterizes one and only one state. But we may actually need to make more than one distinct measurement in order to specify the state fully. One typically assumes that there exists a *complete set of commuting observables*, $\{\hat{a}_i : 1 \leq i \leq k\}$, such that one can label every state uniquely by a list of their simultaneous eigenvalues: i.e.,

$$\mathscr{H} = L^2(\Sigma) = \bigoplus \mathbb{C} \left| a_1 \ldots a_k \right\rangle , \tag{11}$$

with a_i running over the spectrum of $\hat{a}_i$. Of course, the operators $\hat{a}_i$ can be simultaneously diagonalized (as is done above) if and only if they mutually commute. The existence of the CSCO is the postulate that there exists a set of compatible measurements which serve to fully characterize the state space. The Born rule also needs slight generalization: if the eigenspace corresponding to an observation $\hat{a} = a$ is degenerate, the probability will be $\langle\psi| P_a |\psi\rangle$, where P_a is the projection operator onto the eigenspace. Since $\sum_a P_a = 1$, it is immediate that the probabilities of all possible measurements sum to unity.

Thirdly, one needs to be careful: $\mathscr{H}$ is emphatically *not* the space of L^2 functions over the phase space! The issue, of course, is that pairs of coordinates like q and p have a canonical nonzero Poisson bracket, and so fail to commute; one can choose only *one* of such a pair as part of a set of commuting observables. This is, of course, related to the Stone–von Neumann theorem, which says that the algebra of a canonical pair of real variables has a unique unitary representation on $L^2(\mathbb{R})$. The consistent choice of "half" of the phase space variables, in order to construct the Hilbert space, is a source of difficulty in quantization of general symplectic manifolds; the additional data required is sometimes called a *polarization*.

Finally, let us remark on how one forms composite systems in quantum mechanics. Armed with the ideas we have developed, it is easy to see how this should go: If measuring $\hat{a}$ completely characterizes one subsystem, and measuring $\hat{b}$ completely characterizes the other, then a measurement of both of these should characterize their composite. We expect that the two measurements are compatible because each pertains only to one of the subsystems, and it is supposed that the two could in principle be isolated from one another. The joint Hilbert space is therefore the L^2-span of the $|a_\gamma, b_\zeta\rangle$—but this is just the tensor product of $\mathscr{H}_1 = \operatorname{span} |a_\gamma\rangle$ and $\mathscr{H}_2 = \operatorname{span} |b_\zeta\rangle$. (We are passing over issues arising in infinite dimensions in silence; our aim is to give an intuitive idea as to *why* the tensor product is the correct structure to use.)

Measurements on one subsystem that do not affect the other should then be represented by the relevant operator on the subsystem, tensored with the trivial operator on the other; this means that it will act correctly on the basis vectors of a tensor product basis, returning the value of the measurement independent of the state of the other system. The natural dynamics for a noninteracting pair, then, is just the sum of the two Hamiltonians, where each is taken to act in this way on the entire system:

$$H = H_1 \otimes 1 + 1 \otimes H_2. \tag{12}$$

(We will often write this and similar expressions in the abbreviated form $H = H_1 + H_2$ in what follows, the natural maps between spaces of operators being understood implicitly.)

Of course, just as the Hilbert space contained many states that were not eigenstates for an operator, and so (with strange and counterintuitive consequences) didn't correspond to any particular value of the corresponding measurement, the composite Hilbert space $\mathscr{H} = \mathscr{H}_1 \otimes \mathscr{H}_2$ contains many states that are not basis states in the tensor product basis. Moreover, many of these states can't be basis states in *any* tensor product basis—just as rank-one operators span, but do not exhaust, $\operatorname{End}(\mathscr{H}) = \mathscr{H} \otimes \mathscr{H}^\vee$. Physicists refer to such states as *entangled* between the two subsystems, and entanglement remains an inexhaustible source of strange behavior and theoretical interest even in the present day. We will not be able to say much about it here, but we feel it is important to note in closing that entanglement is connected to interactions between subsystems. If the Hamiltonian takes the non-interacting form (12), it is obvious that it can be diagonalized in a tensor-product basis, between the energy eigenstates of each subsystem. Thus, all eigenstates of the composite Hamiltonian have zero entanglement. Conversely, when energy eigenstates are entangled, it follows that there are interactions between the subsystems. There has been great interest in developing quantitative measures of entanglement, and exploring their properties in energy eigenstates of interacting systems, such as the vacuum state of a quantum field theory (see Chapter 5 of [**33**] and references therein).

2. A few words about field theories

At root, a *field theory* is a theory whose dynamical degrees of freedom are fields, and *fields* are quantities (like the electric field, the air pressure, or the temperature) that can be measured independently throughout space. It is common to make further requirements, for example that the field theory be *Lorentz-invariant*, i.e. compatible with the principle of special relativity. But requirements like these are not inherent in the concept, and there are interesting examples where they are not met. Other than the fact that the degrees of freedom are labeled or indexed by points in space, all of the generalities of §1 are equally true for field theories.

We will make one further assumption, though, which we already identified in the very beginning of the article: namely, that the theory exhibits *locality*. In other words, the action functional that encodes the theory's dynamics should be able to be computed in a local fashion. For a typical continuum field theory, this will amount to the statement that

$$S = \int dt\, L = \int dt\, d\mathbf{x}\, \ell[q, \dot{q}],$$

where ℓ, the *Lagrangian density*, depends on the fields $q(\mathbf{x})$ and their derivatives at the same point of space only. Similarly, field theories (just like any other theories) admit a Hamiltonian operator, and locality means that it is likewise the integral over space of a Hamiltonian density:

$$H = \int d\mathbf{x}\, h[q, p].$$

For discretized field theories and spin systems, the concept of locality will be implemented in a slightly different fashion, but its essential meaning is the same.

A further assumption we can make is that ℓ and h depend only on the values at a given point, and not explicitly on the point (the value of $\mathbf{x}$) itself. This is an assumption of *translation invariance* or *homogeneity:* it means that the dynamics at each point in space are the same.

The most basic example satisfying these requirements is a free scalar field theory: its degree of freedom is a single real-valued function $q(\mathbf{x})$, whose dynamics are determined by the functional

$$S = \frac{1}{2}\int dt\, d\mathbf{x}\, \left(\partial^\mu q \partial_\mu q + m^2 q^2\right) \tag{13}$$

After an integration by parts, we can write it as

$$S = \frac{1}{2}\int dt\, d\mathbf{x}\, q \left(-\partial^\mu \partial_\mu + m^2\right) q, \tag{14}$$

and it is a simple exercise to check that the associated Euler–Lagrange equation is the *Klein–Gordon equation:*

$$\left(\frac{\partial^2}{\partial t^2} - \nabla^2 + m^2\right) q(t, \mathbf{x}) = 0. \tag{15}$$

We'll look at this equation in a little more detail, in the simplest possible case: namely, in one spatial dimension, setting the parameter $m^2 = 0$. Then it reduces to the wave equation,

$$\left(\frac{\partial^2}{\partial t^2} - \frac{\partial^2}{\partial x^2}\right) q(t, x) = 0. \tag{16}$$

Of course, solving this equation is a triviality: because it's translation-invariant and time-independent, we can also diagonalize translations. This amounts to asking that $q(t, x)$ is also an eigenfunction of the translation operators d/dx and d/dt, in other words, a plane wave:

$$q(t, x) \propto \exp i(kx - \omega t). \tag{17}$$

The wave equation then becomes an algebraic relation (the *dispersion relation*) on the eigenvalues of these operators: it reads simply $k^2 = \omega^2$, which expresses the fact that waves can move in either direction at a speed independent of their frequency. (Here we've tacitly set this speed to 1. The fact that the speed of light is constant reflects the fact that, in the absence of matter, electromagnetism is a free theory very similar to this one.) The essential meaning of the word "free"[2] is that (15) is linear: the superposition of any two solutions is again a solution, which means that waves can pass through one another and neither is affected by the other's presence. No scattering, or interaction, of any kind takes place.

Let's now imagine discretizing this system: instead of the degree of freedom q being defined for all values of x, we'll imagine that it can be measured only for $x = \alpha a$, where α is an integer labeling lattice sites, and a is a new dimensionful

[2]There is an unfortunate clash of vocabulary here. "Free" objects, in mathematics, tend to be the most general possible examples of their kind, so that all others arise as quotients, or (what is the same) by introducing additional relations. In physics, a "free" theory is one without interactions, i.e., one where the scattering is in some sense the *least* general possible. The author is grateful to B. Knudsen for related discussion that drove this point home.

parameter that measures the lattice spacing. An immediate analogue of (16) is obtained by replacing spatial derivatives by difference operators:

$$\frac{\partial^2}{\partial t^2} q_\alpha(t) + \frac{1}{a^2}\left(2q_\alpha(t) - q_{\alpha+1}(t) - q_{\alpha-1}(t)\right) = 0. \tag{18}$$

This equation is still local, in the sense that a particle only interacts with its nearest neighbors. And it is solved by the same ansatz as (17): we diagonalize time translations and (discrete!) lattice translations, choosing eigenfunctions

$$q_\alpha(t) \propto \exp i(ka\alpha - \omega t). \tag{19}$$

It is then trivial to see that the new dispersion relation is

$$\omega^2 = \frac{2}{a^2}(1 - \cos(kaL)) = \frac{4}{a^2}\sin^2(ka/2). \tag{20}$$

As the reader will have no trouble checking, this reduces precisely to the linear dispersion we found before in the limit $ka \ll 1$; however, it deviates from it when $k \sim 1/a$, i.e., at sufficiently short wavelengths. And this makes sense: short-wavelength waves will be not be represented well by discrete sampling, just as high frequencies on a CD are distorted due to digitization noise. Taking the continuum limit, $a \to 0$, recovers the continuous wave equation we studied previously.

Another important effect is that the range of allowable wavelengths is no longer infinite: the parameter k is now circle-valued, expressing the fact that the Pontryagin dual of $\mathbb{Z}$ is the circle group, and this has the effect of imposing a hard momentum cutoff at $k = \pi/a$. This is often called an *ultraviolet cutoff* (UV) in physics, and eliminates one common source of divergences in field theories. However, some infinities may remain, because we are still dealing with a system with infinitely many degrees of freedom: the parameter $k \in [-\pi/a, \pi/a]$ labeling independent solutions is continuous. To have a system with only finitely many degrees of freedom, we will also have to make the system size finite by imposing boundary conditions. One possible choice would be *periodic* boundary conditions, so that the lattice is a discretization of the circle rather than the entire real line. These would require, for a linear arrangement of N total sites, that $\exp(ikNa) = 1$, i.e. that k is an integer multiple of $2\pi/Na$. (If one wanted to talk pretty, one would say that the translation symmetry is now $\mathbb{Z}/N\mathbb{Z}$, which is Pontryagin self-dual.) Of course, if the momenta of the mode we are interested in is large compared to the scale $2\pi/Na$ on which the discreteness occurs, the finite size doesn't make a substantial change in the behavior of the system. This should be familiar from everyday experience: the air in a room is obviously a system with finitely many degrees of freedom, in which the infrared (IR) regulator is the size of the room and the UV regulator is the intermolecular spacing. But the fact that one can speak words and be understood in such a room attests that the nondispersive wave equation is an exceptionally good model, at least for the wavelengths characteristic of the human voice. Recalling that the speed of sound in air is about 340 m/s, and that the fundamental frequency of the human voice is typically between 85–255 Hz, one sees that the typical wavelengths are about 1–4 m, essentially at the IR scale.

Finite system size is often referred to as an *infrared regulator* in physics, because it removes infinities that have to do with the momentum k becoming infinitesimally small (or, what is the same, the wavelength of the mode becoming infinitely long) rather than large. As we have just seen, imposing both infrared and ultraviolet regulators typically yields a system with only finitely many degrees of freedom, whose behavior closely approximates that of the original system as long as the modes being studied are not close to either cutoff. Furthermore, as our example shows, the infrared cutoff typically takes the form of boundary conditions (periodic or otherwise), which can often be added or removed as desired without affecting the rest of the analysis or the solvability of the system. The limit $N \to \infty$ is referred to by physicists as the thermodynamic limit.

However, there is yet a third type of regulator that can be introduced, that pertains not to the number of degrees of freedom but to the nature of the degrees of freedom themselves. It is to this last approximation (and, with it, at last, to the subject of spin systems) that we turn in the next section.

3. From field theories to spin systems

So far, we've talked about field theories in which the degrees of freedom are functions on space, and we've shown how introducing a lattice discretization, together with a finite system size, can reduce a field theory's dynamics to that of only finitely many degrees of freedom, coupled in a homogeneous and local fashion. But each degree of freedom is still something like a harmonic oscillator; in particular, upon quantization, the Hilbert space of even one degree of freedom is still infinite-dimensional.

We may be interested in replacing this problem by one in which the local Hilbert space is finite-dimensional. This has a number of advantages: First of all, it eliminates any remaining analytic issues. All operators are bounded and trace-class; they're just finite matrices. And one can in principle diagonalize such matrices—the Hamiltonian, for instance—on a computer, although (since the total dimension grows exponentially in the number of sites) this is infeasible in practice. As is usually the case, it's better to think structurally about what makes these matrices special than to attack the problem by brute force. And the relevant matrices are indeed a very special class, due once again to the pervasive notions of *homogeneity* and *locality*.

What do homogeneity and locality mean for spin systems? Well, each of the sites is a subsystem, so it's clear that the total Hilbert space is a tensor product over all the sites:

$$\mathscr{H} = \bigotimes_{\alpha} \mathscr{H}_\alpha \qquad (\mathscr{H}_\alpha \cong \mathbb{C}^d). \tag{21}$$

There's therefore a notion of the *support* of an operator: If it acts nontrivially on the tensor product of Hilbert spaces with α in some set S, extended by the identity on all other sites as we discussed above, then its support is contained in S. (The support is the intersection of all such S.) And if the dynamics are to be homogeneous and local, then we expect the Hamiltonian to be expressible as a sum over sites:

$$H = \sum_{\alpha} h_\alpha, \tag{22}$$

where h_α does not depend explicitly on the parameter α, but rather is constructed in the same way at each site. Moreover, the support of h_α should be "local," in some appropriate sense. One might, for example, demand that it includes α together with only a finite collection of nearby sites. Substantially weaker locality conditions are often considered, though: one might choose to ask only that interactions between separated sites decay exponentially with the separation, or even with a sufficiently fast power law.

We promised that at least some systems satisfying these properties would admit clever techniques of solution, and this is indeed the case. Considering problems of this form opens the door to a wide class of exactly solvable models, often referred to as spin chains or quantum integrable systems. Integrability, of course, is an enormous and active field of research in its own right, and we will not have space to discuss it adequately here; the reader is referred to the literature (for instance, [**3**, **4**, **6**]). We will, however, give one important example of an integrable spin chain in the next section, which will hopefully give both some small idea of the flavor of integrable models and indicate the way in which such models are connected to field theories. In essence, the term means that such models admit reasonably simple closed-form expressions for the spectrum of the Hamiltonian as a function of parameters, and that the exact eigenstates can also be written down.

Why might a truncation to finite-dimensional local Hilbert spaces be reasonable or permissible to make in certain physical systems? Well, as is well-known, the quantum harmonic oscillator has a spectrum of energy eigenstates that are evenly spaced:

$$H|n\rangle = E_n|n\rangle = \hbar\omega\left(n + \frac{1}{2}\right)|n\rangle . \tag{23}$$

If the Hamiltonians h_α of our lattice system consist, for example, of a harmonic oscillator Hamiltonian for the degree of freedom at α, together with terms that weakly couple it to its neighbors, then we know that the $\{|n_\alpha\rangle\}$ (eigenstates of the uncoupled Hamiltonian) at least form a basis of the total Hilbert space. Of course, it won't be an energy eigenbasis anymore, but—since energy is a physically meaningful quantity that is stored locally by the degrees of freedom—we expect that states in which *one* degree of freedom has an enormous amount of energy won't contribute meaningfully to states in which *all* degrees of freedom collectively have much less energy. Thus, one imagines that eigenstates of the coupled system with energy less than some chosen bound can be constructed, at least to a very good approximation, as linear combinations of tensor product states where n_α is less than some cutoff value; and that is precisely the spin system approximation.

To emphasize the central point, this discussion indicates that the passage to a spin-system approximation of a field theory is best interpreted as a cutoff on the energy, or more precisely energy density, of the excitations we will allow ourselves to model. In terms of a field theory of light, we have already imposed a UV and an IR cutoff, that tell us not to think about wavelengths that are too large or too small; we expect this new approximation to work well (at least roughly speaking) only when the *amplitude* of the light, or the number of photons present, is not too arbitrarily large.

Where does the name "spin system" come from? Well, the local Hilbert spaces are finite-dimensional, and the typical physical examples of systems with finite-dimensional Hilbert spaces are precisely spins. By Stone–von Neumann, any continuous degree of freedom has an infinite-dimensional Hilbert space, but the angular momentum algebra (which, again, is just $\mathfrak{su}(2)$) admits a unique unitary representation of every finite dimension. So it is typical in physics to think of these Hilbert spaces as representing spins, whether or not there is actually $SO(3)$ invariance present.

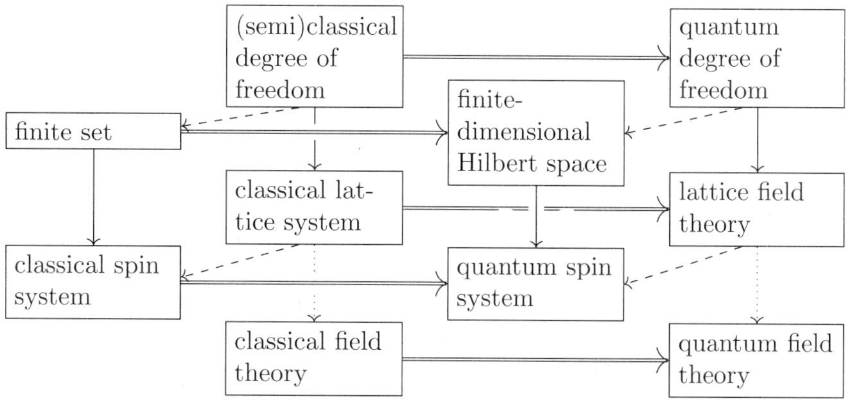

FIGURE 1. A diagram of relations between different types of physical model

What, then, is the classical analogue of a spin degree of freedom? For our purposes, we'll be satisfied with a simple (and of course very incomplete) answer, that the reader can probably guess from our discussion of quantization above. Since, for us, a "spin" is just a finite-dimensional Hilbert space, in which a complete measurement produces only finitely many possible results, the corresponding space of semiclassical states is just a finite set. Of course, the property of a genuine classical phase space that translates to finite-dimensionality of the corresponding Hilbert space is really *compactness;* the finite-dimensional representations of $SU(2)$, for example, can all be thought of via the Borel–Weil–Bott theorem as arising from quantization of the compact phase space $S^2 = P^1(\mathbb{C}) = SU(2)/U(1)$, equipped with different choices of prequantum line bundle. We won't go into this subtlety here, though, and will be content to think of models just based on the set of (quantized!) possible outcomes of a measurement on such a system.

Models made from such finite-set-valued degrees of freedom (like the famous *Ising model*, which we'll discuss more a bit later on) we will term *classical spin systems.* Of course, any notion of classical dynamics as a Hamiltonian flow is out the window; one typically has to interpret such models in the context of statistical mechanics at finite temperature. When the system is in thermal equilibrium with its environment, it effectively occupies each of its possible states s at random with probability $p_s = \exp(-E_s/\tau)$. Here τ is a temperature parameter. We'll talk a bit more about systems of this kind below.

To summarize the discussion so far, and to keep the big picture in view, Figure 1 shows a cartoon of the relationships between the different types of physical systems we've been talking about. The top layer represents different choices of local degrees of freedom; the passage downward, as represented by arrows of the form $\longrightarrow$, represents coupling many such degrees of freedom in a homogeneous, local manner. Dashed arrows (- ->) represent a truncation of the kind we have discussed in this section, of the dimension of the local Hilbert space, and dotted arrows (⋯>) represent passage to a continuum limit. Finally, doubled arrows like $\Rightarrow$ represent quantization.

While we have chosen to represent the arrows with definite directions, we hope it is clear that the indicated relationships are somewhat more bilateral. In the discussion thusfar, we have taken a sort of "top-down" perspective, in which the field theory is viewed as the primordial object, and discretizations of it are approximations made to facilitate calculation or to remove problems with infinities. But we wish to emphasize that the reverse processes are always possible; that is, the discrete systems being considered *behave like* continuum field theories, at least over a wide range of their possible excitations.

The lesson one should draw is that what is truly fundamental is the picture of a field theory as *many* degrees of freedom, coupled together in a way that is parameterized by a geometric space, so that the coupling is *local* and *homogeneous*. And quantum field theory is a universal language for talking about the collective behavior of such many-body systems. Indeed, in many physical applications (like the study of sound waves or magnetic materials), it is the discrete picture that is "fundamental," and the continuum field theory that is a large-scale approximation! The lattice corresponds most closely to reality; field theory is nonetheless an exceptionally good model.

In some sense, the real lesson is that what is "fundamental" is the wrong question to be asking. It is of little physical consequence whether "many" means 10^{23} or ∞, or even what the precise nature of each local degree of freedom is. But let's see how this plays out in another important example.

4. An example: The Heisenberg spin chain and integrability

Following the criteria we've laid out, let's write down the simplest example of a quantum spin system we can, using the ingredients that have appeared in the preceding discussion. (Our discussion largely follows that in Bethe's original paper [**4**], as well as [**6**].) We'll take the local Hilbert space to be two-dimensional; there is then a basis of local (one-site) Hermitian operators consisting of Pauli matrices acting at each site:

$$[s_i^\alpha, s_j^\beta] = i\epsilon_{ijk}\delta^{\alpha\beta}s_k^\alpha. \tag{24}$$

These generate (as an algebra) the entire collection of observables of the system.

The sites will be distributed along a one-dimensional lattice, just as in the discretized wave equation. We'll imagine that the degrees of freedom on each site are quantum spins with spin $S = 1/2$—i.e., two-dimensional Hilbert spaces. The mutual interaction between two such particles is a so-called *exchange interaction*, which tends to align (or anti-align) the spins, and is represented in the Hamiltonian by a term proportional to $-\boldsymbol{s}^\alpha \cdot \boldsymbol{s}^\beta$—i.e., to the dot product of the two spin vectors. For more on the physics of exchange interactions and condensed matter systems more generally, see [**25**].

Furthermore, to ensure locality, we'll assume that the only significant exchange interactions occur between nearest neighbors in the lattice. We can now write a simple local translation-invariant Hamiltonian, satisfying all of the good properties we have outlined above:

$$h_\alpha = -\frac{J}{2}\left(\boldsymbol{s}^{\alpha-1}\cdot\boldsymbol{s}^\alpha + \boldsymbol{s}^\alpha\cdot\boldsymbol{s}^{\alpha+1} - \frac{1}{2}\right) \tag{25}$$

For definiteness, we will assume that $J > 0$; such an exchange interaction is known as ferromagnetic. We can, if we like, also add a term that represents an interaction with an external magnetic field to h_α, such as $B(s_z^\alpha + 1/2)$. The total Hamiltonian then takes the form

$$H = \sum_\alpha h_\alpha = J\left(\frac{N}{4} - \sum_\alpha \boldsymbol{s}^\alpha \cdot \boldsymbol{s}^{\alpha+1}\right) + \left(\frac{N}{2}B + \sum_\alpha \mathbf{B}\cdot \boldsymbol{s}^\alpha\right). \tag{26}$$

Note that $B = |\mathbf{B}|$, and that $\mathbf{B}$ is taken to point along the z-axis. The N-dependent constants are included merely to ensure that the ground-state energy is set to zero; they don't affect the physics, and one can safely ignore them. The Hamiltonian (26) encodes the dynamics of a simple integrable spin system, called the *Heisenberg* or *XXX spin chain*.

If we so desire, we can turn off the interactions by setting $J = 0$. Then the system is just a composite of many noninteracting spins, and so is trivial to solve: the eigenstates are tensor product states of the form

$$|A\rangle = |a_1, \dots, a_d\rangle = |{\downarrow\downarrow} \dots {\downarrow\uparrow\downarrow} \dots\rangle\,. \tag{27}$$

In this notation, A is a subset of $\{\alpha\}$, the set of all lattice sites. It contains the positions of any spins that have been flipped up. The energy of such a state is

$$H\,|A\rangle = B\cdot \#A \cdot |A\rangle\,. \tag{28}$$

To see that the interaction term changes the system meaningfully, we just need to check that the eigenstates of the free problem no longer solve the interacting Hamiltonian. And this is not difficult to do: To see it simply, let's define $s_\pm = s_x \pm i s_y$. These are raising and lowering operators; they satisfy the commutation relations

$$[s_z, s_\pm] = \pm s_\pm, \qquad [s_+, s_-] = 2s_z, \tag{29}$$

which should be familiar from the representation theory of $\mathfrak{su}(2)$. The Hamiltonian can then be rewritten in terms of these operators as

$$H = -J\sum_\alpha \left(s_z^\alpha s_z^{\alpha+1} + \frac{1}{2}\left(s_+^\alpha s_-^{\alpha+1} + s_-^\alpha s_+^{\alpha+1}\right)\right) + B\sum_\alpha s_z^\alpha + E_0, \tag{30}$$

where E_0 denotes the (irrelevant) constant shift terms. When one considers the action of this Hamiltonian on a tensor product state, it is clear that such a state (unless all spins are aligned) cannot possibly be an eigenstate: the terms containing $s_\pm$ represent "hopping" interactions, in which an up-spin moves one site right or left:

$$s_-^\alpha s_+^{\alpha+1}\,|\dots {\downarrow}\overset{\alpha}{\uparrow}{\downarrow}\dots\rangle = |\dots {\downarrow\downarrow\uparrow}\dots\rangle \tag{31}$$

The action of H on $|A\rangle$ thus is a sum of terms in which all elements of A have been perturbed by ± 1. This expression for H also makes it obvious that H commutes with the operator $S_z = \sum_\alpha s_z^\alpha$ that measures the total spin: none of the terms, including the hopping interactions, change the total spin of the state on which they act. As such, while the multi-index A is no longer a good quantum number, its length $\#A$ (which is just the total spin) remains compatible with the interacting Hamiltonian.

Indeed, if we were to set $B = 0$, we could form the total spin operators

$$S_i = \sum_\alpha s_i^\alpha, \tag{32}$$

and *all* of these would then commute with the Hamiltonian. Of course, they do not mutually commute; rather, they satisfy the $\mathfrak{su}(2)$ algebra. But the action of an algebra on $\mathscr{H}$, commuting with the Hamiltonian, is precisely a symmetry in quantum mechanics; this one comes from the obvious $SO(3)$ rotation symmetry of the space in which our spins live. Of course, this symmetry is broken by the application of an external magnetic field, as this picks out a distinguished direction.

At this point, we've satisfied ourselves that H is an interesting, nontrivially interacting Hamiltonian, and we've also discovered that it commutes with (and so can be simultaneously diagonalized with) S_z. We can therefore look for eigenvectors independently in each subspace with a definite value of $\#A$. The subspace $\#A = 0$ is already one-dimensional, so that state $|\emptyset\rangle$ must be an energy eigenvector; in fact, it will be the vacuum state of the system.

The next-easiest place to start is $\#A = 1$. A general state in this subspace takes the form

$$|\psi\rangle = \sum_\alpha c_\alpha |\alpha\rangle, \tag{33}$$

and the eigenvalue equation we are trying to solve becomes

$$H|\psi\rangle = E|\psi\rangle = J\sum_\alpha c_\alpha\Big(|\alpha\rangle - \frac{1}{2}|\alpha+1\rangle - \frac{1}{2}|\alpha-1\rangle\Big) + B\sum_\alpha c_\alpha |\alpha\rangle. \tag{34}$$

A bit of rearrangement of terms puts this into the form of an equation for the c_α's:

$$(E-B)c_\alpha = J\Big(c_\alpha - \frac{1}{2}c_{\alpha+1} - \frac{1}{2}c_{\alpha-1}\Big). \tag{35}$$

This should look surprisingly familiar; it is nothing other than the translation-invariant discrete equation we solved before, for the lattice wave equation! And, of course, this is no coincidence: it is the presence of discrete translation invariance that is responsible for the form of the solution in both cases. The solution, just as it was above, is to pick c_α to be eigenfunctions $\exp(ik\alpha)$ of translation, reducing (35) to a dispersion relation:

$$E = B + J(1-\cos k) = B + 2J\sin^2(k/2). \tag{36}$$

Just as in the classical problem, we could choose to impose boundary conditions to make the problem genuinely finite. (We are then working in a subspace of dimension N, so of course there can be at most N possible eigenvalues of the energy; the allowed values of k must therefore be restricted whenever only finitely many lattice sites are present). A straightforward choice is periodic boundary conditions, which will impose the requirement that $\exp(ikN) = 1$, and thus that k is an integral multiple of $2\pi/N$, exactly as for the discrete wave equation.

At this point, a physicist would say that we've shown that the Heisenberg spin chain admits quasiparticle-like excitations: just as the plane wave solutions in a field theory of electromagnetism correspond to photons of various wavelengths, so the plane-wave solutions we've just found correspond to collective excitations of definite wavelength, which (in this context) are sometimes called *magnons*. Moreover, we've calculated their dispersion relation, which shows how the energy of a magnon is related to its wavelength. If we wanted to, we could write down "magnon creation operators" labeled by momenta:

$$a_k^\dagger = \sum_\alpha \exp(ik\alpha)s_+^\alpha, \tag{37}$$

such that $a_k^\dagger |\emptyset\rangle$ is the "one-magnon" eigenstate with parameter k that we just constructed.

Based on our experience above, we now have some idea of what to expect. In §2, we started with a free, nondispersive model: the continuum wave equation. Discretizing led us to a free, but no longer nondispersive system (the discrete wave equation). Now, we've made a truncation in the local Hilbert-space dimension, and found a similar dispersion relation—even though we have no idea what the "magnon equation of motion" is. (Of course, we could guess it from (37), observing that $\omega \propto k^2$ for long-wavelength modes.)

Is the field theory of magnons still a free theory? If it were, we would have solved it completely; its Hilbert space would be just a Fock space, generated freely by the action of all the creation operators $a_k^\dagger$ on the vacuum. Moreover, the creation operators would all mutually commute with one another, and satisfy commutation relations like

$$[H, a_k^\dagger] \stackrel{?}{=} E_k a_k^\dagger, \tag{38}$$

so that multi-magnon states like $a_k^\dagger a_{k'}^\dagger |\emptyset\rangle$ would be energy eigenstates.

But this doesn't quite work. It is easy to check that, while the magnon creation operators mutually commute, they actually fail to commute with the Hamiltonian in the way specified by (38). One computes that

$$\begin{aligned}
[H, a_k^\dagger] &= \sum_{\alpha,\beta} e^{ik\beta}\Big(-J\Big([s_z^\alpha s_z^{\alpha+1}, s_+^\beta] \\
&\qquad + \frac{1}{2}\big([s_+^\alpha s_-^{\alpha+1}, s_+^\beta] + [s_-^\alpha s_+^{\alpha+1}, s_+^\beta]\big)\Big) + B[s_z^\alpha, s_+^\beta]\Big) \\
&= \sum_{\alpha,\beta} e^{ik\beta}\Big(-J\Big(s_z^\alpha s_+^{\alpha+1}\delta^{\alpha+1,\beta} + s_+^\alpha s_z^{\alpha+1}\delta^{\alpha\beta} \\
&\qquad - s_+^\alpha s_z^{\alpha+1}\delta^{\alpha+1,\beta} - s_z^\alpha s_+^{\alpha+1}\delta^{\alpha\beta}\Big) + B\delta^{\alpha\beta}s_+^\beta\Big) \\
&= Ba_k^\dagger - J\sum_\alpha e^{ik\alpha}\big(s_z^\alpha s_+^{\alpha+1} - s_+^\alpha s_z^{\alpha+1}\big)\big(e^{ik} - 1\big).
\end{aligned} \tag{39}$$

This does not vanish, and so one can't expect that "multiparticle states" created by several successive magnon creation operators are energy eigenstates. Indeed, the only reason that single-particle states are energy eigenstates is that the operator appearing next to J in (39) acts simply on the vacuum. The s_z operators all act by $-1/2$, so that one obtains

$$[H, a_k^\dagger]|\emptyset\rangle = Ba_k^\dagger|\emptyset\rangle + \frac{J}{2}(e^{ik}-1)(e^{-ik}-1)a_k^\dagger|\emptyset\rangle = \big(B + 2J\sin^2(k/2)\big)a_k^\dagger|\emptyset\rangle, \tag{40}$$

recovering the single-particle dispersion relation, as of course one must.

Remarkably, though, one can recover the correct two-magnon eigenstates by a small, but clever, modification of the state $a_k^\dagger a_{k'}^\dagger|\emptyset\rangle$. The form of the resulting solution goes by the name of the *Bethe ansatz*, and plays a key role in the story of integrable systems. One arrives at it by remembering that, in our computation (39), the problematic term (which we want to reduce to be proportional to $a_k^\dagger$) in fact *does* work out correctly, as long as the two spins are separated from one another. The problem only appears when the spins are neighbors.

Let's write out the state as follows:

$$a_k^\dagger a_{k'}^\dagger \left|\emptyset\right\rangle = \sum_{\alpha\neq\beta} e^{i(k\alpha+k'\beta)} \left|\alpha\beta\right\rangle = \sum_{\alpha<\beta} \left(e^{i(k\alpha+k'\beta)} + e^{i(k'\alpha+k\beta)}\right) \left|\alpha\beta\right\rangle . \tag{41}$$

Bethe's insight was to consider the slightly more general state

$$\sum_{\alpha<\beta} \left(e^{i(k\alpha+k'\beta)} + Ae^{i(k'\alpha+k\beta)}\right) \left|\alpha\beta\right\rangle , \tag{42}$$

where A is a constant to be determined. One imagines that it represents an interaction of some sort, that takes place when one magnon passes through the other. By using this ansatz, it is a good exercise to check that (42) gives an eigenstate of the Hamiltonian precisely when

$$A = \frac{\cot(k/2) - \cot(k'/2) - 2i}{\cot(k/2) - \cot(k'/2) + 2i} = \exp i\theta(k, k'). \tag{43}$$

A is thus a pure phase, which is a function only of the two magnon momenta. It encodes magnon-magnon scattering. One is led to the conclusion that the particles are *almost* free: they can pass through one another without exchanging any momentum, but, in doing so, they acquire a phase shift. In the time domain, the meaning of this is that they experience a momentum-dependent time delay.

The energy of the resulting state is just the sum of the energies of the two magnons independently; that is, $E(k) + E(k')$, with E defined using the dispersion relation (36). One might therefore be tempted to say that the interaction has no effect on the energy spectrum whatsoever. But this isn't quite true! When periodic boundary conditions are imposed, the quantization conditions on allowed values of k and k' are altered by the scattering. They become

$$kN = \theta(k', k), \quad k'N = \theta(k, k') \pmod{2\pi}. \tag{44}$$

And this means that the allowed *discrete* spectrum of two-magnon states isn't quite just the set of sums of allowed energies of one-magnon states.

The truly amazing thing is that Bethe's ansatz works to find exact eigenstates when *any* number of quasiparticles are present. One simply writes down the "free-theory" ansatz analogous to (41), writing it as a sum over basis vectors with each coefficient carrying terms from all possible reorderings of the momenta. Then, one adds a phase for each pairwise interaction between particles. The result is an exact multi-magnon state, and the quantization conditions are also simple to intuitively understand. They just state that the wavefunction for each magnon must be single-valued, after accounting for the phase acquired by transporting it past all the others:

$$k_i N = \sum_{j\neq i} \theta(k_j, k_i) \pmod{2\pi}. \tag{45}$$

One can then sum the single-magnon energies for any collection of k_i permitted by the quantizability conditions, and obtain the corresponding energy eigenvalue of the system. Remarkably, the Bethe ansatz produces a *complete* solution of the model, with a pleasing physical interpretation, in terms of simple algebraic constraints (45) on parameters!

It would be easy to fill this volume (and several more besides) with material about integrable systems. So we'll have to make an abrupt exit from the subject at some point, and this is as good a time as any. Given just a bit more space, one would like to discuss how integrable models are reformulated in terms of an *algebraic* version of the Bethe ansatz, which constructs a Lax representation of the Hamiltonian and its conserved quantities, and encodes the scattering data θ in terms of an R-matrix satisfying the famous Yang–Baxter equation. Such R-matrices are closely connected to braid group representations, and therefore to the theory of anyons in two-dimensional systems. The connection continues on to knot invariants, and hence to line operators in three-dimensional TQFT (another theme of ours). But we will have to leave this bridge unbuilt, and refer instead to the extensive literature on the subject [**10**, **28**].

By way of closing, let us just mention that we earlier defined complete integrability as the existence of as many mutually compatible conserved quantities as possible, and suggested that it is often connected to the action of a symmetry algebra on the system (for example, rotational invariance in the hydrogen atom). These ideas hold true for the Heisenberg chain, which admits not just the $\mathfrak{su}(2)$ algebra we defined before, but a much larger symmetry algebra.

We'll see later, in our discussion of the toric code model, that the truly crucial feature is the presence of an algebra of non-local symmetries (which in that case are line operators). Non-local symmetries are also a crucial theme in the study of integrable systems: the additional symmetries of the Heisenberg chain act non-locally, and together constitute an algebra called the Yangian. For example, the quadratic generators of this algebra take the following form:

$$S_z^{(2)} = \sum_{\alpha<\beta} (s_+^\alpha s_-^\beta - s_-^\alpha s_+^\beta), \qquad S_\pm^{(2)} = \sum_{\alpha<\beta} (s_z^\alpha s_\pm^\beta - s_\pm^\alpha s_z^\beta). \tag{46}$$

It's an instructive exercise to show that these operators commute with the Hamiltonian, as long as periodic boundary conditions are imposed. Further commutation relations of these operators generate new, higher-order nonlocal symmetries; the Yangian is the algebra of all of these taken together. It is a deformation of the algebra of polynomials with coefficients in (the universal enveloping algebra of) $\mathfrak{su}(2)$. Further discussion of Yangian symmetry can be found, for example, in [**5**, **21**].

5. Questions to ask about systems, or, what is a phase of matter?

Having spent so long discussing different types of physical models, and the way in which they all arise from field theories by successive processes of discretization or regularization—or, to take the complementary perspective, illustrating the way in which field theory is a universal language for talking about a wide class of discrete and continuous models in a regime of their parameter space, provided that they consist of many similar degrees of freedom coupled in a spatially local and homogeneous way—we will now turn and say a few words about the kind of questions physicists might ask about the behavior of such systems. The kind of thing one would like to do is to make a kind of field guide to such systems, or otherwise understand the "zoology" of possible behaviors they can exhibit. To have some clear ways to start classifying them, one wants a list of well-defined and salient questions (invariants or characteristic features) that can tell them apart.

As we've already emphasized, field theory should primarily be thought of in a physical context as an effective or emergent description, rather than something complete and self-contained. One often doesn't know a precise description of the microscopic physics underlying the particular effective field theory one is studying, and (especially in applications to condensed matter) it is always possible that various impurities and additional interactions are present as perturbations to the system. So what is most often interesting about a field theory is its *infrared* behavior, i.e., how it behaves for modes of sufficiently long wavelength or low energy. Furthermore, one would typically only think of a characteristic feature or behavior of the theory as meaningful if it is robust against small perturbations or variations of the parameters of the system.

A first-order question of this kind would be to simply ask about the spectrum of the Hamiltonian. In general, this isn't a wildly interesting problem in field theories; since particle-like solutions typically exist with variable momenta, the spectrum of the Hamiltonian is usually continuous, at least above a certain value of the energy. But the low-lying spectrum can differ: theories exist in which the continuous spectrum extends all the way to the lowest-energy state. These are termed *gapless* theories. On the other hand, theories may also be *gapped:* that is, the vacuum may be in a discrete part of the spectrum, separated by some amount Δ from any other states, whether continuous or discrete. (A typical example is a field theory in which all particles are massive; the gap is then the mass of the lightest particle.) It is clear that the set of gapped theories is "open," in that an infinitesimal perturbation to a gapped system will result in another gapped system. Put differently, the property of being gapped is robust, in the sense we discussed above.

Of course, in a spin system, the Hilbert space is finite-dimensional, and so no portion of the spectrum can actually be continuous. So we need to interpret these ideas in the context of a thermodynamic limit. That is, we say a system is gapped if the gap persists at a finite value in the limit where the number of lattice sites grows to infinity.

We've already decided that one should consider small variations to the parameters of the Hamiltonian as insignificant from the point of view of the classification we'd like to achieve. This means that two Hamiltonians H_0 and H_1 should be considered equivalent if they're connected by a sufficiently short path H_t in the space of possible Hamiltonians. But what should "sufficiently short" mean? In the context of gapped theories, there's an immediate answer: any such path between gapped Hamiltonians can be subdivided into arbitrarily "short" equivalences between nearby gapped Hamiltonians, *unless* the gap closes at some point along the path. So the natural thing to consider is the set of path components of the collection of gapped Hamiltonians. We call such an equivalence class of gapped systems a *phase*.

How can one characterize phases? Well, the next simple thing one might ask is about a gapped theory is whether or not its vacuum state is unique. As we will discuss, in certain situations the vacuum degeneracy is an invariant of the phase. So one is led to ask about what kinds of degenerate vacua can appear in field theories or spin systems. This, of course, naturally suggests that one should study TQFT—but what kinds of field theories should one expect to have vacuum degeneracies described by interesting TQFTs?

As we mentioned earlier, degeneracies in quantum-mechanical systems are generically not present. It is often the case that the Hamiltonian alone will constitute a complete set of observables; in a family of generic Hamiltonians, one expects

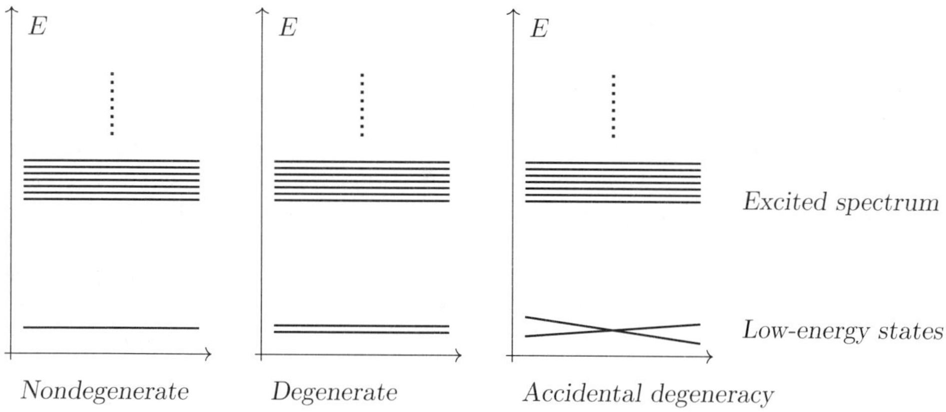

FIGURE 2. Different possible situations in gapped systems. The abscissa indicates a parameter, or possible path of nearby Hamiltonians.

degeneracies in the energy spectrum to occur only on a locus of positive codimension. Such degeneracies are sometimes termed *accidental*; for an illustration, see Fig. 2, at right. So then how can robust ground state degeneracy occur? One mechanism involves the presence of a symmetry and the phenomenon of *spontaneous symmetry breaking*.

As we have discussed above, a symmetry is present in a quantum-mechanical system when the Hamiltonian commutes with the action of some algebra. This may be an algebra of observables, corresponding to the generators of a continuous symmetry, or simply the group algebra of a discrete group of symmetries. Such a situation implies that each eigenspace of the Hamiltonian carries a representation of the algebra. It might happen that the ground state in the thermodynamic limit carries a nontrivial representation; this is termed a spontaneous breaking of the symmetry. The terminology arises because the vacuum fails to be invariant under the symmetry, even though the Hamiltonian that determines it *is* invariant. (In a sense, the solutions to the equations of motion are less symmetric than the equations of motion themselves.) When the representation on the vacuum is of dimension larger than one, spontaneous symmetry breaking results in a vacuum degeneracy.

Note that while in a finite system each eigenstate will generally carry an irreducible representation (absent accidental degeneracy), in the thermodynamic limit several irreducible representations might become degenerate in energy. In particular, the ground state might carry a reducible representation.

The prototypical phenomenon to keep in mind here is something like ferromagnetism in iron. Iron consists of a number of constituent molecules with a spin, so that each produces a magnetic field and interacts with its neighbors in a manner similar to the Heisenberg model discussed above. At high temperatures, the equilibrium state of the iron will be disordered; the interactions will be insufficient to align faraway spins, and the net magnetization will be zero. As the system is cooled, though, the role of interactions becomes more important, and below a particular critical temperature the equilibrium state of the iron will be a permanent magnet.

The spatial rotation symmetry is then spontaneously broken: the magnetization points in a particular direction in space, but the rotation group acts to produce other inequivalent equilibrium states.

One can therefore detect what phase the system is in by measuring the magnetic field: when its magnitude is zero, the system is in the disordered phase (the symmetry is unbroken). Conversely, when it is nonzero, the symmetry is broken and the system is in the spontaneously ordered phase. Observables that detect the phase of the system such as this are termed *order parameters;* it is clear from our discussion that the order parameter necessarily exhibits nonanalytic dependence on the temperature at the phase transition. Landau proposed a theory of phase transitions with precisely these ingredients: a symmetry that becomes spontaneously broken, and a locally measurable order parameter that detects the corresponding phase transition. Although simple, the Landau theory was extremely successful, predicting even the precise kind of nonanalytic behavior one should expect of the order parameter near the phase transition.

A simple but famous semiclassical model of ferromagnetic behavior[3] is the *Ising model*, which we mentioned briefly above. In this model, one imagines that the classical set of states is defined by a collection of spins taking the values ± 1; these interact with their nearest neighbors in a way that favors alignment. The system is then taken to occupy each of its states at random, with probability given by $\exp(-E/\tau)$. One thus sees that there is a competition between order and disorder. States in which all spins are aligned are the energetically most favorable, and so the most probable. On the other hand, there are enormously more states in which some number of spins are misaligned; for example, the degeneracy of a state with half spins up and half down is $\binom{N}{N/2}$, which is an incredibly large number. If the temperature is taken to infinity, all states are equally likely, so that disorder clearly triumphs. On the other hand, at zero temperature, any excitation has zero probability, so that only the ordered state is possible. That there is a sharp phase transition between these two phases in the two-dimensional Ising model is a celebrated result of Peierls (the system was later solved exactly by Onsager).

The equilibrium state in a thermodynamic system like the Ising model can be determined by minimizing the *free energy*, which is defined to be the combination $F = E - \tau S$. Here E is the energy and τ the temperature, as before, while S is the entropy, which measures disorder, or more precisely the number of different degenerate states available to the system. In Landau's theory, one writes the free energy as a function of the order parameter (here, the magnetization ϕ), and postulates that it is both analytic and invariant under the relevant symmetry. The typical model for a phase transition is then of the form

$$F(\phi) = (\tau - \tau_c)\phi^2 + \lambda\phi^4 + \dots, \tag{47}$$

where the dots represent possible higher-order terms in a Taylor series. (Here we are treating the simplest case, where ϕ is valued in $\mathbb{R}$, and the relevant symmetry is $O(1) \cong \mathbb{Z}/2\mathbb{Z}$.) One sees immediately that, even though F is perfectly analytic in ϕ and τ, the equilibrium value ϕ_0 of the order parameter (i.e. the minimum of F)

[3]For a discussion of an analogous quantum-mechanical model, the transverse field quantum Ising model, see, for example, [**25**, chapter 5].

is not: rather,

$$\phi_0 = \begin{cases} \sqrt{(\tau_c - \tau)/\lambda}, & \tau < \tau_c; \\ 0, & \tau > \tau_c. \end{cases} \tag{48}$$

Landau's simple analysis of the phase transition can be refined to include fluctuations of the order parameter; the complete framework is known as the Landau–Ginzburg–Wilson paradigm of classical phase transitions. It is worth noting that the example (47) famously recurs in elementary particle physics, in the context of spontaneously broken gauge symmetry. It is the self-interaction potential of the Higgs boson.

Successful as the Landau–Ginzburg–Wilson theory was, it was eventually realized that it did not provide a perfect classification of phases. Indeed, physicists found examples of systems in which *no* local symmetry breaking or order parameter accounted for the degeneracy. Let us say that two degenerate vacua, $|a\rangle$ and $|b\rangle$, are *locally indistinguishable* if, for any local operator $\hat{A}$,

$$\langle a| \hat{A} |b\rangle \propto \delta_{ab}. \tag{49}$$

That is, $\hat{A}$ acts in the space of vacua by an operator proportional to the identity, and so measuring it does not serve to distinguish one degenerate vacuum from another.

The study of phases with locally indistingushable robust vacuum degeneracy has given rise to the subject of *topological order*, which is more or less the topic of the conference. Physicists have had to go beyond Landau's theory of phases; in doing so, they have taken seriously the idea that the structure that controls the degenerate vacuum sector or deep infrared limit of a topologically ordered system is a nontrivial topological field theory. (For early papers on the subject that are still well worth reading, see [**31**, **32**].) In the coming section, we'll say a bit more about phases, the structures that they have, and what is known about them. Then we'll conclude with an example of a spin system with topological order, that will shed some light on how topologically ordered systems fit into the idea that robust degeneracies only arise from the action of a symmetry. In this instance, that "symmetry" will be related to the presence of *anyons* (particles with unusual exchange statistics). Indeed, topological quantum field theories in spacetime dimension three can all be thought of as arising from anyons [**24**, **29**]. And one can even often construct Hamiltonians whose vacuum sector represents such a TQFT: a construction of Levin and Wen [**20**], generalizing the toric code model, produces such a model from the data of a modular tensor category. (The Levin-Wen construction only gives a subset of topological orders, those with "gappable boundaries," but does produce all parity- and time-reversal-invariant topological phases; see [**19**, **20**] for details, as well as [**15**, **16**] for connections to extended TQFT.)

We'll see why ground state degeneracies are connected to the presence of anyons a bit later on; to foreshadow a bit, the essential idea is that *nonlocal* operators can be constructed from the worldlines of anyons, which act as an algebra and commute with the Hamiltonian. This algebra is then "spontaneously broken," resulting in a nontrivial robust vacuum degeneracy. In some sense, therefore, Landau's paradigm is still intact; one just needs to relax the assumption that the symmetry algebra consists of local operators. Versions of this idea go back to 't Hooft, who introduced line operators as order parameters in gauge theory [**27**]; see [**11**] for a recent discussion of generalized non-local symmetry.

It's also worth mentioning here that there are slightly more general notions of phase that one might consider, defined by cutting down the class of Hamiltonians one is interested in. For instance, one might insist that all Hamiltonians be time-reversal invariant, or respect some other discrete symmetry. Then the relevant question becomes to understand gapped *symmetric* Hamiltonians, regarded as equivalent if they can be connected by a path of gapped symmetric Hamiltonians. Notions like this are termed *symmetry-protected topological order* in the literature, but we won't have space to offer further discussion here; the reader is referred to the other lectures in this volume. We'll also have to give a slight refinement of our definition of a phase, that allows one to change the dimension of the local Hilbert space; we turn to this in the next section.

6. Stacking

A perspective that we've tried to highlight throughout this lecture is the crucial importance of the notion of composition of subsystems in physical theories. Other fundamental ideas, such as locality, only make sense if we know what it means for a system to be composed of subsystems and what it means for two subsystems to be noninteracting.

The reader can likely guess that there's a sense in which two separate field theories (or spin systems), living over the same space (or lattice), can be concatenated as well. This operation is sometimes called "stacking," and it's defined in the obvious way: The new local Hilbert space is the tensor product of the local Hilbert spaces of each system; the new local Hamiltonians h_α are the sums of the previous local Hamiltonians, extended by tensoring with the identity.

This gives a notion of composition on spin systems, making the collection of spin systems into an abelian monoid. How does this play with the notion of equivalence we defined before? Well, it's clear that an equivalence between systems A and A' induces an equivalence between $A \otimes B$ and $A' \otimes B$, and likewise for the other subsystem. So stacking descends to a composition law on phases. One thing to note, though, is that this composition is actually valued in a more fine space that maps down to the space of phases. Any equivalence that varies the parameters of only one subsystem is a path of *interaction-free* Hamiltonians of the composite. There are thus two questions one might ask about a spin system that is the composite of two constituent spin systems: one might ask about the path-components of the set of interaction-free Hamiltonians, or about the path-components of all Hamiltonians (ignoring the decomposition). The first maps to the second, but there is no reason the map must be either injective or surjective: there may be phases that contain no representative noninteracting Hamiltonian, and two noninteracting Hamiltonians may be connected by a path only when interactions are allowed to be turned on.

As the reader can probably guess, analogous considerations imply that phases are classified differently depending on whether or not one considers interacting systems, or just free systems—i.e., whether one considers the path components of the space of free Hamiltonians, or all Hamiltonians. Again, this map is in general neither injective nor surjective. For a recent discussion, see [**13**].

spacetime dim.	2	3	4
bosons	0	$\mathbb{Z}$ $\nu = 8$ integer QHE	0
fermions	$\mathbb{Z}/2\mathbb{Z}$ Majorana wire	$\mathbb{Z}$ $p + ip$ superconductor ($\nu = 1$ QHE represents $2 \in \mathbb{Z}$)	0

FIGURE 3. Groups of invertible phases under stacking in various dimensions.

There's also a sensible physical notion of a trivial or identity phase: we'll say that a Hamiltonian represents the trivial phase if it has a unique ground state, which is a tensor product over all local degrees of freedom. That is, it's of the form

$$|\emptyset\rangle = \bigotimes_{\alpha} |\psi\rangle . \tag{50}$$

Put differently, the local degrees of freedom in a trivial phase don't have any meaningful interaction (or correlation, or entanglement) with one another at all.

It would be nice if the "identity" phase were the identity for our composition operation, but that's not true as we've defined things so far. The essential problem is that, as we emphasized before, the local Hilbert space dimension is not an essential detail or ingredient, and different systems with different values of the local dimension may represent the same physics. But our definition of a phase only allows us to regard systems as equivalent when the local dimensions agree.

To remedy this, let's rectify our definition by making a notion of "stable equivalence:" We'll say that two spin systems are equivalent if they can be connected by a path of Hamiltonians, after either of them is tensored with any number of Hamiltonians representing the trivial phase. The trivial phase is then the identity for stacking by definition.

Now we have a unital abelian monoid of phases, and so we can ask what the group of *invertible* phases is. These are Hamiltonians H representing phases that are nontrivial, but for which a partner Hamiltonian $\bar{H}$ can be found such that $H + \bar{H}$ admits a deformation to the trivial phase. Computing these groups of invertible phases, possibly with additional subtleties such as protecting symmetry groups, is the essential subject of Dan's lectures, and we leave it to him to discuss; a table showing a couple of examples of groups of invertible phases is given in Fig. 3.

7. Topological order and the toric code

As a last invitation, we'll discuss a very famous spin system, due to Alexei Kitaev [**14**], that exhibits a protected ground-state degeneracy *without* a local order parameter. This model goes by the name of the *toric code;* while it will be discussed again elsewhere (e.g. in Dan Freed's lectures in this volume), we feel it is useful to give a physicist's perspective on the model and its features of interest, and to indicate how it fits into the bird's-eye view we have been outlining. Our presentation is slightly nonstandard, but we feel that it is the clearest and simplest way to analyze the system.

The underlying lattice of the toric code is a two-dimensional square lattice, which we'll think of as being bicolored with alternating black and white squares like a chessboard, and oriented diagonally, as shown in Fig. 4. Just as in the Heisenberg model, each vertex of the lattice carries a local Hilbert space, which is taken to be two dimensional. The algebra of observables is therefore again generated (as an algebra) by the one-site Pauli matrices σ_i^α. (As Kitaev points out [**14**], the model also makes sense on irregular lattices, and can be defined on a surface of any genus. We'll only discuss the genus-one case here.)

Consider the following operators, defined to act on the four sites surrounding either a black or a white square:

$$h_w = \prod_{\alpha \in w} \sigma_x^\alpha, \qquad h_b = \prod_{\alpha \in b} \sigma_z^\alpha. \tag{51}$$

(In what follows, the label w will always refer to a chosen white square, and b to a black one.) For our purposes here, it is important that $\sigma_i^2 = 1$, for reasons that will become clear momentarily.

It is obvious that all of the black-square and all of the white-square operators mutually commute. What is slightly less obvious is that in fact *all* of the operators (51) are mutually commuting. The case to check is that of a white square and a black square, adjacent along an edge. The important calculation (which makes a straightforward but healthful exercise) is

$$[\sigma_x^\alpha \sigma_x^\beta, \sigma_z^\alpha \sigma_z^\beta] = 0. \tag{52}$$

From this, it follows that all of the h_w and h_b can be simultaneously diagonalized! Since they square to the identity, each of these operators has possible eigenvalues ± 1; however, because the product over all w of h_w is the identity operator (and similarly for the h_b), there are two constraints on the set of the mutual eigenvalues, which insist that the number of white squares w with $h_w = -1$ is even, and likewise for black squares. (In the end, we will choose periodic boundary conditions.)

The Hamiltonian of the toric code is just a linear combination of all these mutually commuting local Hamiltonians:

$$H = -\sum_w h_w - \sum_b h_b. \tag{53}$$

We have already, in some sense, computed the spectrum of this Hamiltonian: we expect that all the local operators h can be simultaneously diagonalized, and any set of eigenvalues chosen that obey the constraints we mentioned above can be realized. But let's try to get a clearer picture of the behavior of the model, and (in particular) figure out whether or not the ground state is degenerate.

To do this, one should consider defining the following *line operators:*

$$W_\gamma = \prod_{\alpha \in \gamma'} \sigma_x^\alpha, \qquad B_{\gamma'} = \prod_{\alpha \in \gamma} \sigma_z^\alpha. \tag{54}$$

Here, the paths γ and γ' with respect to which the operators are defined, are paths of squares like those taken by chess bishops: they pass over squares of the same color, always moving diagonally. A site α belongs to a path if the colored squares on both sides of it do, i.e., if a chess bishop moving along the path of squares would have passed over that vertex. The picture in Fig. 4 should make this clear. A word of caution: the operators W are defined on paths corresponding to chess bishops on the *black* squares, and B on those on the *white* squares! The reversal occurs

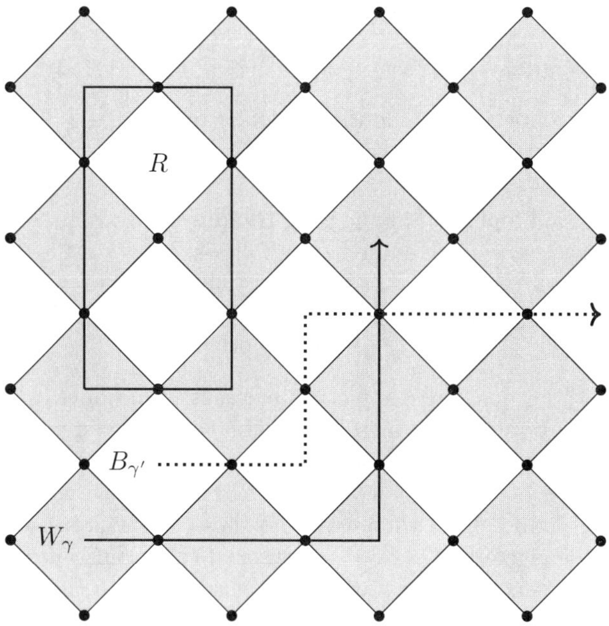

FIGURE 4. A picture of the toric code and its line operators

because the path of a chess bishop on white squares defines a partition of the black squares, and vice versa; a clash in the notation in some place cannot be avoided.[4]

When γ is taken to be the shortest possible closed path, so that it has length four, turns in a square and encloses exactly one white square, it is easy to see that $W_\gamma = h_w$, where w is the enclosed square. W_γ therefore commutes with all local Hamiltonians. Indeed, this generalizes to *any* closed path: since $\sigma_i^2 = 1$, the product of two adjacent operators $h_w h'_w$ is equal to the line operator on the closed path of length six encircling both adjacent squares. (See the region marked R in Fig. 4.) And it is trivial to see that this generalizes to

$$W_{\partial R} = \prod_{w \in R'} h_w, \qquad B_{\partial R'} = \prod_{b \in R} h_b, \tag{55}$$

for any regions R, R' and corresponding nullhomologous boundary curves.

What if the curves are open? Well, the easiest way to understand this case is to look at the commutation relations between different line operators. For this, it's useful to recall that the Pauli matrices also obey anticommutation relations of the form

$$\{\sigma_i^\alpha, \sigma_j^\beta\} = 2\delta^{\alpha\beta}\delta_{ij}. \tag{56}$$

Now, it's easy to see that all the W operators commute with one another, independent of their support (all the local operators they're built out of mutually commute), and likewise for the B operators. But what about a W_γ and a $B_{\gamma'}$? Well, suppose that the paths γ and γ' cross exactly once, at a site α. Then it's

[4]Of course, the color of the bishop doesn't correspond to the color of the squares on which it moves in chess either.

easy to see that

$$W_\gamma B_{\gamma'} = \cdots \sigma_x^\alpha \sigma_z^\alpha \cdots = -\cdots \sigma_z^\alpha \sigma_x^\alpha \cdots = -B_{\gamma'} W_\gamma. \tag{57}$$

More generally, it follows by an identical calculation that

$$W_\gamma B_{\gamma'} = (-)^{\#(\gamma\cap\gamma')} B_{\gamma'} W_\gamma. \tag{58}$$

And, by what we said before (that the local Hamiltonians are in fact particular line operators for closed curves), it immediately follows that

$$W_\gamma h_b = \begin{cases} -h_b W_\gamma, & \gamma \text{ ends at } b \\ +h_b W_\gamma, & \text{otherwise.} \end{cases} \tag{59}$$

In light of these relations, a clear picture emerges: Line operators supported on closed curves are products of local Hamiltonians, and so commute with H and all the h's. Line operators supported on open curves, on the other hand, commute with all Hamiltonians of the same color, and all those of the opposite color *except* those two on the endpoints of the curve. At those two squares, the line operator flips the sign of the eigenvalue of h. It follows that, if (for instance) W_γ acts on the ground state, a new energy eigenstate is produced in which the energy is bumped up by 2.

Energy eigenstates in the toric code, therefore, can be represented by imagining quasiparticles of two species living on the white and black squares. Each particle number is conserved modulo 2, so that we can imagine pairs of particles connected by line operators W and B on open curves. Furthermore, the particles can be moved around freely; the physical state is independent of how the line operators are drawn, and we may deform them however we like.

However, while the physical state does not depend on how the lines are drawn, its *phase* may! And this phase is not insignificant; rather, since one can imagine processes that act on the state by the corresponding phase, it is possible in principle to measure such relative phases—and they can have important consequences. For example, in the usual spin/statistics theorem, it is assumed that the physical state corresponding to a collection of indistinguishable particles is invariant under exchange of particles. However, this does not preclude the possibility that the phase of the wavefunction transforms according to a character of S_n. Since the character needs to be defined for all n, the only meaningful possibilities are the trivial or sign characters; these correspond to bosonic and fermionic particles, and have far-reaching physical consequences, including the Pauli exclusion principle.

In two spatial dimensions, though, S_n is replaced by the braid group B_n, and more interesting possibilities—*anyonic exchange statistics*—are possible. Indeed, we've constructed the phase already in the above: Imagine that there's a white and a black quasiparticle in a region of the lattice, connected by strings to distant partners. It is clear that, if the white quasiparticle is transported once around the black one, the intersection number of the strings changes by exactly one, so that the state acquires a phase -1!

So we've shown that excitations of the toric code model correspond to anyonic quasiparticles, appearing at the end of the line operators W and B. But furthermore—and this is really the punch line—when the spatial topology is nontrivial, so that there are closed curves that do not bound, the corresponding line operators commute with all local Hamiltonians. But they form an algebra that *cannot be represented nontrivially* in any energy eigenspace! If periodic boundary conditions are imposed on the lattice, and X and Y are curves (of either type γ or γ') representing a basis of the first homology, then the relevant algebra is generated by $W_{X,Y}$ and $B_{X,Y}$, subject to the relations

$$\begin{gathered} W_{X,Y}^2 = B_{X,Y}^2 = 1, \\ W_X B_Y = -B_Y W_X, \quad W_Y B_X = -B_X W_Y, \\ [W_X, B_X] = [W_Y, B_Y] = 0. \end{gathered} \tag{60}$$

And this algebra, by a simple analogue of the Stone–von Neumann theorem, admits an essentially unique four-dimensional irreducible representation. (There is no possible trivial representation, because of the presence of the nontrivial exchange phase! To put it differently, we've already demanded that a central element in our algebra be represented by the fixed nontrivial number -1.) Indeed, this algebra is essentially two commuting copies of the quantum torus algebra for $\mathbb{Z}/2\mathbb{Z}$-valued variables, and the quantum torus is itself nothing more than the exponentiated form of the canonical commutation relations $[q,p] = i\hbar$. This family of related constructions sometimes go under the name of *Heisenberg groups* [**30**, for example]; see [**12**] for a detailed discussion of related constructions in physics.

Thus, *every* eigenspace of the toric code has a fourfold degeneracy, coming from the action of the algebra of line operators wrapping nontrivial cycles. The anyonic statistics of the quasiparticles ensure that the phases in that algebra are nontrivial, and therefore that it cannot be trivially represented.

In fact, there is a much more general algebraic structure at work here: a version of the toric code can be defined for *any* finite group G, not just for $\mathbb{Z}/2\mathbb{Z}$. It is a simple spin system whose vacuum sector defines a topological quantum field theory known as *Dijkgraaf–Witten theory* [**8**]; the fields in this theory are just principal G-bundles, and so its path integral reduces to a finite sum. Just as in the case we studied, an algebra of symmetries is responsible for the vacuum degeneracy: indeed, a general Tannaka–Krein-type theorem guarantees that *any* three-dimensional TQFT has such a "quasi-quantum group" symmetry [**22**], and theorems in the converse direction allow one to build a 3d TQFT out of any appropriate symmetry algebra [**1**, **23**].

For gauge theory with finite gauge group G, the relevant algebra is the so-called *twisted Drinfeld double* of G [**9**]; this construction takes the two natural Hopf algebras associated to a finite group (functions on G and their dual algebra), associates a new Hopf algebra structure to their tensor product, and then twists the operations by the cocycle in $H^3(G, U(1))$ that defines the classical (exponentiated) action functional of the theory. The role of these algebras in Dijkgraaf–Witten theory was pointed out in [**7**].

One thus sees that we've returned to the idea that symmetries are responsible for every robust degeneracy in quantum-mechanical systems. Properly understood, a symmetry is nothing other than the action of a collection of operators commuting with the Hamiltonian. In quantum field theory, though, those operators may be local, or have extended support in spacetime [**11**]. Topological order, at least in spatial dimension two, arises when the model admits quasiparticles that admit mutually nontrivial anyonic exchange statistics. One can then consider operators defined by creating quasiparticles, transporting them around one another (or along nontrivial paths in the spatial topology) to obtain a nontrivial transformation of the state, and then annihilating them to return to the vacuum. The resulting algebras of line operators must be represented nontrivially, due to the existence of phases, and the corresponding symmetry is therefore always "spontaneously broken:" One obtains a robust, topology-dependent vacuum degeneracy, which is described by a three-dimensional topological quantum field theory. And the use of an axiomatic picture of topological quantum field theory to study the zoology of possible phases—and, in particular, to compute the groups of invertible phases that we mentioned above—is the subject of Dan's lectures in this volume. We hope we've managed to provide some small window onto why these questions were formulated in the physics literature, how a physicist might begin to think about them, and why they are of such great interest.

Acknowledgments

First and foremost, this paper could not have been written without Max Metlitski, on whose lecture at the Bozeman summer school it is in part based, and who was an invaluable help in producing the final version. The author is also deeply grateful to Ryan Grady and David Ayala, for their hard work organizing such a wonderful meeting and for the invitation to prepare this paper, and to Montana State University, for its warm hospitality during the workshop. Further thanks are due to Johannes Walcher for comments on the draft, to the anonymous referee for careful proofreading and helpful suggestions, and to Alex Turzillo for conversations. The author's work is supported in part by the Deutsche Forschungsgemeinschaft, within the framework of the Exzellenzinitiative an der Universität Heidelberg.

References

[1] Daniel Altschüler and Antoine Coste, *Quasi-quantum groups, knots, three-manifolds, and topological field theory*, Comm. Math. Phys. **150** (1992), no. 1, 83–107. MR1188498

[2] Vladimir I. Arnold, *Mathematical methods of classical mechanics*, Graduate Texts in Mathematics, vol. 60, Springer–Verlag, 1991.

[3] Rodney J. Baxter, *Exactly solved models in statistical mechanics*, Academic Press, Inc. [Harcourt Brace Jovanovich, Publishers], London, 1982. MR690578

[4] Hans Bethe, *Zur Theorie der Metalle: Eigenwerte und Eigenfunktionen der linearen Atomkette*, Zeitschrift für Physik **71** (1931), no. 3-4, 205–226.

[5] Kevin Costello, *Supersymmetric gauge theory and the Yangian*, arXiv preprint arXiv:1303.2632 (2013).

[6] Marius de Leeuw and Constantin Candu, *Spin chains*, unpublished course notes from "Introduction to Integrability," Institute for Theoretical Physics, ETH Zürich (`http://edu.itp.phys.ethz.ch/fs13/int/`), 2013.

[7] R. Dijkgraaf, V. Pasquier, and P. Roche, *Quasi Hopf algebras, group cohomology and orbifold models*, Nuclear Phys. B Proc. Suppl. **18B** (1990), 60–72 (1991), DOI 10.1016/0920-5632(91)90123-V. Recent advances in field theory (Annecy-le-Vieux, 1990). MR1128130

[8] Robbert Dijkgraaf and Edward Witten, *Topological gauge theories and group cohomology*, Comm. Math. Phys. **129** (1990), no. 2, 393–429. MR1048699

[9] V. G. Drinfel′d, *Quantum groups* (Russian, with English summary), Zap. Nauchn. Sem. Leningrad. Otdel. Mat. Inst. Steklov. (LOMI) **155** (1986), no. Differentsial′naya Geometriya, Gruppy Li i Mekh. VIII, 18–49, 193, DOI 10.1007/BF01247086; English transl., J. Soviet Math. **41** (1988), no. 2, 898–915. MR869575

[10] Jürg Fröhlich, *Statistics of fields, the Yang-Baxter equation, and the theory of knots and links*, Nonperturbative quantum field theory (Cargèse, 1987), NATO Adv. Sci. Inst. Ser. B Phys., vol. 185, Plenum, New York, 1988, pp. 71–100. MR1008276

[11] Davide Gaiotto, Anton Kapustin, Nathan Seiberg, and Brian Willett, *Generalized global symmetries*, J. High Energy Phys. **2** (2015), 172, front matter+61. MR3321281

[12] Roger Howe, *Dual pairs in physics: harmonic oscillators, photons, electrons, and singletons*, Applications of group theory in physics and mathematical physics (Chicago, 1982), Lectures in Appl. Math., vol. 21, Amer. Math. Soc., Providence, RI, 1985, pp. 179–207. MR789290

[13] Anton Kapustin, Ryan Thorngren, Alex Turzillo, and Zitao Wang, *Fermionic symmetry protected topological phases and cobordisms*, J. High Energy Phys. **12** (2015), 052, front matter+20pp. MR3464750

[14] A. Yu. Kitaev, *Fault-tolerant quantum computation by anyons*, Ann. Physics **303** (2003), no. 1, 2–30, DOI 10.1016/S0003-4916(02)00018-0. MR1951039

[15] Alexei Kitaev and Liang Kong, *Models for gapped boundaries and domain walls*, Comm. Math. Phys. **313** (2012), no. 2, 351–373, DOI 10.1007/s00220-012-1500-5. MR2942952

[16] L. Kong, *Some universal properties of Levin-Wen models*, XVIIth International Congress on Mathematical Physics, World Sci. Publ., Hackensack, NJ, 2014, pp. 444–455. MR3204497

[17] L. D. Landau and E. M. Lifshitz, *Mechanics*, Course of Theoretical Physics, Vol. 1. Translated from the Russian by J. B. Bell, Pergamon Press, Oxford-London-New York-Paris; Addison-Wesley Publishing Co., Inc., Reading, Mass., 1960. MR0120782

[18] A. J. Leggett, *Lecture 26: the Kitaev models*, unpublished course notes from "Physics 598: Physics in Two Dimensions," U.I.U.C. (`https://courses.physics.illinois.edu/phys598PTD/fa2013/`), 2013.

[19] Michael A. Levin, *Protected edge modes without symmetry*, Physical Review X **3** (2013), no. 2, 021009.

[20] Michael A. Levin and Xiao-Gang Wen, *String-net condensation: A physical mechanism for topological phases*, Physical Review B **71** (2005), no. 4, 045110.

[21] N. J. MacKay, *Introduction to Yangian symmetry in integrable field theory*, Internat. J. Modern Phys. A **20** (2005), no. 30, 7189–7217, DOI 10.1142/S0217751X05022317. MR2192353

[22] Shahn Majid, *Quasi-quantum groups as internal symmetries of topological quantum field theories*, Lett. Math. Phys. **22** (1991), no. 2, 83–90, DOI 10.1007/BF00405171. MR1122044

[23] N. Yu. Reshetikhin and V. G. Turaev, *Ribbon graphs and their invariants derived from quantum groups*, Comm. Math. Phys. **127** (1990), no. 1, 1–26. MR1036112

[24] Eric C. Rowell and Zhenghan Wang, *Mathematics of topological quantum computing*, Bull. Amer. Math. Soc. (N.S.) **55** (2018), no. 2, 183–238, DOI 10.1090/bull/1605. MR3777017

[25] Subir Sachdev, *Quantum phase transitions*, 2nd ed., Cambridge University Press, Cambridge, 2011. MR2789712

[26] Jun John Sakurai, *Modern quantum mechanics*, Addison–Wesley, 1994.

[27] G. 't Hooft, *On the phase transition towards permanent quark confinement*, Nuclear Phys. B **138** (1978), no. 1, 1–25, DOI 10.1016/0550-3213(78)90153-0. MR0489444

[28] V. G. Turaev, *The Yang-Baxter equation and invariants of links*, Invent. Math. **92** (1988), no. 3, 527–553, DOI 10.1007/BF01393746. MR939474

[29] Vladimir G. Turaev, *Quantum invariants of knots and 3-manifolds*, De Gruyter Studies in Mathematics, vol. 18, De Gruyter, Berlin, 2016. Third edition [of MR1292673]. MR3617439

[30] André Weil, *Sur certains groupes d'opérateurs unitaires* (French), Acta Math. **111** (1964), 143–211, DOI 10.1007/BF02391012. MR0165033

[31] Xiao-Gang Wen, *Vacuum degeneracy of chiral spin states in compactified space*, Physical Review B **40** (1989), no. 10, 7387.

[32] X. G. Wen, *Topological orders in rigid states*, Internat. J. Modern Phys. B **4** (1990), no. 2, 239–271, DOI 10.1142/S0217979290000139. MR1043302

[33] Bei Zeng, Xie Chen, Duan-Lu Zhou, and Xiao-Gang Wen, *Quantum information meets quantum matter—from quantum entanglement to topological phase in many-body systems*, arXiv preprint arXiv:1508.02595 (2015).

Mathematisches Institut, Ruprecht-Karls-Universität Heidelberg, 69120 Heidelberg, Germany

Email address: `saberi@mathi.uni-heidelberg.de`

Contemporary Mathematics
Volume **718**, 2018
https://doi.org/10.1090/conm/718/14478

The Arf-Brown TQFT of pin$^-$ surfaces

Arun Debray and Sam Gunningham

Abstract. The Arf-Brown invariant $AB(\Sigma)$ is an 8th root of unity associated to a surface Σ equipped with a Pin$^-$ structure. In this note we investigate a certain fully extended, invertible, topological quantum field theory (TQFT) whose partition function is the Arf-Brown invariant. Our motivation comes from the recent work of Freed-Hopkins on the classification of topological phases, of which the Arf-Brown TQFT provides a nice example of the general theory; physically, it can be thought of as the low energy effective theory of the Majorana chain, or as the anomaly theory of a free fermion in 1 dimension.

Contents

2010 *Mathematics Subject Classification.* Primary 57R56; Secondary 57R15, 81T25.

1. Introduction

As part of a general program to classify and understand topological phases of matter within condensed-matter physics, there is a large body of recent work focusing on the special case of symmetry-protected topological (SPT) phases. This classification question has been studied by many authors in different settings and with many different approaches: for lists of references, see [**GJF17**, §1] and [**FH16**, §9.3]. It is believed that the low-energy physics of SPT phases is often described by invertible topological quantum field theories (TQFTs), which admit a purely mathematical classification, and that the classification of a given class of SPTs often agrees with the classification of the analogous class of invertible TQFTs. At the same time, work on the mathematical theory of invertible TQFTs has understood their classification as a problem in stable homotopy theory [**GMTW09**, **FH16**, **SP17**]. Freed-Hopkins [**FH16**] use this to answer the classification problem across a wide range of dimensions and symmetry types.

In this paper, we explain this perspective on classifying invertible TQFTs and SPT phases in a specific setting, focusing on 2-dimensional theories formulated on manifolds with a pin$^-$ structure. Freed-Hopkins show that the group of deformation classes of 2D invertible pin$^-$ TQFTs is isomorphic to $\mathbb{Z}/8$, and is generated by a TQFT Z_{AB} whose partition function is the Arf-Brown invariant of a pin$^-$ surface, a generalization of the Arf invariant of a spin surface.

In §3, we provide three definitions for the Arf-Brown invariant, and compare each to an analogous definition of the Arf invariant: in §3.1, the original intersection-theoretic description due to Brown [**Bro71**]; in §3.2, an index-theoretic description due to Zhang [**Zha94**, **Zha17**]; and in §3.3, a new description using a twisted Atiyah-Bott-Shapiro map.

Then, in §4, we discuss how the classification of 2D invertible TQFTs reduces to a homotopy-theoretic problem. Our approach follows Freed-Hopkins [**FH16**, §5], but the fact that we're in dimension 2 allows for an explicit description of the 2-categories and homotopy 2-types that enter this argument, which are more complicated in higher dimensions. Moreover, some aspects of the story, such as the choice of a target category and the stable homotopy hypothesis, are understood in dimension 2 but not in higher dimensions. In §4.1, we review some generalities of invertible 2D TQFTs, and in §4.2 discuss the stable homotopy hypothesis in dimension 2, a theorem of Gurski-Johnson-Osorno [**GJO17**]. We use this in §4.3 to classify 2D invertible TQFTs of a given symmetry type valued in the Morita 2-category $\mathsf{sAlg}_\mathbb{C}$ of complex superalgebras. This is similar to a classification theorem of Schommer-Pries [**SP17**, Theorem 7.6] but with a different choice of target; hence in Proposition 4.21, we provide a proof for a folklore theorem identifying the homotopy type of $\mathsf{sAlg}_\mathbb{C}^\times$.

In §5, we apply this to pin$^-$ theories: the homotopy-theoretic approach to invertible TQFTs and the KO-theoretic description of the Arf-Brown invariant combine to define the Arf-Brown TQFT Z_{AB}, which was previously known to exist but hasn't been explicitly studied until now. We discuss what this theory assigns to closed pin$^-$ 0-, 1-, and 2-manifolds and how it relates to the twisted Atiyah-Bott-Shapiro orientation.

Finally, in §6, we discuss a conjectural appearance of the Arf-Brown TQFT in physics, as the low-energy theory of the Majorana chain. In §6.1, we provide some background on SPTs and the low-energy approach to their classification. We then formulate the Majorana chain on an arbitrary compact pin$^-$ 1-manifold in §6.3, and discuss its low-energy TQFT and how it relates to the Arf-Brown TQFT in §6.4. We find that the space of ground states of the Majorana chain depends on a pin$^-$ structure, which is expected, but doesn't appear to have been determined before.

We also provide some preliminaries on Clifford algebras and pin manifolds in §2.1, and on the stable homotopy theory that we use in §2.2.

2. Preliminaries

2.1. Clifford algebras, pin groups, and pin structures. Pin structures are generalizations of spin structures to unoriented vector bundles and manifolds. In this section, we define the pin groups and state a few useful results about them. For proofs and a more detailed exposition, see [**ABS64**].

DEFINITION 2.1. Let k be a field of characteristic not equal to 2, S be a finite set, and $\mathfrak{o}\colon S \to \{\pm 1\}$ be a function. The *Clifford algebra* $C\ell(k, S, \mathfrak{o})$ is defined to be the k-algebra

$$C\ell(k,S,\mathfrak{o}) := T(k[S])/(s^2 = \mathfrak{o}(s), st = -ts \mid s,t \in S, s \neq t), \tag{2.2}$$

where $T(k[S])$ denotes the tensor algebra of the space of functions $S \to k$, and we identify s with the function equal to 1 at s and 0 elsewhere.

For $S := \{1,\ldots,m\} \cup \{-1,\ldots,-n\}$ and $\mathfrak{o}(x) := \operatorname{sign}(x)$, we'll write $C\ell_{m,n}(k) := C\ell(k,S,\mathfrak{o})$, as well as $C\ell_n(k) := C\ell_{n,0}(k)$ and $C\ell_{-n}(k) := C\ell_{0,n}(k)$. If $k = \mathbb{C}$, we'll suppress $\mathbb{C}$ from the notation, e.g. writing $C\ell_{m,n}$, $C\ell_n$, and $C\ell_{-n}$.

The ideal in the quotient in (2.2) contains only even-degree elements of the tensor algebra, so the Clifford algebras are $\mathbb{Z}/2$-graded algebras, or *superalgebras*. If a is a homogeneous element in a $\mathbb{Z}/2$-graded algebra or module, we will let $|a| \in \mathbb{Z}/2$ denote its degree.

LEMMA 2.3 ([**ABS64**, Proposition 1.6]). *Let S_1 and S_2 be finite sets and $\mathfrak{o}_i\colon S_i \to \{\pm 1\}$ be functions. If $\mathfrak{o}\colon S_1 \amalg S_2 \to \{\pm 1\}$ is $\mathfrak{o}_i$ on S_i, then there is a canonical isomorphism*

$$C\ell(k,S_1,\mathfrak{o}_1) \otimes_k C\ell(k,S_2,\mathfrak{o}_2) \cong C\ell(k, S_1 \amalg S_2, \mathfrak{o}).$$

For this to be true, we must use the graded tensor product, whose multiplication contains a sign: if a, b, a', b' are homogeneous elements, then

$$(a \otimes b)\cdot(a' \otimes b') = (-1)^{|b'||a'|} aa' \otimes bb'. \tag{2.4}$$

Let $\alpha \in \operatorname{End}(C\ell(k,S,\mathfrak{o}))$ be the *grading operator*, whose action on a homogeneous element a is multiplication by $(-1)^{|a|}$.

DEFINITION 2.5. The *Clifford group* is

$$\Gamma(k,S,\mathfrak{o}) := \{x \in C\ell(k,S,\mathfrak{o})^\times \mid \alpha(x)yx^{-1} \in k[S] \subset C\ell(k,S,\mathfrak{o}) \text{ for all } y \in k[S]\}.$$

Here we use the canonical map $k[S] \hookrightarrow T(k[S]) \twoheadrightarrow C\ell(k,S,\mathfrak{o})$, which is injective.

DEFINITION 2.6. There is an involution $\beta\colon C\ell(k,S,\mathfrak{o}) \to C\ell(k,S,\mathfrak{o})$ induced from the map $\widetilde{\beta}\colon T(k[S]) \to T(k[S])$ sending a homogeneous element

$$f_1 \otimes \cdots \otimes f_n \mapsto f_n \otimes \cdots \otimes f_1.$$

The *Clifford norm* $N\colon \Gamma(k,S,\mathfrak{o}) \to k^\times$ is defined by $N(x) := \beta(x)\cdot x$.

Definition 2.7. The *pin group* $\mathrm{Pin}(k, S, \mathfrak{o})$ associated to the Clifford algebra in Definition 2.1 is the kernel of the Clifford norm. The *spin group* $\mathrm{Spin}(k, S, \mathfrak{o})$ is the subgroup of $\mathrm{Pin}(k, S, \mathfrak{o})$ which is even in the grading on the Clifford algebra.

We are interested in the case where $k = \mathbb{R}$, so that the pin and spin groups are Lie groups. If we specialize to $C\ell_{\pm n}(\mathbb{R})$, they're compact Lie groups.

Definition 2.8. Let Pin_n^+ denote the pin group associated to $C\ell_n(\mathbb{R})$ and Pin_n^- denote the pin group associated to $C\ell_{-n}(\mathbb{R})$. The corresponding spin groups are canonically isomorphic, so we denote either one by Spin_n.

Proposition 2.9. *Let* $\mathrm{Pin}_n^\pm$ *denote either of* Pin_n^+ *or* Pin_n^-. *Then, there are group extensions*

$$1 \longrightarrow \mathrm{Spin}_n \longrightarrow \mathrm{Pin}_n^\pm \xrightarrow{\pi_0} \mathbb{Z}/2 \longrightarrow 1 \tag{2.10a}$$

$$1 \longrightarrow \mathbb{Z}/2 \longrightarrow \mathrm{Pin}_n^\pm \xrightarrow{\rho} \mathrm{O}_n \longrightarrow 1. \tag{2.10b}$$

Let $\rho\colon H \to G$ be a homomorphism of Lie groups and $\pi\colon P \to M$ be a principal G-bundle. Recall that a *reduction of the structure group* of P to H is data $(\pi'\colon Q \to M, \theta)$ such that

- $\pi'\colon Q \to M$ is a principal H-bundle, and
- $\theta\colon Q \times_H G \to P$ is an isomorphism of principal G-bundles, where H acts on G through ρ.

An equivalence of reductions $(Q_1, \theta_1) \to (Q_2, \theta_2)$ is a map $\psi\colon Q_1 \to Q_2$ intertwining θ_1 and θ_2.

Definition 2.11. If $\rho\colon H \to \mathrm{GL}_n(\mathbb{R})$ is a Lie group homomorphism, an *H-structure* on a vector bundle $E \to X$ is an equivalence class of reductions of the structure group of the principal $\mathrm{GL}_n(\mathbb{R})$-bundle of frames of E to H. If M is a smooth manifold and $E = TM$, this is called a *tangential H-structure* on M; if M is a smooth manifold and E is its stable normal bundle, this is called a *normal H-structure.*

For example, an SO_n-structure is the same thing as an orientation. A *spin structure* on an n-manifold M is a tangential Spin_n structure, and we define pin^+ and pin^- structures analogously.

Remark 2.12. We note that such structures are *stable* in the following sense: a $(\mathrm{s})\mathrm{pin}^\pm$ structure on a vector bundle V is equivalent to a $(\mathrm{s})\mathrm{pin}^\pm$ structure on $V \oplus \underline{\mathbb{R}}$. In particular, a stable framing on a vector bundle (a trivialization of $V \oplus \underline{\mathbb{R}}^N$ for some N) determines a $(\mathrm{s})\mathrm{pin}^\pm$ structure.

Proposition 2.13 ([**KT90**, Lemma 1.3]). *Let $E \to X$ be a vector bundle and $w_n(E) \in H^n(X; \mathbb{Z}/2)$ denote its nth Stiefel-Whitney class.*

- *E admits a spin structure iff $w_1(E) = 0$ and $w_2(E) = 0$.*
- *E admits a pin^+ structure iff $w_2(E) = 0$.*
- *E admits a pin^- structure iff $w_2(E) + w_1(E)^2 = 0$.*

In all cases, if E admits one of these structures, the set of such structures (in the spin case, with fixed orientation) on E is an $H^1(X; \mathbb{Z}/2)$-torsor.

Corollary 2.14. *Let M be a closed manifold of dimension at most 2. Then, M has a pin^- structure, and has a spin structure if and only if it is orientable. If* $\dim M = 2$, *then M has a pin^+ structure iff its Euler characteristic is even.*

REMARK 2.15. There are a few facts about pin structures which might be surprising to a reader who has only studied spin manifolds. A tangential spin structure is equivalent data to a normal spin structure, but this is false for pin structures: a tangential pin$^+$ structure is equivalent to a normal pin$^-$ structure, and vice versa. This is discussed in [**KT90**, §1], and will be relevant in our homotopical approach to 2D pin$^-$ TQFTs.

The product of spin manifolds has an induced spin structure, but using Proposition 2.13 one can write down pin$^-$ manifolds whose product doesn't have a pin$^-$ structure; the same phenomenon occurs for pin$^+$ structure. This means that the pin$^+$ and pin$^-$ cobordism groups are not rings, though they are modules over the spin cobordism ring.

PROPOSITION 2.16 ([**KT90**, §2]). *Let Ω_n^H denote the cobordism group of n-manifolds with tangential H-structure. Then:*

(1) *There are isomorphisms $\Omega_1^{\mathrm{Spin}} \cong \mathbb{Z}/2$ and $\Omega_1^{\mathrm{Pin}^-} \cong \mathbb{Z}/2$, and the forgetful map $\Omega_1^{\mathrm{Spin}} \to \Omega_1^{\mathrm{Pin}^-}$ is an isomorphism. Both are generated by the circle with structure induced by its Lie group framing.*
(2) *There are isomorphisms $\Omega_2^{\mathrm{Spin}} \cong \mathbb{Z}/2$ and $\Omega_2^{\mathrm{Pin}^-} \cong \mathbb{Z}/8$ which identify the forgetful map $\Omega_2^{\mathrm{Spin}} \to \Omega_2^{\mathrm{Pin}^-}$ with the map sending $1 \mapsto 4$. The torus with spin structure induced from its Lie group framing generates Ω_2^{Spin}, and $\mathbb{RP}^2$ (with either of its two pin$^-$ structures) generates $\Omega_2^{\mathrm{Pin}^-}$.*

In particular, the two isomorphism classes of pin$^-$ circles aren't cobordant: one bounds and the other doesn't. We denote the bounding pin$^-$ circle by S_b^1, and the nonbounding pin$^-$ circle by S_{nb}^1. This applies *mutatis mutandis* to the two spin circles.

2.2. Homotopy theory. We follow the conventions in [**BC18**]. If the reader is unfamiliar with spectra and the stable homotopy category we recommend they first read Section 2 of loc. cit. Here, we briefly recall some notation, basic definitions, and examples.

2.2.1. *The (stable) homotopy category.*

- The unstable homotopy category is denoted $h\mathcal{S}$. This category receives a map from the category $\mathcal{T}op$ of topological spaces and continuous maps which takes weak equivalences in $\mathcal{T}op$ to isomorphisms in $h\mathcal{S}$; by abuse of notation we will denote the image of a topological space in $h\mathcal{S}$ by the same symbol, and refer to the objects of $h\mathcal{S}$ as spaces.
- We write $[X, Y]$ for the space of morphisms in $h\mathcal{S}$ between spaces X and Y; if we choose "nice enough" representatives for X and Y (for example CW-complexes), then this is given by the set of homotopy classes of maps between X and Y.
- We denote by $h\mathcal{S}p$ the stable homotopy category (also known as the homotopy category of spectra). This category receives a functor from the category of prespectra which takes weak equivalences of prespectra to isomorphisms in $h\mathcal{S}p$. As in the unstable case, we will refer to objects of $h\mathcal{S}p$ simply as spectra, and use the same symbol for a prespectrum and its corresponding object in $h\mathcal{S}p$.
- Given a pair of spectra E, F, we write $[E, F]$ for the set of morphisms $\mathrm{Hom}_{h\mathcal{S}p}(E, F)$, and $[E, F]_n$ for $[\Sigma^n E, F]$; if E and F are "nice enough"

(for example, if they are *CW-spectra* – prespectra such that each space is a CW complex, the structure maps are cellular inclusions, and the adjoints of the structure maps are homeomorphisms) then $[E, F]$ is given by homotopy classes of maps between spectra. There are natural abelian group structures on $[E, F]$ and $[E, F]_n$.

- The category hSp carries a symmetric monoidal structure $\wedge$, called the smash product. There is also a mapping object (internal hom) $\mathrm{Map}(E, F)$ whose homotopy groups are $\pi_n(\mathrm{Map}(E, F)) = [E, F]_n$.

2.2.2. *Examples of spectra.*

EXAMPLE 2.17. Given a pointed space X, we have the *suspension spectrum* $\Sigma^\infty X$, which may be presented by a prespectrum whose nth space is $\Sigma^n X$. A special case of this construction is the *sphere spectrum* $\mathbb{S} := \Sigma^\infty S^0$, which is the unit object for the smash product.

Given a spectrum E and a space X, we write $E^n(X)$ for the *E-cohomology group* $[\Sigma^\infty X, E]_n$. Similarly, we write $E_n(X)$ for the *E-homology group* $\pi_n(\Sigma^\infty X \wedge E)$. These are examples of *generalized cohomology* (resp. *homology*) *theories*: they satisfy all of the Eilenberg-Steenrod axioms except the dimension axiom.

REMARK 2.18. The Brown representability theorem [**Bro62**] states that any generalized cohomology theory h^* (resp. generalized homology theory h_*) arises from a spectrum E in this manner; we say that h^* (resp. h_*) is *represented by E*.

EXAMPLE 2.19. Ordinary cohomology with coefficients in an abelian group A is represented by the *Eilenberg-MacLane spectrum HA*. This may be modeled as a spectrum whose nth space is the Eilenberg-MacLane space $K(n, A)$ for $n > 0$, and whose nonpositive spaces are trivial.

Complex K-theory is a generalized cohomology theory represented by a spectrum denoted KU. Similarly, real K-theory is represented by a spectrum KO.

A spectrum is called *connective* if it has trivial negative homotopy groups. Given a connective spectrum E, its zeroth space[1] has the structure of an infinite loop space. In fact, the homotopy theory of connective spectra is equivalent to that of infinite loop spaces: given an infinite loop space, by definition it has a sequence of deloopings which form the spaces in the corresponding spectrum. See Adams [**Ada78**] for more on this correspondence.

2.2.3. *Thom spaces and Thom spectra.* Given a space X and a vector bundle $V \to X$, one may form the *Thom spectrum* $\tau(V)$, which is the suspension spectrum of the Thom space. Thom spectra satisfy the property

$$\tau(V \oplus W) \simeq \tau(V) \wedge \tau(W).$$

We extend this definition to include *virtual* vector bundles (formal differences of vector bundles). For example, there is a Thom spectrum $\tau(-V)$. Explicitly, one can find another bundle W such that $V \oplus W \simeq \underline{\mathbb{R}}^N$ for some N; then, $\tau(-V) \simeq \Sigma^{-N}\tau(W)$.

The classifying space BO_n of the nth orthogonal group carries a universal vector bundle $\gamma_n := EO_n \times_{\mathrm{O}_n} \mathbb{R}^n \to BO_n$. We denote by MO_n the Thom spectrum $\tau(\gamma_n)$; taking the colimit, we have a spectrum MO, represented by a prespectrum whose

[1]Here it is essential that one considers a spectrum rather than just a prespectrum (i.e. the adjoints of the structure maps are homeomorphisms).

nth space is the Thom space of γ_n. One defines analogous spectra MH_n and MH for any family of groups H_n with compatible homomorphisms $H_n \to \mathrm{O}_n$ (such as SO_n, Spin_n, and $\mathrm{Pin}_n^\pm$).

The *Madsen-Tillmann* spectrum MTO_n is defined as the Thom spectrum of the virtual bundle $-\gamma_n$. There are maps $\Sigma^n MTO_n \to \Sigma^{n+1} MTO_{n+1}$, and the direct limit is denoted MTO. One defines MTH_n and MTH analogously for groups H_n as above.

The reason we care about Thom spectra is that their homotopy groups compute cobordism.

THEOREM 2.20 (Thom [**Tho54**], Pontrjagin [**Pon55**]). *Let $\rho_n\colon H_n \to \mathrm{O}_n$ be a compatible family of Lie group homomorphisms. Then there are isomorphisms*

- $\pi_n(MTH) \cong \Omega_n^H$*, the cobordism group of n-manifolds with tangential H_n-structure, and*
- $\pi_n(MH) \cong \Omega_n^{\nu H}$*, the cobordism group of n-manifolds with normal H_n-structure.*

Given a tangential structure $\rho_n\colon H_n \to \mathrm{O}_n$, there is a *perpendicular* tangential structure $\rho_n^\perp\colon H_n^\perp \to \mathrm{O}_n$ such that a tangential H_n-structure is equivalent data to a normal $H_n^\perp$-structure [**FH16**, §7.1]. In this case, there's a weak equivalence $MH^\perp \simeq MTH$. For example, $\mathrm{O}^\perp = \mathrm{O}$, $\mathrm{Spin}^\perp = \mathrm{Spin}$, and $(\mathrm{Pin}^\pm)^\perp = \mathrm{Pin}^\mp$, so there are weak equivalences $MO \simeq MTO$, $MSpin \simeq MTSpin$, $MPin^+ \simeq MTPin^-$, and $MPin^- \simeq MTPin^+$.

Thom spectra are also useful for understanding duality in hSp.

THEOREM 2.21 (Atiyah [**Ati61**]). *Let M be a closed manifold. Then $\Sigma^\infty M$ is dualizable in hSp, and its dual is the Thom spectrum of the stable normal bundle ν of M.*

We won't say very much about duality, but we note in particular that if B is a spectrum with dual $B^\vee$, then for any spectra A and C there is a weak equivalence

$$\mathrm{Map}(A \wedge B, C) \simeq \mathrm{Map}(A, B^\vee \wedge C). \tag{2.22}$$

For more on duality, see [**Ada74**, §III.5].

3. The Arf-Brown invariant of a pin$^-$ surface

In this section, we give various constructions of the Arf-Brown invariant of a pin$^-$ surface: intersection theoretic in §3.1, index-theoretic in §3.2, and KO-theoretic in §3.3.

3.1. Intersection-theoretic descriptions of the invariants. The Arf invariant of a spin surface and the Arf-Brown invariant of a pin$^-$ surface are complete cobordism invariants defined using intersection theory.

3.1.1. *The Arf invariant of a spin surface.* Let Σ be a closed oriented surface. If $x, y \in H_1(\Sigma; \mathbb{Z}/2)$, then the mod 2 intersection number $I_2(x, y) \in \mathbb{Z}/2$ is defined by choosing smooth, transverse representative curves for x and y and computing the number of points mod 2 in their intersection. This does not depend on the choice of representatives and defines a non-degenerate bilinear pairing

$$I_2\colon H_1(\Sigma; \mathbb{Z}/2) \otimes H_1(\Sigma; \mathbb{Z}/2) \to \mathbb{Z}/2.$$

A $\mathbb{Z}/2\mathbb{Z}$*-quadratic enhancement* of I_2 is a quadratic form on $H_1(\Sigma;\mathbb{Z}/2)$ whose induced bilinear form is I_2. Explicitly, this is a function

$$q\colon H_1(\Sigma;\mathbb{Z}/2) \to \mathbb{Z}/2$$

such that for all $x, y \in H_1(\Sigma;\mathbb{Z}/2)$,

$$q(x+y) = q(x) + q(y) + I_2(x,y). \tag{3.1}$$

The set of $\mathbb{Z}/2\mathbb{Z}$-quadratic enhancements of I_2 is an $H^1(\Sigma;\mathbb{Z}/2)$-torsor: given a $\gamma \in H^1(\Sigma;\mathbb{Z}/2)$ and a quadratic enhancement $q\colon H_1(\Sigma;\mathbb{Z}/2) \to \mathbb{Z}/2$, the function $q_\gamma(x) := q(x) + \gamma(x)$ is again a quadratic enhancement.

We have the following relationship between spin structures and quadratic enhancements of the intersection form.

THEOREM 3.2 ([**Joh80, Ati71**]). *There is an isomorphism of $H^1(\Sigma;\mathbb{Z}/2)$-torsors between the set of $\mathbb{Z}/2\mathbb{Z}$-quadratic enhancements of I_2 and isomorphism classes of spin structures on Σ.*

REMARK 3.3. Given a spin structure on Σ, the associated quadratic form is easy to describe: it takes a homology class represented by an embedded circle to either 0 or 1 depending on whether the induced spin structure on the circle is bounding or non-bounding.

DEFINITION 3.4. Given a spin surface Σ with corresponding quadratic form q, the Arf invariant of q may be defined as follows. If $\{e_i, f_i\}$ is a basis of $H_1(\Sigma;\mathbb{Z}/2)$ which is symplectic with respect to the intersection form, then

$$\mathrm{Arf}(\Sigma) := \sum_i q(e_i)q(f_i) \in \mathbb{Z}/2. \tag{3.5}$$

THEOREM 3.6 ([**KT90**]). *The Arf invariant is a spin bordism invariant, and defines an isomorphism*

$$\mathrm{Arf} : \Omega_2^{\mathrm{Spin}} \to \mathbb{Z}/2. \tag{3.7}$$

EXAMPLE 3.8. Let $T = S^1 \times S^1$ denote the torus with spin structure afforded by the Lie group framing. Consider the symplectic basis $\{e, f\}$ for $H_1(T;\mathbb{Z}/2)$ corresponding to the embedded circles $S^1 \times \{1\}$ and $\{1\} \times S^1$. As each circle carries the non-bounding spin structure, the associated quadratic form q takes the values $q(e) = q(f) = 1$. Thus the Arf invariant is $1 \in \mathbb{Z}/2$, and hence T is a generator for the spin bordism group.

3.1.2. *The Arf-Brown invariant of a pin$^-$ surface.* Now suppose Σ is any closed surface (not necessarily oriented). Then $H_1(\Sigma;\mathbb{Z}/2)$ still carries a non-degenerate intersection form I_2, although $H_1(\Sigma;\mathbb{Z}/2)$ may be odd dimensional, and will not admit a symplectic basis in general. In this case, one must consider the following notion:

DEFINITION 3.9. A $\mathbb{Z}/4$*-quadratic enhancement of the intersection form* on Σ is a function

$$q\colon H_1(\Sigma;\mathbb{Z}/2) \to \mathbb{Z}/4 \tag{3.10}$$

such that for all $x, y \in H_1(\Sigma;\mathbb{Z}/2)$,

$$q(x+y) = q(x) + q(y) + 2\cdot I_2(x,y), \tag{3.11}$$

where $(2\cdot)\colon \mathbb{Z}/2 \hookrightarrow \mathbb{Z}/4$ is inclusion.

As with $\mathbb{Z}/2$-quadratic enhancements, the set of $\mathbb{Z}/4$-quadratic enhancements is an $H^1(\Sigma;\mathbb{Z}/2)$-torsor: given a $\gamma \in H^1(\Sigma;\mathbb{Z}/2)$ and a quadratic enhancement $q\colon H_1(\Sigma;\mathbb{Z}/2) \to \mathbb{Z}/4$, the function $q_\gamma(x) := q(x) + 2\cdot\gamma(x)$ is again a $\mathbb{Z}/4$-quadratic enhancement.

THEOREM 3.12 ([**KT90**]). *For any closed surface Σ, there is an isomorphism of $H^1(\Sigma;\mathbb{Z}/2)$-torsors from the set of pin$^-$ structures on Σ to the set of $\mathbb{Z}/4$-quadratic enhancements of the intersection form on Σ.*

DEFINITION 3.13 ([**Bro71, KT90**]). Let Σ be a pin$^-$ surface and let $q\colon H_1(\Sigma;\mathbb{Z}/2) \to \mathbb{Z}/4$ be its associated quadratic enhancement. The *Arf-Brown invariant* of Σ is the unit complex number

$$AB(\Sigma) := \frac{1}{\sqrt{|H_1(\Sigma;\mathbb{Z}/2)|}} \sum_{x\in H_1(\Sigma;\mathbb{Z}/2)} \exp\left(\frac{2\pi i q(x)}{4}\right). \tag{3.14}$$

This is sometimes called the *Kervaire invariant* or the *Arf-Brown-Kervaire invariant.*

THEOREM 3.15 ([**Bro71, KT90**]). *The Arf-Brown invariant $AB(\Sigma)$ is a pin$^-$ bordism invariant, and defines an isomorphism*

$$AB : \Omega_2^{\mathrm{Pin}^-} \to \mu_8 \cong \mathbb{Z}/8 \tag{3.16}$$

where μ_8 denotes the group of eighth roots of unity.

EXAMPLE 3.17. Let us compute the value of the Arf-Brown invariant for the pin$^-$ structures on $\mathbb{RP}^2$. In that case $H_1(\mathbb{RP}^2;\mathbb{Z}/2) \cong \mathbb{Z}/2$, and there are two quadratic enhancements of the intersection form which take the image of the non-zero homology class to either 1 or 3 mod 4. In the first case, we see that the Arf-Brown invariant is $\exp(\frac{2\pi i}{8})$, and in the second $\exp(\frac{-2\pi i}{8})$. It follows that either structure gives a generator for pin$^-$ bordism.

If Σ is an oriented surface, then a $\mathbb{Z}/4$-quadratic enhancement is necessarily valued in the even elements of $\mathbb{Z}/4$, and thus recovers the $\mathbb{Z}/2$-quadratic enhancement corresponding to a spin structure. Moreover, the Arf-Brown invariant of such a quadratic enhancement is the (exponentiated) Arf invariant of the corresponding quadratic form.

3.2. Index-theoretic description of the invariants. The Arf(-Brown) invariant of a spin (or pin$^-$) surface admits an alternative description in terms of Dirac operators acting on sections of (s)pinor bundles. In the spin case, the Arf invariant corresponds to the mod 2 index or Atiyah invariant of a spin Riemann surface – the mod 2 dimension of the space of holomorphic sections of a theta-characteristic. In the pin$^-$ case, the Arf-Brown invariant may be interpreted as the reduced η-invariant of a twisted Dirac operator as defined and studied by Zhang [**Zha94, Zha17**].

3.2.1. *The Atiyah invariant of a spin surface.* Let Σ be a closed surface equipped with a Riemannian metric and a spin structure.[2] Then Σ carries a graded spinor bundle

$$S_\Sigma = P_{\mathrm{Spin}_2} \times_{\mathrm{Spin}_2} C\ell_{-2}(\mathbb{R}) \tag{3.18}$$

[2]Here, we consider a Riemannian metric to induce a *negative* definite quadratic form on the fibers of T^*_Σ.

with a left action of the bundle of Clifford algebras $C\ell(T^*_\Sigma)$ and a commuting right action of the constant algebra $C\ell_{-2}(\mathbb{R})$. This bundle splits as a sum of its graded components $S^0_\Sigma \oplus S^1_\Sigma$, where each S^i_Σ carries a fiberwise action of $C\ell^0_{-2}(\mathbb{R}) \cong \mathbb{C}$ and thus may be considered as a complex line bundle.

There is a Dirac operator

$$D^+_\Sigma : C^\infty(S^0_\Sigma) \to C^\infty(S^1_\Sigma)$$

given by composing the canonical connection operator

$$\nabla : C^\infty(S_\Sigma) \to C^\infty(T^*_\Sigma \otimes S_\Sigma)$$

with the action of sections of T^*_Σ via Clifford multiplication.

DEFINITION 3.19. The *Atiyah invariant* of the spin surface Σ is

$$\dim \ker(D^+_\Sigma) \bmod 2 \in \mathbb{Z}/2.$$

REMARK 3.20. The Riemannian metric and orientation on Σ determine a complex structure, and the even spinor bundle S^0_Σ defines a square root of the holomorphic cotangent bundle (such a square root is known as a *theta characteristic*). The Dirac operator is identified with the $\overline{\partial}$ operator defining the holomorphic structure on S^0_Σ. Thus the Atiyah invariant of Σ is the mod 2 dimension of the space of holomorphic sections of S^+_Σ.

PROPOSITION 3.21 ([**Joh80**]). *Given a closed spin surface Σ, the Atiyah invariant and Arf invariant coincide.*

3.2.2. *Reduced η-invariant of a pin$^-$ surface.* Given a pin$^-$ surface Σ, the pinor bundle

$$S_\Sigma = P_{\mathrm{Pin}_2^-} \times_{\mathrm{Pin}_2^-} C\ell_{-2}(\mathbb{R})$$

still makes sense, though it doesn't carry a natural $\mathbb{Z}/2$-grading. However, as Clifford multiplication is not Pin_2^--equivariant, the formula for the Dirac operator now defines a map on sections

$$D_\Sigma : C^\infty(S_\Sigma) \to C^\infty(S_\Sigma \otimes \delta),$$

where δ is the orientation bundle.

To get an operator acting on sections of the same bundle, we may apply the following trick (which is spelled out in [**Sto88**] in the analogous case of a pin$^+$ manifold of dimension 4 mod 8). The left regular action of $C\ell_{-2}(\mathbb{R}) \cong \mathbb{H}$ on itself extends to an action of $C\ell_{-3}(\mathbb{R}) \cong \mathbb{H} \oplus \mathbb{H}$; now compose the operator D_Σ with the action of $e_3 \in C\ell_{-3}(\mathbb{R})$ to get a self-adjoint operator $\widetilde{D}_\Sigma$ on sections S_Σ called the *twisted Dirac operator.*

The twisted Dirac operator has an associated η-function, defined for $s \in \mathbb{C}$ with $\mathrm{Re}(s) \gg 0$ by the formula

$$\eta_{\widetilde{D}}(s) = \sum_{\lambda \neq 0} \mathrm{sign}(\lambda) \dim E(\lambda) |\lambda|^{-s}$$

where $E(\lambda)$ is the eigenspace with eigenvalue λ. This function admits a meromorphic extension to $s = 0$, and thus we may define the *reduced η-invariant*:

$$\overline{\eta}(\widetilde{D}) = \frac{\dim \ker(\widetilde{D}) + \eta_{\widetilde{D}}(0)}{2} \bmod 2\mathbb{Z} \in \mathbb{R}/2\mathbb{Z}.$$

One can check (see [**Zha17**, Proposition 2.7]) that if $\Sigma = \partial M$ is the boundary of a pin$^-$ 3-manifold M, the reduced eta invariant of Σ is zero (i.e. the non-reduced invariant is an even integer). Thus for any pin$^-$ surface Σ, the above quantity is a well-defined cobordism invariant (and in particular is independent of the choice of metric used to define it).

PROPOSITION 3.22 ([**Zha94, Zha17**]). *Given a pin$^-$ surface Σ, the reduced η-invariant $\overline{\eta}(\widetilde{D}_\Sigma)$ is an element of $\mathbb{Z}[\frac{1}{4}]/2\mathbb{Z}$, and agrees with the Arf-Brown invariant of Σ under the isomorphism given by the exponential map $\mathbb{Z}[\frac{1}{4}]/2\mathbb{Z} \cong \mu_8 \subseteq \mathbb{C}^\times$.*

REMARK 3.23. In the spin case, the contribution from the η-invariant vanishes (as the spectrum is symmetric), and we are just left with half the dimension of $\ker(D_\Sigma)$ (or equivalently the dimension of $\ker(D_\Sigma^+)$) mod 2.

3.3. KO-theoretic descriptions of the invariants. Here we explain how the analytic index-theoretic invariants of the previous section may be expressed topologically in terms of pushforwards in (twisted) KO-theory.

3.3.1. *The Atiyah-Bott-Shapiro orientation and pushforward maps in KO-theory.* Let $\pi : V \to X$ be a rank-k real vector bundle equipped with a spin structure, and let $\tau(V)$ denote its Thom space. The Clifford module construction of Atiyah-Bott-Shapiro [**ABS64**] determines a Thom isomorphism

$$KO^*(X) \xrightarrow{\cong} \widetilde{KO}^{*+k}(\tau(V)). \tag{3.24}$$

This isomorphism is given by multiplication by a Thom class $u_V \in \widetilde{KO}^n(\tau(V))$, which may be described as follows. The spin structure on V determines a graded spinor bundle $S_V = S_V^0 \oplus S_V^1$ (see (3.18)), which carries a left action of the Clifford bundle $C\ell(V)$. Pulling S_V back to the total space of V, we obtain a pair of bundles together with a homomorphism

$$\pi^*(S_V^1) \to \pi^*(S_V^0)$$

given by Clifford multiplication, which is an isomorphism away from the zero section. This defines an element of $KO(V, V - \{0\}) \cong \widetilde{KO}(\tau(V))$ which is the required Thom class.

Using the Thom isomorphism, we can define a pushforward for an n-dimensional spin manifold M. Choose an embedding $M \to \mathbb{R}^N$ for some large N, and let $\nu \to M$ be the normal bundle, which has rank $N - n$. Using the tubular neighborhood theorem to embed $\nu \hookrightarrow \mathbb{R}^N$, consider the Pontrjagin-Thom collapse map

$$PT_\nu : S^N = \mathbb{R}^N \cup \{\infty\} \to \tau(\nu)$$

which takes the complement of the tubular neighborhood in S^N to the basepoint of the Thom space.

DEFINITION 3.25. The *pushforward map in KO-theory* for X is the composition

$$\pi_!^M : KO^*(M) \xrightarrow[(3.24)]{} \widetilde{KO}^{*+N-n}(\tau(\nu)) \xrightarrow{PT_\nu^*} \widetilde{KO}^{*+N-n}(S^N) \xrightarrow{s} KO^{*-n}(\mathrm{pt}), \tag{3.26}$$

where s is the suspension isomorphism.

One may check that this invariant does not depend on the choice of embedding for large enough N. Moreover, by considering the appropriate modification for manifolds with boundary, one can check that the association of $\pi_!^M(1_M) \in KO^{-n}(\mathrm{pt})$ to a closed spin n-manifold is a cobordism invariant.

REMARK 3.27. Another perspective on the pushforward in KO-theory is that a closed spin n-manifold M carries a fundamental class $[M] \in KO_n(M)$, Spanier-Whitehead dual to the Thom class in $u_M \in \widetilde{KO}^{-n}(\Sigma^{-N}\tau(\nu))$. We can then pushforward to get a class in $KO_n(\mathrm{pt}) = KO^{-n}(\mathrm{pt})$.

The collection of Thom classes given by the Atiyah-Bott-Shapiro orientation may be succinctly formulated as a morphism of spectra (in fact of E_∞-ring spectra [**Joa04, AHR10**])

$$\widehat{A}\colon \mathit{MSpin} \to \mathit{KO}. \tag{3.28}$$

The Pontrjagin-Thom construction can then be interpreted as a homomorphism (in fact, isomorphism) from the spin cobordism groups Ω_n^{Spin} to the homotopy groups $\pi_n(\mathit{MSpin})$, as in Theorem 2.20; the induced map on homotopy groups

$$\Omega_n^{\mathrm{Spin}} \cong \pi_n(\mathit{MSpin}) \to \pi_n(\mathit{KO}) = KO^{-n}(\mathrm{pt})$$

takes the class of a closed spin n-manifold M to the pushforward $\pi_!^M(1_M)$.

With the pushforward in hand, we can give a description for the Arf invariant of a spin surface which is simpler, if less intuitive, than the one given in the previous section.

PROPOSITION 3.29 ([**Ati71**]). *Let Σ be a spin surface. The Atiyah/Arf invariant of Σ is the pushforward of* 1*:*

$$A'(\Sigma) := \pi_!^\Sigma(1) \in KO^{-2}(\mathrm{pt}) \cong \mathbb{Z}/2. \tag{3.30}$$

REMARK 3.31. From this perspective, the Arf invariant is an example of a generalized characteristic number, specifically a *KO-Pontrjagin number* as constructed by Anderson-Brown-Peterson [**ABP66**].

3.3.2. *Twisted Thom isomorphism for pin⁻ manifolds.* Next we discuss the generalization of this invariant to pin^- surfaces. One issue is that pin^- vector bundles are not oriented for KO, so it is not immediately clear how to define a pushforward.

The key idea is the observation that a $\mathrm{pin}^\pm$ structure on a vector bundle V is equivalent to a spin structure on the virtual vector bundle $V \mp \det(V)$ (see [**KT90**]). In particular, we have a Thom class

$$u_{V\mp\det(V)} \in KO^n(\tau(V \mp \det(V)))$$

Now if M is a closed pin^- n-manifold with an embedding in $\mathbb{R}^N$ for $N \gg 0$, its normal bundle ν is equipped with a pin^+ structure, so we have a corresponding Thom class

$$u_M \in KO^{N-n-1}(\tau(\nu - \delta)) = KO^{N-n}(\tau(\nu + 1 - \delta).$$

Alternatively, by Spanier-Whitehead duality, there is a fundamental class

$$[M] \in KO_n(\tau(\delta - 1)).$$

These ideas are best understood from the perspective of twisted KO-theory.

DEFINITION 3.32. Given a space X equipped with a map $w : X \to BO_1$ (which we may think of as classifying either a double cover X^w or a real line bundle L_w), we define the *twisted KO-cohomology*

$$KO^{w+n}(X) := KO^n(\tau(L_w - 1)).$$

Similarly, we have the *twisted KO-homology*

$$KO_{w+n}(X) := KO_n(\tau(1 - L_w)).$$

REMARK 3.33. This construction is an example of the notion of a twisted generalized cohomology theory. This and other examples can be found in [**FHT11**] (e.g. see Example 2.28) and [**ABG10**].

3.3.3. *Pushforward in twisted KO-homology.* Using this language, we may now give a KO-theoretic construction of the Arf-Brown (or reduced η-invariant) of a pin$^-$ surface Σ. Let

$$w_1 : \Sigma \to BO_1$$

denote the classifying map of the orientation line bundle. Then consider the push-forward

$$w_{1!}([\Sigma]) \in KO_{2+w}(BO_1). \tag{3.34}$$

The group

$$KO_{2+w}(BO_1) = KO_2(\Sigma^{-1}MO_1) \cong KO_3(\mathbb{RP}^\infty)$$

is isomorphic to the direct limit of cyclic 2-groups $\mathbb{Z}/2^\infty$, or equivalently, the collection of all 2^nth roots of unity in $\mathbb{C}^\times$. One may also apply the above construction for the connective theory ko, in which case (see [**BG10**])

$$ko_{2+w}(BO_1) \cong \mathbb{Z}/8$$

and thus the class $w_{1!}([\Sigma])$ may be interpreted as an 8th root of unity via the exponential map.

PROPOSITION 3.35. *The assignment $M \mapsto w_{1!}([M])$ defines an isomorphism*

$$\Omega_2^{\mathrm{Pin}^-} \xrightarrow{\sim} ko_{2+w}(BO_1)$$

and thus after a suitable choice of isomorphism $ko_{2+w}(BO_1) \simeq \mu_8$, we may identify $w_{1!}([M])$ with the Arf-Brown invariant.

REMARK 3.36. The advantage of this perspective is that it naturally occurs as the induced map on homotopy groups of a map of spectra. Namely, the observation from §3.3.2 relating pin$^-$ structures and spin structures leads to the identification

$$MTPin^- \simeq MPin^+ \simeq MSpin \wedge \Sigma^{-1}MO_1. \tag{3.37}$$

Thus smashing the Atiyah-Bott-Shapiro orientation with the factor $\Sigma^{-1}MO_1$ leads to a map of spectra

$$MTPin^- \to ko \wedge \Sigma^{-1}MO_1. \tag{3.38}$$

The class $w_{1!}([M])$ as defined in (3.34) is precisely the image of the class in $\Omega_2^{\mathrm{Pin}^-}$ defined by M.

PROOF OF PROPOSITION 3.35. The morphism $\Omega_2^{\mathrm{Pin}^-} \to ko_{2+w}(BO_1)$ given by $M \mapsto w_{1!}([M])$ is just the induced morphism on the homotopy groups π_2 associated to the map of spectra (3.38). To check that we get an isomorphism on π_2, note that the original Atiyah-Bott-Shapiro map (3.37) is 8-connected (see [**ABP67**]) and thus it remains so after taking the smash product with the connective spectrum $\Sigma^{-1}MO_1$, in particular inducing an isomorphism on π_2 as required. □

REMARK 3.39. One may define an invariant in $\mathbb{Z}/8$ associated to any vector bundle on Σ by first considering the twisted Poincaré duality isomorphism

$$KO^0(\Sigma) \to KO_{2+w_1}(\Sigma),$$

then taking the pushforward in twisted KO-homology as before.

We now mention a few closely related KO-theoretic constructions of the Arf-Brown invariant that appear in the literature.

3.3.4. *Zhang's construction.* Given a pin$^-$ surface Σ, the classifying map of the orientation double cover

$$w_1 : \Sigma \to BO_1 \simeq \mathbb{RP}^\infty$$

is homotopic to a map which factors through the two skeleton $\mathbb{RP}^2$. Let $\widetilde{w_1}$ denote the corresponding map $\Sigma \to \mathbb{RP}^2$. The stable normal bundle of $\widetilde{w_1}$ is a virtual vector bundle of rank 0 which carries a spin structure (after choosing a pin$^-$ structure on $\mathbb{RP}^2$), and thus there is a well-defined pushforward map

$$KO^*(\Sigma) \to KO^*(\mathbb{RP}^2).$$

In [**Zha17**], Zhang defines the topological index of a vector bundle V on Σ as the image of $f_!V$ under a certain homomorphism

$$KO^0(\mathbb{RP}^2) \cong \mathbb{Z} \oplus \mathbb{Z}/4 \to \mathbb{Z}[\frac{1}{4}]/2 \cong \mathbb{Z}/8.$$

It is shown in loc cit. that the topological index agrees with the reduced η-invariant of a twisted Dirac operator on V. In particular, the topological index of the trivial bundle agrees with the Arf-Brown invariant.

One can check that Zhang's construction of the topological index agrees with the one in Remark 3.39 (first observe that in each case, the index factors through $KO^0(\mathbb{RP}^2)$, then check that the morphism to $\mathbb{Z}/8$ is the same).

3.3.5. *Distler-Freed-Moore construction.* Distler-Freed-Moore [**DFM10**] give a slightly different KO-theoretic construction. In order to state it, we will need to consider two modifications of the cohomology theory KO: first, we consider the Postnikov truncation of the connective cover $R := KO\langle 0,1,2,3,4\rangle$; then we take R-cohomology with coefficients in $\mathbb{R}/\mathbb{Z}$ (this is represented by a spectrum $R(\mathbb{R}/\mathbb{Z})$ which fits in to an exact triangle $R \to R \wedge H\mathbb{R} \to R(\mathbb{R}/\mathbb{Z})$). Moreover, the authors consider a twisted version of this theory, as explained above.

It is shown in [**DFM10**] that $R^{w_1-2}(BO_1;\mathbb{R}/\mathbb{Z})$ is cyclic of order 8. Given a closed pin$^-$ surface Σ, the Arf-Brown invariant is obtained as the image of a generator under the morphisms

$$R^{w_1-2}(BO_1;\mathbb{R}/\mathbb{Z}) \xrightarrow{w_1^*} R^{w_1-2}(\Sigma;\mathbb{R}/\mathbb{Z}) \xrightarrow{\pi_!^\Sigma} R^{-4}(\mathrm{pt};\mathbb{R}/\mathbb{Z}) \cong \mathbb{R}/\mathbb{Z} \xrightarrow{\exp} \mathbb{C}^\times.$$

4. Invertible TQFTs via stable homotopy theory

In this section, we give a brief exposition of TQFTs leading to the classification of invertible TQFTs in terms of homotopy theory. For concreteness we will focus on the case of 2-dimensional TQFTs. The notion of fully extended TQFT requires some ideas from higher category theory, which we will discuss only at an informal level. However, when considering invertible theories, we will see that the higher categorical aspects may be reformulated in the (perhaps) more familiar terms of stable homotopy theory.

4.1. What is a (2D) invertible TQFT?.

4.1.1. *Atiyah-Segal Axioms.* In [**Ati88**] (see also [**Seg88**]) an n-dimensional TQFT is axiomatized as a symmetric monoidal functor

$$Z : \mathsf{Bord}_{n,n-1} \to \mathsf{Vect}_{\mathbb{C}}.$$

The source category has objects given by closed $(n-1)$-manifolds, and morphisms are cobordisms between them. Thus a TQFT Z will assign a vector space $Z(Y^{n-1})$ to a closed $(n-1)$-manifold, and a linear map

$$Z(X) : Z(Y_1) \to Z(Y_2)$$

whenever X is a cobordism between Y_1 and Y_2. Moreover, this assignment is compatible with the symmetric monoidal structures on each side: disjoint union of manifolds is taken to tensor product of vector spaces. In particular, the empty $(n-1)$-manifold $\emptyset^{n-1}$ is the unit object for the symmetric monoidal structure on $\mathsf{Bord}_{n,n-1}$, and thus we can identify $Z(\emptyset^{n-1})$ with the trivial line $\mathbb{C}$. Given a closed n-manifold, X^n, we may interpret it as a cobordism between empty $(n-1)$-manifolds, and thus we obtain a number $Z(X) \in \mathbb{C}$; this is referred to as the *partition function* of the theory.

In order to study the Arf-Brown TQFT, we will need a number of variations, extensions, and simplifications of these axioms.

4.1.2. *Fully extended TQFTs.* A 2-dimensional TQFT as defined above may be thought of as an assignment of an invariant $Z(X) \in \mathbb{C}$ to every closed 2-manifold X, which satisfies a certain locality property. Namely, $Z(X)$ may be computed by decomposing X in to pairs of pants and discs. We will be interested in *fully extended TQFTs* which satisfy a stronger form of locality, allowing the partition function to be computed by decomposing X in to arbitrarily small pieces (for example, a triangulation).

One way to make this precise involves replacing the ordinary category $\mathsf{Bord}_{2,1}$ with a certain 2-category Bord_2. In this paper, we will use the term 2-category to denote what some authors would call a *weak* 2-*category*, or *bicategory*. We will treat the subject in an informal and expository manner—a 2-category should have objects, 1-morphisms, and 2-morphisms which may be composed in various ways; for further details, see [**Bén67, SP17**]. In our case, the objects of Bord_2 are now closed 0-manifolds, 1-morphisms are 1-dimensional cobordisms, and 2-morphisms are diffeomorphism classes of certain 2-manifolds with corners, interpreted as cobordisms between 1-dimensional cobordisms.

Similarly we must replace the category $\mathsf{Vect}_{\mathbb{C}}$ with an appropriate 2-category of coefficients for the theory. One choice is the Morita 2-category $\mathsf{Alg}_{\mathbb{C}}$, whose objects are algebras, 1-morphisms are bimodules, and 2-morphisms are bimodule homomorphisms. As in the usual Atiyah-Segal axioms, these 2-categories carry symmetric monoidal structures (for more details on symmetric monoidal 2-categories, see [**Shu10**]). A fully extended 2D TQFT may be formulated as a symmetric monoidal functor

$$Z : \mathsf{Bord}_2 \to \mathsf{Alg}_{\mathbb{C}}.$$

Note that the original categories $\mathsf{Bord}_{1,2}$ and $\mathsf{Vect}_{\mathbb{C}}$ sit as the endomorphism categories of the unit objects in Bord_2 and $\mathsf{Alg}_{\mathbb{C}}$ respectively, and thus a fully extended TQFT gives rise a TQFT as in Atiyah and Segal's original definition.

4.1.3. *Tangential structure.* As the Arf-Brown invariant is an invariant of pin$^-$ surfaces, to define the Arf-Brown TQFT, we must consider a variant of the bordism category in which all the manifolds are equipped with a pin$^-$ structure. More generally, for any Lie group H equipped with a homomorphism $H \to O_2$, there is a 2-category Bord_2^H defined as above, but now all manifolds are equipped with H-structures.[3]

REMARK 4.1 (Structure groups). In the cases of interest, the group H is usually part of a family H_n, $n \in \mathbb{Z}_{>0}$, equipped with maps $H_n \to \mathrm{O}_n$. For example, we could take H_n to be SO_n, Spin_n, $\mathrm{Pin}_n^{\pm}$, or O_n itself. These examples all have the additional property that an H_n-structure on a vector bundle V is equivalent to an H_{n+1}-structure on $V \oplus \underline{\mathbb{R}}$; thus we may think of an H_n structure as an H-structure on the corresponding stable vector bundle (where $H = \varinjlim H_n$). In that case, we will generally write Bord_2^H instead of $\mathsf{Bord}_2^{H_2}$. We will also consider the framed case $H_n = 1$; however, note that a framing on an n-dimensional bundle is not determined by a stable framing in general.

REMARK 4.2. Given a symmetric monoidal 2-category $\mathcal{C}$, we say that $\mathcal{C}$ *has duals* if every object of $\mathcal{C}$ admits a dual with respect to the monoidal structure and every 1-morphism of $\mathcal{C}$ admits an adjoint. The bordism 2-categories $\mathsf{Bord}_2^{H_2}$ all have duals in this sense. Moreover, one version of the cobordism hypothesis [**Lur09**] states that the framed bordism category Bord_2^{fr} (obtained by taking $H_2 = 1$ above) is universal for this property, in the following sense: given any symmetric monoidal 2-category with duals $\mathcal{C}$, symmetric monoidal functors $\mathsf{Bord}_2^{fr} \to \mathcal{C}$ are in correspondence with objects of $\mathcal{C}$ (where the correspondence assigns to a functor, the value of the framed 0-manifold pt_+).

4.1.4. *Gradings.* To allow for more interesting theories we can also enlarge the coefficient category by considering algebras and bimodules with a $\mathbb{Z}/2\mathbb{Z}$-grading (and morphisms compatible with this grading). This gives rise to a 2-category $\mathsf{sAlg}_{\mathbb{C}}$ (the 2-category of superalgebras), which is equipped with a symmetric monoidal structure incorporating the Koszul rule of signs. Later we will see that this choice of target category is universal in a certain sense (see Corollary 4.22).

[3]One defines an H-structure on a 0- or 1-manifold by taking the direct sum of the tangent bundle with a trivial bundle of appropriate rank (so the total rank is two). Or better, one can consider the manifolds appearing in the bordism category as being equipped with 2-dimensional collars. Making this precise in the smooth category takes some care; see [**SP17**] for a careful treatment in the 2-dimensional case.

4.1.5. *Examples.*

EXAMPLE 4.3 (Euler Theories). There is a TQFT Z_1 defined on unoriented manifolds which assigns only identity objects and morphisms in $\mathsf{sAlg}_{\mathbb{C}}$. In particular, the partition function takes the constant value 1. Slightly more interesting is the *Euler theory* Z_λ, associated to a complex number $\lambda \in \mathbb{C}^\times$. This agrees with the trivial theory on 0 and 1-manifolds, but assigns the number $\lambda^{\chi(X)}$ (considered as a linear map between trivial lines) to any 2-dimensional cobordism.

EXAMPLE 4.4. [2-dimensional Dijkgraaf-Witten Theory [**DW90**, **FQ93**, **FHLT10**]] Given a finite group G, there is a 2D oriented TQFT whose partition function on a closed surface is the weighted sum

$$Z_G(\Sigma) = \sum_{[P]} \frac{1}{|\operatorname{Aut}(P)|}.$$

This theory assigns the group algebra of G to a point, and the space of class functions to a circle.

EXAMPLE 4.5 (The Arf-Brown Theory). We will see in 5.0.1 that the Arf-Brown invariant is the partition function of a 2D fully extended TQFT. In other words, there is a symmetric monoidal functor

$$Z_{AB} : \mathsf{Bord}_2^{\mathrm{Pin}^-} \to \mathsf{sAlg}_{\mathbb{C}}$$

such that for a pin$^-$ surface X, $Z_{AB}(X) = AB(X)$. The Arf-Brown theory takes the following values on lower dimensional manifolds:

- For a bounding pin$^-$ circle S^1_b, $Z_{AB}(S^1_b) \cong \mathbb{C}$, an even line.
- For a non-bounding pin$^-$ circle S^1_{nb}, $Z_{AB}(S^1_{nb}) \cong \mathbb{C}[1]$, an odd line.
- We have $Z_{AB}(\mathrm{pt}) \cong C\ell_1$, the first Clifford algebra.

4.1.6. *Invertible theories.* The Euler theories and the Arf-Brown theory have the property that every value of the functor Z is invertible (either as an object with respect to the monoidal structure, or as a 1- or 2-morphism) in the symmetric monoidal 2-category $\mathsf{sAlg}_{\mathbb{C}}$ (for example, Dijkgraaf-Witten theory does not have this property unless $G = 1$). Such TQFTs are called *invertible.*

When dealing with invertible theories we may restrict attention to the following class of 2-categories:

DEFINITION 4.6. A 2-category is called a *2-groupoid* if every 1-morphism and 2-morphism is invertible. A symmetric monoidal 2-category $\mathcal{C}$ is called a *Picard 2-groupoid* if it is a 2-groupoid and in addition every object is invertible with respect to the monoidal structure.

Note that if a symmetric monoidal 2-groupoid $\mathcal{C}$ has duals, then it is necessarily a Picard 2-groupoid, i.e. the duals of objects must be inverses.

Given a symmetric monoidal 2-category $\mathcal{C}$, we may consider the maximal subcategory $\mathcal{C}^\times$ which is a Picard 2-groupoid (i.e. throw out any non-invertible objects and morphisms). A TQFT

$$Z : \mathsf{Bord}_2^{H_2} \to \mathsf{sAlg}_{\mathbb{C}}$$

is invertible if and only if Z factors through $\mathsf{sAlg}_{\mathbb{C}}^\times \to \mathsf{sAlg}_{\mathbb{C}}$.

There is another way to associate a 2-groupoid ${}^{\mathrm{loc}}\mathcal{C}$ to a 2-category $\mathcal{C}$ by formally inverting any non-invertible 1 and 2-morphisms. This procedure is left adjoint to the inclusion of 2-groupoids in to 2-categories (the maximal 2-groupoid is right adjoint). In other words, any functor from $\mathcal{C}$ to a 2-groupoid will factor uniquely through ${}^{\mathrm{loc}}\mathcal{C}$. Moreover, if $\mathcal{C}$ is symmetric monoidal and has duals, then ${}^{\mathrm{loc}}\mathcal{C}$ will be a Picard 2-groupoid. We note the following upshot: the data of an invertible (fully extended) H_2-TQFT is given by a functor of Picard 2-groupoids

$$Z : {}^{\mathrm{loc}}\mathsf{Bord}_2^{H_2} \to \mathsf{sAlg}_{\mathbb{C}}^{\times}.$$

4.2. The homotopy hypothesis and stable 2-types. Now let us explain how an invertible TQFT may be reformulated in terms of maps in the stable homotopy category.

4.2.1. *The fundamental 2-groupoid and Postnikov truncations.* To begin with let us recall the following 2-category associated to a space.

EXAMPLE 4.7. Let X be a topological space. The *fundamental 2-groupoid* $\pi_{\leq 2}(X)$ of X is given as follows:

- the objects are points of X,
- the 1-morphisms are paths between points, and
- the 2-morphisms are given by homotopy classes of homotopies between paths.

To understand what information about a space the fundamental 2-groupoid captures, let us recall the following:

DEFINITION 4.8. We say that a topological space X is a *homotopy n-type* if the homotopy groups $\pi_i(X)$ are non-zero only if $i = 0, 1, \dots n$. Given any space X, there is a Postnikov fibration

$$f_{\leq n} : X \to X_{\leq n}$$

where $X_{\leq n}$ is an n-type and $f_{\leq n}$ induces an isomorphism on π_i for $i = 0, 1, \dots n$. We refer to $X_{\leq n}$ as the *(Postnikov) n-truncation* of X, or simply the *n-type* of X.

4.2.2. *Unstable homotopy hypothesis.* The idea of the homotopy hypothesis (formulated by Grothendieck in "Pursuing Stacks") is that the homotopy theory of n-types should be modeled by n-groupoids. There are many forms the homotopy hypothesis might take, depending on what flavor of higher category theory one considers. In some cases, the result is almost tautological (if ones uses an inherently homotopical theory of higher category), and in others it is false (if one uses a too strict a higher category theory).

Here's one form of the homotopy hypothesis in the case $n = 2$.[4]

THEOREM 4.9 (2-dimensional homotopy hypothesis [**Noo07**, Prop. 9.8]). *The assignment from a homotopy 2-type to its fundamental 2-groupoid defines an equivalence of categories between the homotopy category of 2-types and the category of 2-groupoids, with morphisms equivalence classes of 2-functors.*

The inverse functor is given by the *classifying space* of a 2-groupoid. In general, the classifying space of a 2-category assigns a space $|\mathcal{C}|$ to a 2-category $\mathcal{C}$ (see [**Dus02**]), constructed as the geometric realization of a certain simplicial set—the *nerve* of $\mathcal{C}$.

[4]We thank the referee for finding a reference for this theorem.

REMARK 4.10. The classifying space of a 2-category $\mathcal{C}$ is weakly equivalent to the classifying space of its localization ${}^{\mathrm{loc}}\mathcal{C}$ (in fact, one may construct the localization by taking the fundamental 2-groupoid of its classifying space).

4.2.3. *Stable homotopy hypothesis.* A symmetric monoidal structure on a 2-category $\mathcal{C}$ induces a binary operation on the classifying space $|\mathcal{C}|$. This operation is commutative and associative up to homotopy; more precisely, it can be shown that $|\mathcal{C}|$ carries an E_∞ structure. If $\mathcal{C}$ has duals (for example, if it is a Picard 2-groupoid), then every object has an inverse up to homotopy, and the E_∞ structure is said to be grouplike.

Foundational results in homotopy theory (see [**Ada78**, §2.3] and the references therein, or [**Lur17**, §5.1.3] for a modern approach) state that a grouplike E_∞-structure on a space X is equivalent to an infinite loop space structure on X. Equivalently, $X = \Omega^\infty E$ may be identified as the zeroth space in a connective spectrum E (the other spaces in the spectrum are given by iterated deloopings or loop spaces of X).

DEFINITION 4.11. A *stable n-type* is a connective spectrum E such that $\Omega^\infty E$ is a homotopy n-type.

Thus we arrive at the *stable homotopy hypothesis*, which states that there is an equivalence between:

- The homotopy category stable 2-types.
- The category of Picard 2-groupoids with equivalence classes of symmetric monoidal functors.

Thanks to recent work of [**GJO17**] this is now a precisely formulated theorem.

4.3. Classifying invertible TQFTs up to isomorphism. One consequence of the stable homotopy hypothesis is that we may encode a 2D H_2-TQFT as a morphism of stable 2-types:

$$|Z| : |\mathsf{Bord}_2^{H_2}| \to |\mathsf{sAlg}_{\mathbb{C}}^{\times}|.$$

Thus to classify isomorphism classes of invertible TQFTs, we should classify homotopy classes of maps between stable 2-types as above.

4.3.1. *Stable Postnikov data.* Let us first unpack the case of a stable 1-type, which according to the appropriate version of the homotopy hypothesis, is equivalent data to a Picard groupoid [**JO12**, Theorem 1.5].

A stable 1-type E may be succinctly encoded in terms of the following data:

- a pair of abelian groups $A = \pi_0 E$ and $B = \pi_1 E$, and
- a homomorphism $k : A/2A \to B$.

The homomorphism k encodes the k-invariant

$$HA \to \Sigma^2 HB$$

which defines the Postnikov tower of E.

REMARK 4.12 ([**GJOS17**, §3]). The homomorphism k may also be identified with the action of the generator $\eta \in \pi_1(\mathbb{S}) \cong \mathbb{Z}/2$ on the homotopy groups of E (see Remark 4.19).

PROPOSITION 4.13 ([**JO12**]). *Given a Picard 1-groupoid $\mathcal{C}$, we recover the above data as follows:*

- $\pi_0(|\mathcal{C}|)$ *is the set of equivalence classes of objects in $\mathcal{C}$, with group structure coming from the symmetric monoidal structure.*
- $\pi_1(|\mathcal{C}|)$ *is the set of automorphisms of the unit object in $\mathcal{C}$.*
- *Given an object $a \in \mathcal{C}$, the element $k(a) \in \pi_1(|\mathcal{C}|)$ may be identified with the image of the symmetry map $\sigma \in \mathrm{Aut}(a \otimes a)$ under the equivalence $\mathrm{Aut}(1_{\mathcal{C}}) \simeq \mathrm{Aut}(a \otimes a)$ induced by the functor $- \otimes (a \otimes a)$.*

Similarly, a stable 2-type may be described by the following data:

$$\begin{array}{ccccc} \Sigma^2 HC & \xrightarrow{i_2} & E = E\langle 0,1,2\rangle & & \\ & & \downarrow & & \\ \Sigma^1 HB & \xrightarrow{i_1} & E\langle 0,1\rangle & \xrightarrow{k_1} & \Sigma^3 HC \\ & & \downarrow & & \\ \Sigma HA & \xrightarrow{i_0} & E\langle 0\rangle & \xrightarrow{k_0} & \Sigma^2 HB \end{array}$$

As we have seen considering stable 1-types, the maps $k_0 i_0$ and $k_1 i_1$ are classified by homomorphisms

$$A/2A \to B \tag{4.14}$$

and

$$B/2B \to C. \tag{4.15}$$

The map i_0 is an equivalence, so k_0 is determined by $k_0 i_0$. However, these data do not fix the homotopy class of k_1 itself and thus do not fix the homotopy type of E in general.[5]

4.3.2. *Bordism and Madsen-Tillmann Spectra.* Given a Lie group H_n with a map $H_n \to \mathrm{O}_n$, Galatius-Madsen-Tillmann-Weiss [**GMTW09**] define the Madsen-Tillmann spectrum MTH_n as the Thom spectrum of the virtual vector bundle $-\gamma_n$ over BH_n (where γ_n is pulled back from the tautological vector bundle on $B\mathrm{O}_n$). Building on the fundamental work of [**GMTW09, MW07**], Schommer-Pries [**SP17**] shows that the classifying space of $\mathsf{Bord}_2^{H_2}$ is given by the Postnikov 2-truncation of $\Omega^{\infty-2} MTH_2$.[6]

Let us also consider the following variant of the bordism 2-category. Now suppose H_n is a family of groups as in Remark 4.1. Associated to such data we have the *stable bordism 2-category*, a symmetric monoidal 2-category $\mathsf{Bord}_{2,st}^{H}$ defined in a similar way to $\mathsf{Bord}_2^{H_2}$, except the manifolds are now equipped with a H-structure on their stabilized tangent bundles (the direct limit of the sequence obtained by taking iterated direct sums with the trivial vector bundle), and crucially that 2-morphisms are given by 2-cobordisms modulo the relation given by 3-cobordisms (as opposed to diffeomorphism as before).

[5]For example, consider the Postnikov truncation $ku_{\leq 2}$ of connective complex K-theory. We have that $\pi_1(ku) = 0$, so both (4.14) and (4.15) are necessarily zero; however, there is a non-zero k-invariant $H\mathbb{Z} \to \Sigma^3 H\mathbb{Z}$.

[6]More generally, in [**SP17**] it is shown that the classifying space of the (∞, n)-category $\mathsf{Bord}_n^{H_n}$ is $\Sigma^n MTH_n$.

Remarkably, the stable bordism 2-category is already a Picard 2-groupoid (the duals in the usual cobordism 2-category are inverses modulo cobordism). Its classifying space is the stable 2-type associated to the spectrum MTH, the direct limit of spectra $\Sigma^n MTH_n$, which represents the cohomology theory defined by H-cobordism of manifolds: $\pi_i(MTH) = \Omega_i^H$ consists of smooth closed i-manifolds, with a H-structure on the stable tangent bundle, up to H-cobordism.

EXAMPLE 4.16. In the case $H = 1$, we have the *framed* stable 2D-bordism category, $\mathsf{Bord}_{2,st}^{fr}$. Pontrjagin-Thom theory identifies the classifying space of this category with the stable 2-type of the sphere spectrum $\mathbb{S}$. The homotopy groups are given by

$$\pi_0(\mathbb{S}) \cong \mathbb{Z}, \pi_1(\mathbb{S}) \cong \mathbb{Z}/2, \pi_2(\mathbb{S}) \cong \mathbb{Z}/2.$$

The generator η of $\pi_1(\mathbb{S})$ is represented by the stably framed manifold S^1 with its Lie group framing, and the generator η^2 of $\pi_2(\mathbb{S})$ is represented by $S^1 \times S^1$, also with its Lie group framing (see Remark 4.19).

Let H_n be a family of groups as in Remark 4.1. There is a canonical functor

$$\mathsf{Bord}_2^{H_2} \to \mathsf{Bord}_{2,st}^{H}.$$

DEFINITION 4.17. An H_2-TQFT

$$Z : \mathsf{Bord}_2^{H_2} \to \mathsf{sAlg}_{\mathbb{C}}$$

is called *stable* if it factors through $\mathsf{Bord}_{2,st}^{H}$.

EXAMPLE 4.18. We have described the homotopy groups of

$$|\mathsf{Bord}_{2,st}^{\mathrm{Pin}^-}| \simeq MTPin^-\langle 0,1,2\rangle$$

in Proposition 2.16; in particular, recall that $\pi_2(MTPin^-) = \Omega_2^{\mathrm{Pin}^-} \simeq \mathbb{Z}/8$. By the results of [**SP17**, **GMTW09**], there is an equivalence:

$$|\mathsf{Bord}_2^{\mathrm{Pin}^-}| \simeq \Sigma^2 MTPin_2^-\langle 0,1,2\rangle.$$

According to [**RW14**], we have $\pi_2(\Sigma^2 MTPin_2^-) \cong \mathbb{Z} \oplus \mathbb{Z}/4$, and the map

$$\mathsf{Bord}_2^{\mathrm{Pin}^-} \to \mathsf{Bord}_{2,st}^{\mathrm{Pin}^-}$$

induces the map $\mathbb{Z}\oplus\mathbb{Z}/4 \to \mathbb{Z}/8$ which takes (a, b) to $a+2b \mod 8$. The projection on to the first factor is represented by the Euler characteristic. In particular, we see that the Euler theory Z_λ of Example 4.3 is stable if and only if $\lambda = \pm 1$.

REMARK 4.19. The Thom spectra MO, $MSpin$, and $\mathbb{S}$ carry natural E_∞-ring structures; geometrically, the ring operations corresponds to direct product of manifolds. Moreover, any spectrum is a module spectrum for the sphere spectrum in a unique way (just as any any abelian group is a module for the integers). In particular, the graded ring $\bigoplus_i \pi_i(\mathbb{S})$ acts on $\bigoplus_i \pi_i(E)$ for any spectrum E. In the case E is a bordism spectrum then this action may be understood in terms of direct products of appropriately structured manifolds. Note however, that $MPin^\pm$ are not ring spectra – only module spectra for $MSpin$.

4.3.3. *Brown-Comenetz duality and invertible superalgebras.* We define a spectrum $I\mathbb{C}^\times$, which is a variant of the Brown-Comenetz dual of the sphere spectrum [**BC76**] (in the original construction $\mathbb{Q}/\mathbb{Z}$ was used in place of $\mathbb{C}^\times$). Abstractly, the spectrum $I\mathbb{C}^\times$ may be defined up to homotopy in terms of its corresponding cohomology theory: for a spectrum E we have

$$(I\mathbb{C}^\times)^i(E) = \pi_i(E)^\vee = \mathrm{Hom}(\pi_i(E), \mathbb{C}^\times),$$

where we write $A^\vee$ to denote the Pontrjagin dual of an abelian group A. By construction, we have

$$\pi_i(I\mathbb{C}^\times) = (I\mathbb{C}^\times)^{-i}(\mathbb{S}) = \pi_{-i}(\mathbb{S})^\vee$$

(in particular, $I\mathbb{C}^\times$ has vanishing homotopy groups in positive degree).

Consider the shifted spectrum $\Sigma^k I\mathbb{C}^\times$. This is characterized by the following universal property among spectra E:

$$[E, \Sigma^k I\mathbb{C}^\times] \cong \mathrm{Hom}(\pi_k(E), \mathbb{C}^\times) = \pi_k(E)^\vee.$$

In particular, a character $c_k \in \pi_k(E)^\vee$ determines morphisms for each positive integer i:

$$c_{k-i} : \pi_{k-i}(E) \to \pi_{k-i}(\Sigma^k I\mathbb{C}^\times) = \pi_i(\mathbb{S})^\vee.$$

Unwinding the definitions, we see that these morphisms are computed by the action of $\pi_*(\mathbb{S})$ on $\pi_*(E)$:

LEMMA 4.20. *Given E and $c_k \in \pi_k(E)^\vee$ as above, we have*

$$c_{k-i}(x)(\xi) = c_k(\xi \cdot x)$$

for all $x \in \pi_{k-i}(E)$, $\xi \in \pi_i(\mathbb{S})$.

The following key result identifies the Picard 2-groupoid corresponding to $\Sigma^2 I\mathbb{C}^\times$.

PROPOSITION 4.21. *The stable 2-type $|\mathsf{sAlg}_{\mathbb{C}}^\times|$ is equivalent to the connective cover of $\Sigma^2 I\mathbb{C}^\times$.*

We give a proof of this result in 4.3.5 below.

As an immediate consequence of Proposition 4.21 we obtain the following notable result:

COROLLARY 4.22. *Any character $c : \pi_2(MTH_2) \to \mathbb{C}^\times$ arises as the partition function of a unique invertible TQFT*

$$Z_c \colon \mathsf{Bord}_2^{H_2} \to \mathsf{sAlg}_{\mathbb{C}}.$$

EXAMPLE 4.23. Continuing with Example 4.18, we see that isomorphism classes of invertible pin$^-$ TQFTs are given by

$$(\mathbb{Z} \oplus \mathbb{Z}/4)^\vee = \mathbb{C}^\times \times \mu_4.$$

On the other hand, the stable theories are represented by the cyclic subgroup generated by (ζ_8, ζ_4) (where ζ_4, resp. ζ_8, denotes a primitive 4th, resp. 8th, root of unity).

4.3.4. *Deformation classes of theories and the Freed-Hopkins classification.* The natural computation from the perspective of topological phases is not to do with *isomorphism classes* of theories, but rather *deformation classes.* Informally this may be understood in terms of replacing the 2-category $\mathsf{sAlg}_{\mathbb{C}}$ with an appropriate topological 2-category in which $\mathbb{C}$ is considered with its continuous topology. More precisely, one considers the *Anderson dual* spectrum $I\mathbb{Z}$ which may be defined as the homotopy fiber of the map $H\mathbb{C} \to I\mathbb{C}^\times$ given by the exponential map. In particular, there is a map

$$|\mathsf{sAlg}_{\mathbb{C}}^\times| \simeq \Sigma^2 I\mathbb{C}^\times \to \Sigma^3 I\mathbb{Z},$$

which can be understood as passing from the discrete to the continuous topology on $\mathbb{C}$. Deformation classes of 2-dimensional invertible theories may then be computed in terms of homotopy classes of maps in to $\Sigma^3 I\mathbb{Z}$.

In fact, the relevant computation for the classification of symmetry protected phases concerns the deformation classes of reflection positive theories. Freed-Hopkins define an invertible TQFT with *reflection structure* to be one whose associated map of spectra is equivariant with respect to certain involutions on the domain and codomain [**FH16**, Ansatz 7.13], and a TQFT with *reflection positivity structure* to be one with reflection structure and a trivialization of an associated map [**FH16**, Definition 8.15]. Reflection structure implies that *isomorphism* classes of reflection positive invertible TQFTs are classified by homotopy classes of $\mathbb{Z}/2$-equivariant maps of Borel $\mathbb{Z}/2$-equivariant spectra, but Freed-Hopkins prove [**FH16**, Theorem 8.20] that *deformation* classes of reflection-positive invertible TQFTs can be classified in terms of nonequivariant maps of spectra, and that up to deformation, a reflection-positive theory can always be represented by a stable theory. In the case of interest in this paper (2-dimensional pin$^-$ theories) deformation classes of reflection positive theories are in natural bijection with isomorphism classes of stable theories: both groups are cyclic of order 8.

4.3.5. *Proof of* 4.21. By the defining property of $I\mathbb{C}^\times$, there is a unique homotopy class of maps of spectra

$$c : |\mathsf{sAlg}_{\mathbb{C}}^\times| \to \Sigma^2 I\mathbb{C}^\times$$

which induces the identity map

$$c_2 = \pi_2(c) : \mathbb{C}^\times = \pi_2(|\mathsf{sAlg}_{\mathbb{C}}^\times|) \to \pi_2(\Sigma^2 I\mathbb{C}^\times) = \pi_0(\mathbb{S})^\vee = \mathbb{C}^\times.$$

To prove the proposition, we must check that the morphism c_2 induces isomorphisms

$$c_1 : \mathbb{Z}/2\mathbb{Z} \cong \pi_1(|\mathsf{sAlg}_{\mathbb{C}}^\times|) \to \pi_1(\Sigma^2 I\mathbb{C}^\times) = \pi_1(\mathbb{S})^\vee \cong \mu_2$$

and

$$c_0 : \mathbb{Z}/2\mathbb{Z} \cong \pi_0(|\mathsf{sAlg}_{\mathbb{C}}^\times|) \to \pi_0(\Sigma^2 I\mathbb{C}^\times) = \pi_2(\mathbb{S})^\vee \cong \mu_2.$$

By Lemma 4.20, to understand the maps c_0 and c_1, we must compute the action of $\pi_1(\mathbb{S})$ and $\pi_2(\mathbb{S})$ on $\pi_*(|\mathsf{sAlg}_{\mathbb{C}}^\times|)$.

First consider the generator $\eta \in \pi_1(\mathbb{S}) \cong \mathbb{Z}/2\mathbb{Z}$. Recall from Remark 4.12 that the action of η on the homotopy groups of the classifying space of a Picard groupoid is given by the formula in Proposition 4.13 (which also encodes the unique k-invariant of the stable 1-type). Consider the Picard groupoid $\mathsf{sVect}_{\mathbb{C}}^\times$, whose classifying space is weakly equivalent to $\Omega|\mathsf{sAlg}_{\mathbb{C}}^\times|$. As explained in Proposition 4.13, the action of η is given by the homomorphism

$$\pi_1(|\mathsf{sAlg}_{\mathbb{C}}^{\times}|) \otimes \mathbb{Z}/2\mathbb{Z} = \pi_0(|\mathsf{sVect}_{\mathbb{C}}^{\times}|) \otimes \mathbb{Z}/2\mathbb{Z} \cong \mathbb{Z}/2\mathbb{Z} \to \pi_1(|\mathsf{sVect}_{\mathbb{C}}^{\times}|) = \mathbb{C}^\times$$

which takes the generator (represented by an odd line) to the number $-1 \in \mathbb{C}^\times$. It follows that c_1 applied to the class of on odd line gives the unique non-trivial character of $\pi_1(\mathbb{S}) \cong \mathbb{Z}/2\mathbb{Z}$, and thus c_1 is an isomorphism as required.

Now $\pi_2(\mathbb{S}) \cong \mathbb{Z}/2\mathbb{Z}$ is generated by η^2. Thus, the action of η^2 is a composite:

$$\pi_0(|\mathsf{sAlg}_{\mathbb{C}}^{\times}|) \xrightarrow{\eta} \pi_1(|\mathsf{sAlg}_{\mathbb{C}}^{\times}|) \xrightarrow{\eta} \pi_2(|\mathsf{sAlg}_{\mathbb{C}}^{\times}|).$$

Thus it remains to compute the map between π_0 and π_1. For this, we consider the Picard groupoid representing the $\langle 0, 1\rangle$ Postnikov truncation of $|\mathsf{sAlg}_{\mathbb{C}}^{\times}|$; its objects are Morita invertible complex superalgebras, and morphisms are *isomorphism classes* of invertible bimodules. The action of η is computed via the same method as before using the symmetry isomorphism. One sees that the generator of $\pi_0(|\mathsf{sAlg}_{\mathbb{C}}^{\times}|)$ (represented by the Clifford algebra $C\ell_1$) is taken by η to the non-trivial class $\pi_1(|\mathsf{sAlg}_{\mathbb{C}}^{\times}|)$ represented by the odd line.

Putting everything together we observe that c_0 takes the class of $C\ell_1$ to the unique non-trivial character of $\pi_2(\mathbb{S})$, and thus is an isomorphism as required.

5. The Arf-Brown TQFT

5.0.1. *The Arf-Brown Theory.* In particular, recall the Arf-Brown invariant

$$AB : \Omega_2^{\mathrm{Pin}^-} \to \mu_8 \subseteq \mathbb{C}^\times.$$

We denote by

$$Z_{AB} : \mathsf{Bord}_2^{\mathrm{Pin}^-} \to \mathsf{sAlg}_{\mathbb{C}}$$

the unique TQFT with partition function given by AB (see Corollary 4.22).

PROPOSITION 5.1. *The pin⁻ Arf-Brown TQFT*

$$Z_{AB} : \mathsf{Bord}_2^{\mathrm{Pin}^-} \to \mathsf{sAlg}_{\mathbb{C}}$$

assigns the following invariants:

- *To a pin⁻ point, Z_{AB} assigns the first Clifford algebra $C\ell_1$.*
- *To a bounding pin⁻ circle, Z_{AB} assigns an even line $\mathbb{C}$.*
- *To a non-bounding pin⁻ circle, Z_{AB} assigns an odd line $\mathbb{C}[1]$.*

PROOF. We use Lemma 4.20 to compute the values of Z_{AB} on closed manifolds in terms of the action of $\pi_*(\mathbb{S})$ on $\pi_*(MTPin^-)$.

Recall that the homotopy groups of $\mathbb{S}$ (respectively $MTPin^-$) are given by cobordism classes of stably framed manifolds (respectively, pin⁻-manifolds). As usual, let η denote the generator of $\pi_1(\mathbb{S})$, which is represented by the cobordism class of the circle S^1 with its Lie group framing (which induces the non-bounding pin⁻ structure). Thus the operation "multiplication by η" on $\pi_*(MTPin^-)$ may be understood as direct product with the pin⁻ manifold S^1_{nb}.

We have [**KT90**] that the class of S^1_{nb} is a generator of $\pi_1(MTPin^-) \cong \mathbb{Z}/2\mathbb{Z}$. The class of $S^1_{nb} \times S^1_{nb}$ is the unique element of order 2 in $\pi_2(MTPin^-)$, and its Arf-Brown invariant is $-1 \in \mathbb{C}^\times$.

It follows that Z_{AB} takes the non-bounding circle to the unique non-trivial character of $\pi_1(\mathbb{S})$ (which takes the class η represented by S^1_{nb} to -1). Similarly, Z_{AB} takes the pin⁻ point to the unique non-trivial character of $\pi_2(\mathbb{S})$ (which takes the class η^2 represented by $S^1_{nb} \times S^1_{nb}$ to -1), as required. □

For a different construction of the nonextended Arf-Brown TQFT, see [**Yon18**, §4.2].

REMARK 5.2. The Arf-Brown theory gives rise to an invertible spin TQFT as the composite

$$Z_A : \mathsf{Bord}_2^{\mathrm{Spin}} \to \mathsf{Bord}_2^{\mathrm{Pin}^-} \to \mathsf{sAlg}_{\mathbb{C}}$$

which assigns the Arf invariant to a closed Spin surface. This TQFT, called the *Arf theory*, was studied extensively in [**Gun16**].

5.0.2. *The Atiyah-Bott-Shapiro orientation revisited.* First let us recall from Section §3.3.1 that the Arf invariant of a closed Spin surface Σ may be constructed as a pushforward in ko-theory:

$$\pi_!^{\Sigma} 1_{\Sigma} \in ko^{-2}(\mathrm{pt}) \cong \mathbb{Z}/2\mathbb{Z}.$$

As explained in §3.3.1, the pushforward map in ko-theory is constructed using the Atiyah-Bott-Shapiro orientation of ko, which may be encoded as a map of spectra

$$\widehat{A} : MTSpin \to ko.$$

In fact (as shown in [**Gun16**]) the entire Arf theory factors naturally through the Atiyah-Bott-Shapiro orientation:[7]

$$Z_A : |\mathsf{Bord}_2^{\mathrm{Spin}}| \to MTSpin \xrightarrow{\widehat{A}} ko \xrightarrow{Cliff} |\mathsf{sAlg}_{\mathbb{C}}^{\times}|.$$

As explained in §3.3.2, the Arf-Brown invariant may also be interpreted as a pushforward in (twisted) KO-theory. One may reinterpret this as arising from the following twisted form of the Atiyah-Bott-Shapiro orientation:

$$MTPin^- \simeq MTSpin \wedge \Sigma^{-1} MO_1 \xrightarrow{\widehat{A}\wedge \mathrm{id}} ko \wedge \Sigma^{-1} MO_1. \tag{5.3}$$

The factor $\Sigma^{-1}MO_1$ is the Thom spectrum of the virtual vector bundle over BO_1 corresponding to the representation sphere $S^{\sigma-1}$.

Note that the twisted Atiyah-Bott-Shapiro orientation induces an isomorphism

$$\pi_2(MTPin^-) \cong \pi_2(ko \wedge \Sigma^{-1}MO_1).$$

It follows that the Arf-Brown theory factors through the twisted ABS map (5.3):

$$|\mathsf{Bord}_2^{\mathrm{Pin}-}| \to MTPin^- \to ko \wedge \Sigma^{-1}MO_1 \to |\mathsf{sAlg}_{\mathbb{C}}^{\times}|.$$

REMARK 5.4. Freed-Hopkins [**FH16**] define an involution on $\Sigma^2 MTPin_2^-$ obtained by taking a pin$^-$ structure to its opposite (obtained by tensoring with the orientation double cover), and an involution on $|\mathsf{sAlg}_{\mathbb{C}}^{\times}|$ induced by complex conjugation. The homotopy fixed point spectra are $(\Sigma^2 MTPin_2^-)^{h\mathbb{Z}/2} \simeq \Sigma^2 MTSpin_2$ and $|\mathsf{sAlg}_{\mathbb{C}}^{\times}|^{h\mathbb{Z}/2} \simeq |\mathsf{sAlg}_{\mathbb{R}}^{\times}|$. A 2D invertible pin$^-$ TQFT is said to have *reflection structure* if the map of spectra $\Sigma^2 MTPin_2^- \to |\mathsf{sAlg}_{\mathbb{C}}^{\times}|$ it defines is equivariant with respect to these involutions [**FH16**] (see also [**JF17**]).

The Arf-Brown theory naturally carries a reflection structure, which follows from the arguments of Freed-Hopkins [**FH16**]. This is also hinted at by physics: in §6 we discuss how the $\mathbb{Z}/8$ classification of deformation classes of 2D invertible pin$^-$ TQFTs is conjecturally linked to the $\mathbb{Z}/8$ classification of 2D pin$^-$ SPT phases.

[7]The map *Cliff* is so called because it takes an element of $ko^0(\mathrm{pt})$, represented by a finite dimensional vector space V, to the (complex) Clifford algebra associated to a positive definite inner product on V.

Some physics-based classifications of these SPTs [**FK11**, **GJF17**] are rooted in real superalgebra, tying the $\mathbb{Z}/8$ classification to the 8-fold periodicity of Morita equivalence classes of real Clifford algebras.

6. The time-reversal-invariant Majorana chain

The Arf-Brown TQFT is believed to arise in physics as part of the classification of topological phases of matter. In this section, we discuss one of its conjectural appearances, as the low-energy theory of the Majorana chain with time-reversal symmetry, and some background on this occurrence.

6.1. Symmetry-protected topological phases. Condensed-matter theorists are interested in classifying topological phases of matter: given a dimension and a collection of symmetries to be preserved (called the *symmetry type*), what physical systems can occur, and what kind of data is needed to specify one up to a suitable notion of equivalence? This is a difficult problem in general, but can be simplified by restricting to nice subclasses of phases.

This problem is complicated by the lack of a mathematical definition of a topological phase. Nonetheless, arguments from physics suggest some properties that a definition will have: for example, given two topological phases with the same dimension and symmetry type, it should be possible to formulate them both on the same ambient space but with no interactions between them, creating another phase. This commutative monoid-like operation is called *stacking*.

DEFINITION 6.1. A *symmetry-protected topological (SPT) phase* is a topological phase of matter which is invertible under stacking: after stacking with some other phase, it's equivalent to the trivial phase.[8]

Though this isn't a mathematical definition, it tells us that equivalence classes of SPTs should form an abelian group. The computation of this abelian group given a dimension and symmetry type has been the subject of considerable recent research activity at the interface of topology and physics (for a long list of references, see [**GJF17**, §1]).

REMARK 6.2. The original definition in physics of a symmetry-protected phase is one which is inequivalent to the trivial theory when its symmetry type is considered, but which is equivalent in the absence of symmetry. According to this definition, the trivial phase is not an SPT, so the group structure is lost. Our interest in the group structure motivates us to allow the trivial phase.

To classify SPTs, one generally needs a model for phases of matter and equivalences between them.[9] Lattice models are a common choice: roughly speaking, an n-dimensional lattice model is a way of assigning to any closed n-manifold M with a simplicial structure the following data:

- a complex vector space $\mathcal{H}$ determined by local combinatorial data on M, called the *state space*; and
- a self-adjoint operator $H\colon \mathcal{H} \to \mathcal{H}$ also determined by local combinatorial data, called the *Hamiltonian*.

[8]We haven't provided a definition of the trivial phase, and the definition will depend on one's model for topological phases. But it should have no interesting physics, and its partition functions on closed manifolds should all be equal to 1.

[9]We note, however, the existence of model-independent approaches [**GJF17**, **Xio17**, **XA18**].

A lattice model is *gapped* if there is an $\varepsilon > 0$ such that as the simplicial structure is refined on any closed n-manifold M, the difference between the two smallest eigenvalues of H is greater than ε. Two gapped lattice models are equivalent if one can be deformed into the other through local deformations of $\mathcal{H}$ and H that preserve a gap in $\operatorname{Spec} H$.

The symmetry type corresponds to a choice of tangential structure on M, expressed in terms of the simplicial structure. Here are some examples.

- The default symmetry type fixes an orientation on M, expressed through a consistent local orientation of its simplices.
- A phase has *time-reversal symmetry* if we can choose M to be unoriented. In this case one doesn't need orientations on simplices. Alternatively, because lattice models are built from local data, one can formulate the model on a simplicial disc, together with an explicit action of reflection on $\mathcal{H}$; this is how time-reversal symmetry is implemented for the Majorana chain.
- There is a notion of a *fermionic SPT* which is believed to correspond to spin structure; see §6.2 below.
- Given a finite group G, an *internal G-symmetry* corresponds to the data of a principal G-bundle $P \to M$. This can be formulated as a function from the 1-simplices of M to G encoding the monodromy of P or by placing a simplicial structure on P itself [**OMD16**].

These symmetries may interact in nontrivial ways: for example, there are two ways to implement time-reversal symmetry in fermionic phases, corresponding to pin$^+$ and pin$^-$ structures on M.

REMARK 6.3. The above is not a rigorous mathematical definition of topological phases of matter. Providing a rigorous framework for this classification problem is a significant open problem in this field.

See [**Sab18**] for more about lattice models.

There are many approaches to classifying SPTs. We will use a low-energy limit approach, because it reduces modulo a conjecture to a completely mathematical problem, the classification of TQFTs.

DEFINITION 6.4. Given a gapped lattice model with Hamiltonian H, its space of *ground states* on a closed manifold M is the eigenspace for the smallest eigenvalue of H.

In examples, this depends on the underlying manifold but not its triangulation, behaving like a topological field theory. Conjecturally, it *is* (part of) a topological field theory:

ANSATZ 6.5. Given a d-dimensional lattice model with symmetry type H_d, there is a fully extended, reflection positive $(d+1)$-dimensional TQFT[10] Z with the same symmetry type, called the *low-energy (effective) field theory* of the lattice model, whose deformation class can be determined from the data of the lattice model, and such that

(1) if N is a closed d-manifold, $Z(N)$ is isomorphic to the space of ground states of the lattice model on N;

[10] In general one must also allow TQFTs tensored with an invertible, non-topological theory; see [**FH16**, §5.4]. However, this will not come into play in this paper.

(2) if $\varphi\colon N \to N$ is a diffeomorphism and N_φ denotes its mapping torus, there is a well-defined action of φ on the ground states of the lattice model on N, and $Z(N_\varphi)$ is the trace of this action.

In addition,

(3) deformation-equivalent lattice models should have deformation-equivalent low-energy effective field theories, and
(4) if S_0 and S_1 are lattice models with low-energy theories Z_0 and Z_1, respectively, the low-energy theory of $S_0 \otimes S_1$ should be $Z_0 \otimes Z_1$.

It is believed that the map sending a lattice model to its low-energy theory is surjective onto the set of deformation classes of fully extended, reflection positive $(d+1)$-dimensional H_d-TQFTs.

For discussion of this prediction, see [**FH16**, **Gai17**, **RW18**]; for discussion of reflection positivity in the invertible case, see [**FH16**, §8.1.3]. For the rest of this section, we assume Ansatz 6.5.

Ansatz 6.5 implies in particular that the group of equivalence classes of d-dimensional SPTs with a given symmetry type is isomorphic to the group of deformation classes of reflection positive invertible $(d+1)$-dimensional TQFTs with the same symmetry type, a fact which Freed-Hopkins [**FH16**, §9.3] use to classify fermionic SPTs. This approach to classifying SPTs is also undertaken in [**Cam17**, **PWY17**, **SSR17a**, **DT18**].

6.1.1. *Context for the Majorana chain.* We now specialize to the group of 2D pin^- SPTs, which is believed to be isomorphic to $\mathbb{Z}/8$.[11] This can be proven assuming Ansatz 6.5, as in [**FH16**, (9.7.8)]: we saw in §4.3.4 that the group of deformation classes of 2D reflection positive invertible pin^- TQFTs is $[MTPin^-, \Sigma^3 I\mathbb{Z}] \cong \mathbb{Z}/8$. Other approaches to this $\mathbb{Z}/8$ classification can be found in [**GW14**, **KTTW15**, **BWHV17**, **CSRL17**, **GJF17**, **SSR17a**].

The Majorana chain is a 2D fermionic SPT phase with time-reversal symmetry making it into a pin^- phase, and several physical arguments have shown that it's the generator of the $\mathbb{Z}/8$ of such phases.[12] The Majorana chain was originally studied by Kitaev [**Kit01**] with an eye towards applications in quantum computing, then given time-reversal symmetry by Fidkowski-Kitaev [**FK10**] and Turner-Pollmann-Berg [**TPB11**], who observed that it generated a $\mathbb{Z}/8$ of SPTs. Therefore, Ansatz 6.5 implies that its low-energy field theory is a tensor product of an odd number of copies of the Arf-Brown theory. In what follows, we will formulate the Majorana chain on a pin^- 1-manifold and study its low-energy behavior.

REMARK 6.6. There's an additional way in which the Arf-Brown theory is expected to arise in physics. Though we won't discuss it in detail, we'll point the interested reader to some references.

Associated to a free fermion theory in dimension d is its anomaly theory, a $(d+1)$-dimensional invertible field theory of the same symmetry type. The group of equivalence classes of 1-dimensional free fermion theories with pin^- symmetry is conjecturally isomorphic to $\mathbb{Z}$ with the Majorana chain as a generator, and its

[11]From a mathematical perspective, because mathematical definitions for SPTs haven't been written down, this isn't yet a theorem. It's expected that once the definitions are in place, it will be true.

[12]Similarly, this is not yet a mathematical theorem because we don't have a mathematical definition of an SPT.

anomaly theory is conjecturally the Arf-Brown TQFT. For general free fermion systems, this conjecture is due to Freed-Hopkins [**FH16**, §9.2.5]; Witten [**Wit16**, §5] provides a physical argument specifically for the time-reversal symmetric Majorana chain.

These two appearances of the Arf-Brown TQFT from the Majorana chain are believed to be related: one can regard a free system as an interacting system with the same dimension and symmetry type, defining a group homomorphism from equivalence classes of free fermion theories to SPTs. For 2D pin$^-$ theories, this is believed to be the quotient map $\mathbb{Z} \to \mathbb{Z}/8$, a surprising fact first noticed by Fidkowski-Kitaev [**FK10**,**FK11**] and Turner-Pollmann-Berg [**TPB11**], and argued a different way by You-Wang-Oon-Xu [**YWOX14**]. Freed-Hopkins [**FH16**, §§9.2, 9.3] provide a conjecture describing this homomorphism in general, then study it in several specific cases, including 2D pin$^-$ theories.

6.2. Combinatorial spin and pin structures. The Majorana chain is a fermionic SPT. This is believed to correspond to building the state space and Hamiltonian using superalgebra. In relativistic quantum field theory, the spin-statistics theorem implies such a system should be formulated on spin manifolds, but in the condensed-matter setting, the theorem doesn't apply. Nonetheless, it appears that spin structures are the correct setting for fermionic phase of matter, in that the data of a fermionic phase of matter depends on a choice of spin structure on the underlying manifold in examples [**GK16**].

Time-reversal symmetry can act on fermionic phases in two ways: by squaring to 1 or by squaring to the grading operator. The former is believed to correspond to a pin$^-$ structure on the underlying manifold, and the latter to a pin$^+$ structure [**KTTW15**].

The Majorana chain admits a time-reversal symmetry T squaring to 1, so to formulate the Majorana chain on a compact 1-manifold M with this symmetry, we must choose a pin$^-$ structure on M and encode it in the data of the lattice somehow. In general, this is somewhat tricky: for spin structures, this was solved by Cimasoni-Reshetikhin [**CR07**] in dimension 2 and Budney [**Bud13**] in all dimensions, and for *pin*$^-$ structures, the 2-dimensional case is solved by Cimasoni [**Cim09**], but appears to be open in higher dimensions. Since we're only studying 1-manifolds, we can use an explicit, simpler construction: there are two pin$^-$ structures on a closed interval relative to a fixed pin$^-$ structure on its boundary, so we will fix pin$^-$ structures on the vertices of M and use the data of the class of the pin$^-$ structure on the edge.

Fix, once and for all, a pin$^-$ point pt.

DEFINITION 6.7. Let I be a closed interval and $\partial I = \{a, b\}$. Fix a pin$^-$ structure on ∂I and pin$^-$ isomorphisms pt $\cong a$ and pt $\cong b$. A *relative pin*$^-$ *structure* on I is a pin$^-$ structure on I which restricts to the specified pin$^-$ structure on ∂I. We consider two relative pin$^-$ structures on I equivalent if there's a pin$^-$ diffeomorphism between them covering the identity and respecting this data, i.e. it restricts to the identity on ∂I and intertwines the pin$^-$ diffeomorphisms with pt.

The pin$^-$ diffeomorphisms $a \cong \mathrm{pt} \cong b$ define a pin$^-$ structure on $I/\partial I$, and sending $I \mapsto I/\partial I$ defines an isomorphism from the set of equivalence classes of relative pin$^-$ structures on I to the set of diffeomorphism classes of pin$^-$ structures on S^1; let I_0 be a relative pin$^-$ structure on I which maps to S^1_{nb} and I_1 be one which maps to S^1_b.

LEMMA 6.8. *Concatenation defines a group structure on the equivalence classes of relative pin$^-$ intervals. This group is isomorphic to $\mathbb{Z}/2$ and the generator is I_0.*

PROOF. Because every 1-manifold M can be stably framed relative to a fixed stable framing on the boundary, we may define the pin$^-$ structures on I_0 and I_1 as those induced by framings. Specifically, I_0 is induced by the trivial framing (the restriction of the usual framing on $\mathbb{R}$ to $[0,1]$), and I_1 is induced by the nontrivial framing. Two concatenated copies of this framing are equivalent to the trivial framing when the endpoints are fixed (see [**DSPS13**, Remark 1.3.1]) and concatenating with the trivial framing does not change the equivalence class of framing on an interval, giving the claimed group structure. □

Let M be a compact pin$^-$ 1-manifold with a simplicial structure, and let $\Delta^i(M)$ denote its set of i-simplices. For each $v \in \Delta^0(M)$, fix a pin$^-$ isomorphism $v \cong \mathrm{pt}$. Since the groupoid of pin$^-$ structures on a point is equivalent to $\bullet/(\mathbb{Z}/2)$, an isomorphism with pt is a choice. For each $e \in \Delta^1(M)$, the pin$^-$ structure on M defines a relative pin$^-$ structure on $\overline{e}$. Thus $e \cong I_j$ for some $j \in \{0,1\}$; define $t(e) := j$. From the function $t\colon \Delta^1(M) \to \mathbb{Z}/2$ one can recover the pin$^-$ structure on M up to isomorphism. We will call t the *combinatorial pin$^-$ data* of M.

LEMMA 6.9. *Let M be a pin$^-$ circle with a simplicial structure and m be the number of edges of e with $t(e) = 1$.*

- *If m is odd, $M \cong S^1_b$.*
- *If m is even, $M \cong S^1_{nb}$.*

PROOF. Fix a vertex $v \in M$. Using the group law from Lemma 6.8, we can concatenate adjacent intervals for all vertices except v, resulting in a simplicial structure on M with a single vertex at v and a single edge e with $t(e) = m \bmod 2$. The result then follows from the definition of t. □

6.3. Defining the Majorana chain. Let M be a compact pin$^-$ 1-manifold with a simplicial structure. Associated to each vertex $v \in \Delta^0(M)$, we associate a trivialized odd line $\mathbb{C}_v^{0|1}$ and define the local state space $\mathcal{H}_v := \Lambda(\mathbb{C}_v)$. The state space for the Majorana chain on M is

$$\mathcal{H} := \bigotimes_{v\in\Delta^0(M)} \mathcal{H}_v. \tag{6.10}$$

Let F denote the space of functions $\Delta^0(M) \to \mathbb{C}$, regarded as a purely odd vector space. Then $\mathcal{H} \cong \Lambda^*(F)$, and hence $\mathcal{H}$ is generated by the δ-functions δ_v for $v \in \Delta^0(M)$, where each δ_v is odd.

DEFINITION 6.11. Let $v \in \Delta^0(M)$.

- The *annihilation operator* associated to v, denoted ι_v, is the interior product with δ_v.
- The *creation operator* associated to v, denoted ε_v, is the exterior product with δ_v.
- The *Majorana operators* associated to v are

$$c_v := \varepsilon_v + \iota_v$$
$$d_v := \varepsilon_v - \iota_v.$$

REMARK 6.12. The notation for the Majorana operators in [**Kit01, FK11**] corresponds to ours as follows: after ordering the vertices $v_1, \ldots, v_n$ on an interval in the direction defined by the orientation, their c_{2j-1} is our c_{v_j}, and their c_{2j} is i times our d_{v_j}. In some papers, the Majorana chain is instead called the *Majorana wire* or *Kitaev chain.*

To define the Hamiltonian, we must orient M. This is a bit surprising, because the Majorana chain admits a time-reversal symmetry and therefore ought to make sense on a pin$^-$ manifold without using the fact that all 1-manifolds are orientable, but if we vary the orientation on M, we obtain a different Hamiltonian. We expect that the eigenspaces for the Hamiltonian, as subspaces of $\mathcal{H}$, end up not depending on the choice of orientation, but verifying this would require a different approach: ours uses a choice of orientation to construct a Clifford module isomorphic to $\mathcal{H}$, but without choosing an isomorphism, making it difficult to track the dependence on orientation.

The Hamiltonian for the Majorana chain is a sum of local terms for each edge. Fix an orientation on M, so that each edge e has an induced orientation; we write $\partial e = v - w$ to mean $\partial e = \{v, w\}$, and that, in the induced orientation on the boundary, v is the positively oriented boundary point and w is the negatively oriented one. For each $v \in \Delta^0(M)$, choose a pin$^-$ isomorphism $v \cong \mathrm{pt}$, and let $t \colon \Delta^1(M) \to \mathbb{Z}/2$ be the induced combinatorial pin$^-$ data. Then, the Hamiltonian on M is

$$H := \frac{1}{2} \sum_{\substack{e \in \Delta^1(M) \\ \partial e = v - w}} (-1)^{t(e)} c_v d_w. \tag{6.13}$$

Time-reversal symmetry acts on $\mathcal{H}$ as complex conjugation; since c_v and d_w are real, this commutes with the Hamiltonian, so the Majorana chain admits a time-reversal symmetry squaring to 1.

REMARK 6.14. In physics, a Majorana fermion is a fermion which is its own antiparticle, meaning that its creation and annihilation operators coincide. Because the Clifford relations imply $c_v^2 = 1$ and $(i \cdot d_v)^2 = 1$, these operators can be interpreted as creating up to two Majorana fermions located at v. The Hamiltonian (6.13) is expressing a relationship between Majorana fermions at adjacent vertices: if $\partial e = v - w$, then the Hamiltonian specifies that low-energy states must have a relationship between the Majorana fermions corresponding to c_v and $i \cdot d_w$.

Because it would be interesting to observe a Majorana fermion, the Majorana chain has been studied experimentally [**MZF$^+$12, DYH$^+$12, DRM$^+$12, RLF12, FVHM$^+$13**]. To our knowledge, however, these experiments have not considered the behavior of the Majorana chain under stacking or time-reversal symmetry.

6.4. The low-energy TQFT. We'd like to use Ansatz 6.5 to determine the deformation class of the low-energy theory Z of the Majorana chain, but it doesn't tell us everything. For example, neither pin$^-$ structure on $\mathbb{RP}^2$ is bordant to a disjoint union of mapping tori, so we won't be able to calculate $Z(\mathbb{RP}^2)$. Nonetheless, Ansatz 6.5 tells us we can compute the state space of any closed 1-manifold and the partition functions of all pin$^-$ tori and Klein bottles. In particular, we'll find that $Z(S^1_{nb})$ is an odd line, which is enough to imply that Z is one of the four generators of the $\mathbb{Z}/8$ of deformation classes of reflection positive 2D pin$^-$ invertible field theories.[13]

Let $\pi\colon M' \to M$ be the orientation double cover, and give M' the simplicial structure which makes π a simplicial map. The orientation of M induces an orientation of the 0-skeleton of M', M'_0, which is a compact oriented 0-manifold, so this orientation defines a function $\mathfrak{o}\colon M'_0 \to \{\pm 1\}$ sending a positively oriented point to 1 and a negatively oriented point to -1.

Let $n := |\Delta^0(M)|$.

LEMMA 6.15. *The algebra generated by c_v and d_v is canonically isomorphic to $C\ell(\pi^{-1}(v), \mathfrak{o})$ and isomorphic to $C\ell_{1,1}$. The algebra generated by all Majorana operators is canonically isomorphic to $C\ell(M'_0, \mathfrak{o})$, and noncanonically isomorphic to $C\ell_{n,n}$.*

PROOF. If $V \in \Delta^0(M)$, let v_+, (resp. v_-) be the positively (resp. negatively) oriented preimage of v. We define the maps $\langle c_v, d_v\rangle \to C\ell(\pi^{-1}(v), \mathfrak{o})$ and $\langle c_w, d_w \mid w \in \Delta^0(V)\rangle \to C\ell(M'_0, \mathfrak{o})$ to send $c_v \mapsto v_+$ and $d_v \mapsto v_-$. For this to define an isomorphism of algebras, one must check the defining relations of the Clifford algebra: $c_v^2 = 1$, $d_v^2 = -1$, $[c_v, d_v] = -1$, and if $v \neq w$, $[c_v, c_w] = [d_v, d_w] = [c_v, d_w] = -1$. These follow directly from the definition of the Majorana operators.

Since $\mathfrak{o}|_{\pi^{-1}(v)}$ sends $v_+ \mapsto 1$ and $v_- \mapsto -1$, $C\ell(\pi^{-1}(v), \mathfrak{o}) \cong C\ell_{1,1}$ and $C\ell(M'_0, \mathfrak{o}) \cong C\ell_{n,n}$, the latter after choosing an ordering of the vertices of v. $\square$

Let M be a spin circle with a simplicial structure, and let $t\colon \Delta^1(M) \to \mathbb{Z}/2$ be the combinatorial data associated to it. Let $n := |\Delta^0(M)|$; then, the state space $\mathcal{H}$ is a $\mathbb{Z}/2$-graded $C\ell(M'_0, \mathfrak{o})$-module.

THEOREM 6.16 ([**ABS64**, §5]). *Up to isomorphism, $C\ell(M'_0, \mathfrak{o})$ has a single irreducible module M, which is 2^n-dimensional. Up to even isomorphism, $C\ell(M'_0, \mathfrak{o})$ has two irreducible supermodules, both isomorphic to M after forgetting the $\mathbb{Z}/2$-grading, and they are parity changes of each other.*

Since $\dim \mathcal{H} = 2^n$, then $\mathcal{H}$ is one of the two irreducible $C\ell(M'_0, \mathfrak{o})$-supermodules. The Hamiltonian acts on $\mathcal{H}$ as an element of $C\ell(M'_0, \mathfrak{o})$, since it's a sum of products of Clifford generators. Thus, to compute its spectrum, it suffices to compute the action of $H \in C\ell(M'_0, \mathfrak{o})$ on any irreducible $C\ell(M'_0, \mathfrak{o})$-module A. To determine the parity of the space of ground states, we need to know whether $\mathcal{H}$ is graded isomorphic to A or ΠA, which we will do by fixing a grading operator $\varepsilon \in C\ell(M'_0, \mathfrak{o})$ and comparing its action on $\mathcal{H}$ and on A.

LEMMA 6.17. *There is a unique isomorphism of superalgebras $\varphi\colon C\ell_{1,1} \xrightarrow{\cong} \mathrm{End}(\mathbb{C}^{1|1})$ sending*

$$v_+ \mapsto \begin{pmatrix} 0 & 1 \\ 1 & 0 \end{pmatrix} \qquad v_- \mapsto \begin{pmatrix} 0 & 1 \\ -1 & 0 \end{pmatrix}. \tag{6.18}$$

PROOF. One can directly verify that $\varphi(v_\pm)$ are odd, $\varphi(v_\pm)^2 = \pm I$, and $\varphi(v_+)$ anticommutes with $\varphi(v_-)$. Uniqueness follows because v_+ and v_- generate $C\ell_{1,1}$. $\square$

[13] Since Ω_2^{Spin} and $\Omega_2^{\mathrm{Pin}^+}$ are generated by mapping tori, this ambiguity does not appear for 2D spin and pin$^+$ phases. For general symmetry types, however, this is not the case, and additional work is needed to uniquely determine the low-energy field theory of an SPT. This perspective is taken up by Shiozaki-Shapourian-Ryu [**SSR17b**].

Thus, $g := \varphi(v_+v_-)$ is a grading operator on $\mathbb{C}^{1|1}$.

Let $e \in \Delta^1(M)$ with $\partial e = v - w$. Since ∂e is an oriented 0-manifold, it comes with a function $\mathfrak{o}_e \colon \partial e \to \{\pm 1\}$; the algebra generated by c_v and d_w is canonically isomorphic to $C\ell(\partial e, \mathfrak{o}_e)$, which is isomorphic to $C\ell_{1,1}$. Let $C\ell(\partial e, \mathfrak{o}_e)$ act on a $\mathbb{C}^{1|1}$ through the isomorphism from Lemma 6.17, and call it $\mathbb{C}_e^{1|1}$. Then there's a canonical isomorphism

$$C\ell(M_0', \mathfrak{o}) \cong \bigotimes_{e\in\Delta^1(M)} C\ell(\partial e, \mathfrak{o}_e), \tag{6.19}$$

so $C\ell(M_0', \mathfrak{o})$ acts on

$$A := \bigotimes_{e\in\Delta^1(M)} \mathbb{C}_e^{1|1}, \tag{6.20}$$

making A into a graded $C\ell(M_0', \mathfrak{o})$-module. Since $\dim_{\mathbb{C}} A = 2^n$, A must be irreducible.

Let m be the number of edges e of M with $t(e) = 1$.

PROPOSITION 6.21. *Let $V \subset A$ denote the eigenspace for the smallest eigenvalue of H acting on A. Then V is one-dimensional, and has parity $n - m$* mod 2.

PROOF. Let $a \in A$ be a pure tensor of homogeneous elements of $\mathbb{C}_e^{1|1}$ as e ranges over the edges of M, and let $|a|$ denote its degree. For any $e \in \Delta^1(M)$, we let $|a|_e$ be 1 if the component of a from $\mathbb{C}_e^{1|1}$ is odd, and 0 otherwise.

If $\partial e = v - w$, then $c_v d_w$ acts by the grading operator g_e on $\mathbb{C}_e^{1|1}$, and therefore the action of H is

$$H = \frac{1}{2}\sum_{e\in\Delta^1(M)} (-1)^{t(e)} \mathrm{id}\otimes\cdots\otimes\mathrm{id}\otimes g_e\otimes\mathrm{id}\otimes\cdots\otimes\mathrm{id}. \tag{6.22}$$

It suffices to describe the action of H on pure tensors of homogeneous elements, so let a be such a tensor. If $t(e) = 0$ for all edges e, then H differs from $(n/2)\cdot\mathrm{id}$ on a by subtracting 1 for each odd component of a. Therefore H acts on a as $(n-2|a|)/2$, in which case the ground states are the top-degree vectors, with eigenvalue $-n/2$.

More generally, if e is an edge with $t(e) = 1$, it contributes $-g_e$ to H instead of g_e. This change is equivalent to multiplying by $2(2|a|_e - 1)$, so if $e_1, \ldots, e_m$ are the edges with $t(e_i) = 1$, then the action of the Hamiltonian on a is

$$H \cdot a = \big(n - 2k + 2(2|a|_{e_1} - 1) + \cdots + 2(2|a|_{e_m} - 1)\big)\frac{a}{2}. \tag{6.23}$$

The a which minimize the eigenvalue are those whose component in $\mathbb{C}_e^{1|1}$ is odd if $t(e) = 0$ and even if $t(e) = 1$; these form a one-dimensional vector space with parity $n - m$. □

PROPOSITION 6.24. *Let*

$$\varepsilon := \prod_{v\in\Delta^0(M)} d_v c_v \in C\ell(M_0', \mathfrak{o}).$$

(The Clifford relations imply this doesn't depend on the order of the vertices in the product.) Then

- *on $\mathcal{H}$, ε acts on a homogeneous degree-k element by multiplication by $(-1)^{n-k}$, and*

- *on A, ε acts on a homogeneous degree-k element by multiplication by $(-1)^{k-1}$.*

PROOF. On $\mathcal{H}$, ε acts as

$$(\varepsilon\cdot) = \prod_{v\in\Delta^0(M)} (\varepsilon_v - \iota_v)(\varepsilon_v + \iota_v) = \prod_v (\varepsilon_v\iota_v - \iota_v\varepsilon_v). \tag{6.25}$$

It suffices to understand how this acts on pure wedges $\omega = \lambda\delta_{v_{i_1}} \wedge \cdots \wedge \delta_{v_{i_\ell}}$. On ω, $\varepsilon_v\iota_v - \iota_v\varepsilon_v$ acts by the identity if $v = v_{i_j}$ for some j, and by -1 otherwise. Therefore $\varepsilon \cdot \omega = (-1)^{n-k}\omega$.

To compute the action of ε on A, we rearrange it into a more convenient form. Choose a $v_1 \in \Delta^0(M)$, and let $v_2, \dots, v_n$ be the vertices encountered in order as one traverses the positively oriented path around M starting at v_1. Thus for each i, there's an edge e_i with $\partial e_i = v_{i+1 \bmod n} - v_i$. Then,

$$\varepsilon = d_{v_1}c_{v_1} \cdots d_{v_n}c_{v_n} = (-1)^n c_{v_1}d_{v_1} \cdots c_{v_n}d_{v_n}. \tag{6.26}$$

Since this string has n letters, reversing it is a permutation of parity $(-1)^n$:

$$= d_{v_n}c_{v_n} \cdots d_{v_1}c_{v_1}. \tag{6.27}$$

Finally, we commute c_{v_1} past the remaining $2n-1$ operators:

$$= -\underbrace{c_{v_1}d_{v_n}}_{g_n} c_{v_n} \cdots \underbrace{c_3d_2}_{g_2}\underbrace{c_2d_1}_{g_1}. \tag{6.28}$$

Therefore ε acts by -1 times the usual grading operator on $\mathbb{C}^{n|n}$ (i.e. the one which is -1 on odd states). □

COROLLARY 6.29. *As graded $C\ell(M_0', \mathfrak{o})$-modules, $\mathcal{H} \cong \Pi^{n-1}A$, so the ground states of the Majorana chain on M are*

- *an even line if m is odd (so $M \cong S^1_b$), and*
- *an odd line if m is even (so $M \cong S^1_{nb}$).*

PROOF. In Proposition 6.21, we saw that the ground states of H acting on A have parity $n-m$ mod 2, but by Proposition 6.24 the difference in the parities of $\mathcal{H}$ and A is $n-1$ mod 2. Hence the ground state space of H acting on $\mathcal{H}$ has parity $n-m-(n-1) = m-1$. □

The parity of the ground states on S^1_b and S^1_{nb} is calculated in a different way by Shapourian-Shiozaki-Ryu [**SSR17a**, Appendix D].

COROLLARY 6.30. *Assuming Ansatz 6.5, the low-energy TQFT Z of the Majorana chain is a generator of the $\mathbb{Z}/8$ of deformation classes of reflection positive pin$^-$ invertible field theories. In particular, its deformation class is an odd multiple of the class of the Arf-Brown theory.*

PROOF. By a result of Schommer-Pries [**SP15**, Theorem 11.1], we know Z is invertible, since there is a pin$^-$ structure on S^2 and $Z(S^1_b)$ and $Z(S^1_{nb})$ are both invertible in $\mathsf{sVect}_{\mathbb{C}}$. Since Z_{AB} generates the $\mathbb{Z}/8$ of deformation classes of reflection positive 2D pin$^-$ invertible TQFTs, Z is deformation equivalent to $(Z_{AB})^{\otimes k}$ for some k, and is a generator iff k is odd.

Because $Z_{AB}(S^1_{nb})$ is an odd line, then $(Z_{AB})^{\otimes k}(S^1_{nb})$ has the same parity as k. Since $Z(S^1_{nb})$ is odd, then k is odd. □

We can also study the Majorana chain on pin$^-$ 1-manifolds with boundary, though again the Hamiltonian depends on an orientation. Kitaev [**Kit01**] found that the space of ground states on an interval I is two-dimensional; from the low-energy perspective, this follows from the fact that for any choice of pin$^-$ structure on I, $Z(I)$ is isomorphic to $C\ell_1$ as a $(C\ell_1, C\ell_1)$-bimodule. We can also see this directly from the lattice.

Suppose $n := |\Delta^0(I)|$. Orient I and let $\partial I = v - w$. Then, c_w and d_v do not appear in the Hamiltonian on I. Since each term in H is $\pm 1/2$ times two Clifford generators not equal to c_w or d_v, both c_w and d_v commute with H, and therefore the algebra they generate, isomorphic to $C\ell_{1,1}$, acts on all eigenspaces of H. In particular, if V denotes the ground states of H, V is a $C\ell_{1,1}$-module, and by Theorem 6.16 is determined up to isomorphism by its dimension, which is even.

We can identify it with $C\ell_1$ in a manner similar to the proof of Proposition 6.21: define A in the same manner as above, except that we pretend there's an extra edge e_∂ joining v and w, so A is a $C\ell(I_0', \mathfrak{o})$-module, where I_0' is the 0-skeleton of the orientation double cover $I' \to I$ and $\mathfrak{o}\colon I_0' \to \{\pm 1\}$ is induced from the orientation as before. If H_{S^1} denotes the Hamiltonian from (6.22) (for the circle), then our Hamiltonian is

$$H = H_{S^1} - \mathrm{id} \otimes \cdots \otimes \mathrm{id} \otimes g_{e_\partial}, \tag{6.31}$$

(where $t(e_\partial) := 0$), whose action on a pure tensor of homogeneous elements $a \in A$ is

$$H \cdot a = \big(n - 2k + 2(2|a|_{e_1} - 1) + \cdots + 2(2|a|_{e_m} - 1) + |a|_{e_\partial}\big)\frac{a}{2}. \tag{6.32}$$

Thus the ground state is two-dimensional, spanned by a pure tensor whose components are odd for all edges with $t(e) = 0$ and even otherwise, and a pure tensor whose components are odd for all edges with $t(e) = 0$ except e_∂, and even otherwise. Since $C\ell_1$ is the unique two-dimensional irreducible (ungraded) $C\ell_{1,1}$-representation up to isomorphism, the space of ground states on I is isomorphic to either $C\ell_1$ or $\Pi C\ell_1$. An argument similar to Proposition 6.24 shows that we get the former. Finally, to match the left $C\ell_{1,1}$-module description of the space of ground states with the $(C\ell_1, C\ell_1)$-bimodule description of $Z(I)$, recall that a left $C\ell_{-1}$-action on a module M is equivalent data to a right $C\ell_1$-action on M, which implies the space of ground states on I is $C\ell_1$ as a $(C\ell_1, C\ell_1)$-bimodule, in accordance with the calculation using the low-energy TQFT.

Acknowledgments

We thank Dan Freed for many helpful conversations and comments on the first draft, the referee for several helpful comments on the first draft, and the organizers of the NSF-CBMS conference, David Ayala and Ryan Grady.

References

[ABG10] Matthew Ando, Andrew J. Blumberg, and David Gepner, *Twists of K-theory and TMF*, Superstrings, geometry, topology, and C^*-algebras, Proc. Sympos. Pure Math., vol. 81, Amer. Math. Soc., Providence, RI, 2010, pp. 27–63, DOI 10.1090/pspum/081/2681757. MR2681757

[ABP66] D. W. Anderson, E. H. Brown Jr., and F. P. Peterson, SU-*cobordism*, KO-*characteristic numbers, and the Kervaire invariant*, Ann. of Math. (2) **83** (1966), 54–67, DOI 10.2307/1970470. MR0189043

[ABP67] D. W. Anderson, E. H. Brown Jr., and F. P. Peterson, *The structure of the Spin cobordism ring*, Ann. of Math. (2) **86** (1967), 271–298, DOI 10.2307/1970690. MR0219077

[ABS64] M.F. Atiyah, R. Bott, and A. Shapiro, *Clifford modules*, Topology **3** (1964), no. Supplement 1, 3–38, `https://ac.els-cdn.com/0040938364900035/1-s2.0-0040938364900035-main.pdf?_tid=4564ce36-f342-11e7-b652-00000aab0f27&acdnat=1515285341_9261e93c966630f1828dd0c2caa00f14`.

[Ada74] J. F. Adams, *Stable homotopy and generalised homology*, University of Chicago Press, Chicago, Ill.-London, 1974. Chicago Lectures in Mathematics. MR0402720

[Ada78] John Frank Adams, *Infinite loop spaces*, Annals of Mathematics Studies, vol. 90, Princeton University Press, Princeton, N.J.; University of Tokyo Press, Tokyo, 1978. MR505692

[AHR10] Matthew Ando, Michael J. Hopkins, and Charles Rezk, *Multplicative orientations of KO-theory and the spectrum of topological modular forms*, `https://faculty.math.illinois.edu/~mando/papers/koandtmf.pdf`.

[Ati61] M. F. Atiyah, *Thom complexes*, Proc. London Math. Soc. (3) **11** (1961), 291–310, DOI 10.1112/plms/s3-11.1.291. MR0131880

[Ati71] Michael F. Atiyah, *Riemann surfaces and spin structures*, Ann. Sci. École Norm. Sup. (4) **4** (1971), 47–62. MR0286136

[Ati88] Michael Atiyah, *Topological quantum field theories*, Inst. Hautes Études Sci. Publ. Math. **68** (1988), 175–186 (1989). MR1001453

[BC76] Edgar H. Brown Jr. and Michael Comenetz, *Pontrjagin duality for generalized homology and cohomology theories*, Amer. J. Math. **98** (1976), no. 1, 1–27, DOI 10.2307/2373610. MR0405403

[BC18] Agnès Beaudry and Jonathan Campbell, *A guide for computing stable homotopy groups*, `https://arxiv.org/pdf/1801.07530.pdf`.

[Bén67] Jean Bénabou, *Introduction to bicategories*, Reports of the Midwest Category Seminar, Springer, Berlin, 1967, pp. 1–77. MR0220789

[BG10] Robert R. Bruner and J. P. C. Greenlees, *Connective real K-theory of finite groups*, Mathematical Surveys and Monographs, vol. 169, American Mathematical Society, Providence, RI, 2010. MR2723113

[Bro62] Edgar H. Brown Jr., *Cohomology theories*, Ann. of Math. (2) **75** (1962), 467–484, DOI 10.2307/1970209. MR0138104

[Bro71] Edgar H. Brown, Jr., *The Kervaire invariant of a manifold*, Algebraic Topology (Arunas Liulevicius, ed.), Proceedings of symposia in pure mathematics, no. 22, American Mathematical Society, 1971, pp. 65–71.

[Bud13] Ryan Budney, *Combinatorial spin structures on triangulated manifolds*, Algebr. Geom. Topol. **18** (2018), no. 3, 1259–1279, DOI 10.2140/agt.2018.18.1259. MR3784005

[BWHV17] Nick Bultinck, Dominic J. Williamson, Jutho Haegeman, and Frank Verstraete, *Fermionic matrix product states and one-dimensional topological phases*, Phys. Rev. B **95** (2017), no. 7, 075108, `https://arxiv.org/pdf/1610.07849.pdf`.

[Cam17] Jonathan Campbell, *Homotopy theoretic classification of symmetry protected phases*, `https://arxiv.org/pdf/1708.04264.pdf`.

[Cim09] David Cimasoni, *Dimers on graphs in non-orientable surfaces*, Lett. Math. Phys. **87** (2009), no. 1-2, 149–179, DOI 10.1007/s11005-009-0299-2. MR2480651

[CR07] David Cimasoni and Nicolai Reshetikhin, *Dimers on surface graphs and spin structures. I*, Comm. Math. Phys. **275** (2007), no. 1, 187–208, DOI 10.1007/s00220-007-0302-7. MR2335773

[CSRL17] Gil Young Cho, Ken Shiozaki, Shinsei Ryu, and Andreas W. W. Ludwig, *Relationship between symmetry protected topological phases and boundary conformal field theories via the entanglement spectrum*, J. Phys. A **50** (2017), no. 30, 304002, 45. MR3673461

[DFM10] Jacques Distler, Daniel S. Freed, and Gregory W. Moore, *Spin structures and superstrings*, Surveys in Differential Geometry (Tomasz Mrowka and Shing-Tung Yau, eds.), vol. 15, International Press, 2010, `http://intlpress.com/site/pub/files/_fulltext/journals/sdg/2010/0015/0001/SDG-2010-0015-0001-a004.pdf`, pp. 99–130.

[DRM+12] Anindya Das, Yuval Ronen, Yonatan Most, Yuval Oreg, Moty Heiblum, and Hadas Shtrikman, *Zero-bias peaks and splitting in an Al-InAs nanowire topological superconductor as a signature of Majorana fermions*, Nature Physics **8** (2012), 887–895, https://www.nature.com/nphys/journal/v8/n12/pdf/nphys2479.pdf.

[DSPS13] Christopher L. Douglas, Christopher Schommer-Pries, and Noah Snyder, *Dualizable tensor categories*, https://arxiv.org/pdf/1312.7188.pdf.

[DT18] Clement Delcamp and Apoorv Tiwari, *From gauge to higher gauge models of topological phases*, https://arxiv.org/pdf/1802.10104.pdf.

[Dus02] John W. Duskin, *Simplicial matrices and the nerves of weak n-categories. I. Nerves of bicategories*, Theory Appl. Categ. **9** (2001/02), 198–308. CT2000 Conference (Como). MR1897816

[DW90] Robbert Dijkgraaf and Edward Witten, *Topological gauge theories and group cohomology*, Comm. Math. Phys. **129** (1990), no. 2, 393–429. MR1048699

[DYH+12] M. T. Deng, C. L. Yu, G. Y. Huang, M. Larsson, P. Caroff, and H. Q. Xu, *Anomalous zero-bias conductance peak in a Nb-InSb nanowire-Nb hybrid device*, Nano Letters **12** (2012), no. 12, 6414–6419, http://pubs.acs.org/doi/pdfplus/10.1021/nl303758w.

[FH16] Daniel S. Freed and Michael J. Hopkins, *Reflection positivity and invertible topological phases*, https://arxiv.org/pdf/1604.06527.pdf.

[FHLT10] Daniel S. Freed, Michael J. Hopkins, Jacob Lurie, and Constantin Teleman, *Topological quantum field theories from compact Lie groups*, A celebration of the mathematical legacy of Raoul Bott, CRM Proc. Lecture Notes, vol. 50, Amer. Math. Soc., Providence, RI, 2010, pp. 367–403. MR2648901

[FHT11] Daniel S. Freed, Michael J. Hopkins, and Constantin Teleman, *Loop groups and twisted K-theory I*, J. Topol. **4** (2011), no. 4, 737–798, DOI 10.1112/jtopol/jtr019. MR2860342

[FK10] Lukasz Fidkowski and Alexei Kitaev, *Effects of interactions on the topological classification of free fermion systems*, Phys. Rev. B **81** (2010), no. 13, 134509, https://arxiv.org/pdf/0904.2197.pdf.

[FK11] Lukasz Fidkowski and Alexei Kitaev, *Topological phases of fermions in one dimension*, Phys. Rev. B **83** (2011), no. 7, 075103, https://arxiv.org/pdf/1008.4138.pdf.

[FQ93] Daniel S. Freed and Frank Quinn, *Chern-Simons theory with finite gauge group*, Comm. Math. Phys. **156** (1993), no. 3, 435–472. MR1240583

[FVHM+13] A. D. K. Finck, D. J. Van Harlingen, P. K. Mohseni, K. Jung, and X. Li, *Anomalous modulation of a zero-bias peak in a hybrid nanowire-superconductor device*, Phys. Rev. Lett. **110** (2013), no. 12, 126406, https://journals.aps.org/prl/pdf/10.1103/PhysRevLett.110.126406.

[Gai17] Davide Gaiotto, *Gapped phases of matter vs. topological field theories*, Conference talk in Hopf Algebras in Kitaev's Quantum Double Models. http://pirsa.org/17070066., 2017.

[GJF17] Davide Gaiotto and Theo Johnson-Freyd, *Symmetry protected topological phases and generalized cohomology*, https://arxiv.org/abs/1712.07950.

[GJO17] Nick Gurski, Niles Johnson, and Angélica M. Osorno, *The 2-dimensional stable homotopy hypothesis*, https://arxiv.org/pdf/1712.07218.pdf.

[GJOS17] Nick Gurski, Niles Johnson, Angélica M. Osorno, and Marc Stephan, *Stable Postnikov data of Picard 2-categories*, Algebr. Geom. Topol. **17** (2017), no. 5, 2763–2806, DOI 10.2140/agt.2017.17.2763. MR3704242

[GK16] Davide Gaiotto and Anton Kapustin, *Spin TQFTs and fermionic phases of matter*, International Journal of Modern Physics A **31** (2016), no. 28n29, 1645044, https://arxiv.org/pdf/1505.05856.pdf.

[GMTW09] Søren Galatius, Ulrike Tillmann, Ib Madsen, and Michael Weiss, *The homotopy type of the cobordism category*, Acta Math. **202** (2009), no. 2, 195–239, DOI 10.1007/s11511-009-0036-9. MR2506750

[Gun16] Sam Gunningham, *Spin Hurwitz numbers and topological quantum field theory*, Geom. Topol. **20** (2016), no. 4, 1859–1907, DOI 10.2140/gt.2016.20.1859. MR3548460

[GW14] Zheng-Cheng Gu and Xiao-Gang Wen, *Symmetry-protected topological orders for interacting fermions: Fermionic topological nonlinear σ models and a special group* *supercohomology theory*, Phys. Rev. B **90** (2014), no. 11, 115141, https://arxiv.org/pdf/1201.2648.pdf.

[JF17] Theo Johnson-Freyd, *Spin, statistics, orientations, unitarity*, Algebr. Geom. Topol. **17** (2017), no. 2, 917–956, DOI 10.2140/agt.2017.17.917. MR3623677

[JO12] Niles Johnson and Angélica M. Osorno, *Modeling stable one-types*, Theory Appl. Categ. **26** (2012), No. 20, 520–537. MR2981952

[Joa04] Michael Joachim, *Higher coherences for equivariant K-theory*, Structured ring spectra, London Math. Soc. Lecture Note Ser., vol. 315, Cambridge Univ. Press, Cambridge, 2004, pp. 87–114, DOI 10.1017/CBO9780511529955.006. MR2122155

[Joh80] Dennis Johnson, *Spin structures and quadratic forms on surfaces*, J. London Math. Soc. (2) **22** (1980), no. 2, 365–373, DOI 10.1112/jlms/s2-22.2.365. MR588283

[Kit01] A. Yu. Kitaev, *Unpaired Majorana fermions in quantum wires*, Physics-Uspekhi **44** (2001), no. 10S, 131, https://arxiv.org/pdf/cond-mat/0010440.pdf.

[KT90] R. C. Kirby and L. R. Taylor, Pin *structures on low-dimensional manifolds*, Geometry of low-dimensional manifolds, 2 (Durham, 1989), London Math. Soc. Lecture Note Ser., vol. 151, Cambridge Univ. Press, Cambridge, 1990, pp. 177–242. MR1171915

[KTTW15] Anton Kapustin, Ryan Thorngren, Alex Turzillo, and Zitao Wang, *Fermionic symmetry protected topological phases and cobordisms*, J. High Energy Phys. **12** (2015), 052, front matter+20pp. MR3464750

[Lur09] Jacob Lurie, *On the classification of topological field theories*, Current developments in mathematics, 2008, Int. Press, Somerville, MA, 2009, pp. 129–280. MR2555928

[Lur17] Jacob Lurie, *Higher algebra*, http://www.math.harvard.edu/~lurie/papers/HA.pdf.

[MW07] Ib Madsen and Michael Weiss, *The stable moduli space of Riemann surfaces: Mumford's conjecture*, Ann. of Math. (2) **165** (2007), no. 3, 843–941, DOI 10.4007/annals.2007.165.843. MR2335797

[MZF+12] V. Mourik, K. Zuo, S. M. Frolov, S. R. Plissard, E. P. A. M. Bakkers, and L. P. Kouwenhoven, *Signatures of Majorana fermions in hybrid superconductor-semiconductor nanowire devices*, Science **336** (2012), no. 6084, 1003–1007, https://arxiv.org/pdf/1204.2792.pdf.

[Noo07] Behrang Noohi, *Notes on 2-groupoids, 2-groups and crossed modules*, Homology Homotopy Appl. **9** (2007), no. 1, 75–106. MR2280287

[OMD16] L. Ortiz and M. A. Martin-Delgado, *A bilayer double semion model with symmetry-enriched topological order*, Ann. Physics **375** (2016), 193–226, DOI 10.1016/j.aop.2016.10.008. MR3577243

[Pon55] L. S. Pontryagin, *Gladkie mnogoobraziya i ih primeneniya v teorii gomotopiĭ* (Russian), Trudy Mat. Inst. im. Steklov. no. 45, Izdat. Akad. Nauk SSSR, Moscow, 1955. MR0071767

[PWY17] Pavel Putrov, Juven Wang, and Shing-Tung Yau, *Braiding statistics and link invariants of bosonic/fermionic topological quantum matter in* $2+1$ *and* $3+1$ *dimensions*, Ann. Physics **384** (2017), 254–287, DOI 10.1016/j.aop.2017.06.019. MR3692208

[RLF12] Leonid P. Rokhinson, Xinyu Liu, and Jacek K. Furdyna, *The fractional a.c. Josephson effect in a semiconductor-superconductor nanowire as a signature of Majorana particles*, Nature Physics **8** (2012), 795–799, https://www.nature.com/nphys/journal/v8/n11/pdf/nphys2429.pdf.

[RW14] Oscar Randal-Williams, *Homology of the moduli spaces and mapping class groups of framed, r-Spin and Pin surfaces*, J. Topol. **7** (2014), no. 1, 155–186, DOI 10.1112/jtopol/jtt029. MR3180616

[RW18] Eric C. Rowell and Zhenghan Wang, *Mathematics of topological quantum computing*, Bull. Amer. Math. Soc. (N.S.) **55** (2018), no. 2, 183–238, DOI 10.1090/bull/1605. MR3777017

[Sab18] Ingmar Saberi, *An introduction to spin systems for mathematicians*, https://arxiv.org/pdf/1801.07270.pdf.

[Seg88] G. B. Segal, *The definition of conformal field theory*, Differential geometrical methods in theoretical physics (Como, 1987), NATO Adv. Sci. Inst. Ser. C Math. Phys. Sci., vol. 250, Kluwer Acad. Publ., Dordrecht, 1988, pp. 165–171. MR981378

[Shu10] Mike Shulman, *Constructing symmetric monoidal bicategories*, https://arxiv.org/pdf/1004.0993v1.pdf.

[SP15] Christopher Schommer-Pries, *Tori detect invertibility of topological field theories*, https://arxiv.org/pdf/1511.01772.pdf.

[SP17] Christopher Schommer-Pries, *Invertible topological field theories*, https://arxiv.org/pdf/1712.08029.pdf.

[SSR17a] Hassan Shapourian, Ken Shiozaki, and Shinsei Ryu, *Many-body topological invariants for fermionic symmetry-protected topological phases*, Phys. Rev. Lett. **118** (2017), no. 21, 216402, 6, DOI 10.1103/PhysRevLett.118.216402. MR3693816

[SSR17b] Ken Shiozaki, Hassan Shapourian, and Shinsei Ryu, *Many-body topological invariants in fermionic symmetry-protected topological phases: Cases of point group symmetries*, Phys. Rev. B **95** (2017), 205139, https://arxiv.org/pdf/1609.05970.pdf.

[Sto88] Stephan Stolz, *Exotic structures on 4-manifolds detected by spectral invariants*, Invent. Math. **94** (1988), no. 1, 147–162, DOI 10.1007/BF01394348. MR958593

[Tho54] René Thom, *Quelques propriétés globales des variétés differentiables*, Ph.D. thesis, University of Paris, 1954, http://www.maths.ed.ac.uk/~aar/papers/thomcob.pdf, pp. 17–86.

[TPB11] Ari M. Turner, Frank Pollmann, and Erez Berg, *Topological phases of one-dimensional fermions: An entanglement point of view*, Phys. Rev. B **83** (2011), no. 7, 075102, https://arxiv.org/pdf/1008.4346.pdf.

[Wit16] Edward Witten, *Fermion path integrals and topological phases*, Rev. Mod. Phys. **88** (2016), no. 3, 035001, https://arxiv.org/pdf/1508.04715.pdf.

[XA18] Charles Zhaoxi Xiong and A. Alexandradinata, *Organizing symmetry-protected topological phases by layering and symmetry reduction: A minimalist perspective*, Phys. Rev. B **97** (2018), 115153, https://arxiv.org/pdf/1709.06998.pdf.

[Xio17] Charles Zhaoxi Xiong, *Minimalist approach to the classification of symmetry protected topological phases*, https://arxiv.org/pdf/1701.00004.pdf.

[Yon18] Kazuya Yonekura, *On the cobordism classification of symmetry protected topological phases*, http://arxiv.org/pdf/1803.10796.pdf.

[YWOX14] Yi-Zhuang You, Zhong Wang, Jeremy Oon, and Cenke Xu, *Topological number and fermion Green's function for strongly interacting topological superconductors*, Phys. Rev. B **90** (2014), no. 6, 060502, https://arxiv.org/pdf/1403.4938.pdf.

[Zha94] Wei Ping Zhang, *Circle bundles, adiabatic limits of η-invariants and Rokhlin congruences* (English, with English and French summaries), Ann. Inst. Fourier (Grenoble) **44** (1994), no. 1, 249–270. MR1262887

[Zha17] WeiPing Zhang, *A* mod 2 *index theorem for* pin$^-$ *manifolds*, Sci. China Math. **60** (2017), no. 9, 1615–1632, DOI 10.1007/s11425-016-9040-7. MR3689189

Department of Mathematics, University of Texas, Austin, Texas 78712
Email address: a.debray@math.utexas.edu

Department of Mathematics, University of Texas, Austin, Texas 78712
Email address: gunningham@math.utexas.edu

Contemporary Mathematics
Volume **718**, 2018
https://doi.org/10.1090/conm/718/14476

A guide for computing stable homotopy groups

Agnès Beaudry and Jonathan A. Campbell

Abstract. This paper contains an overview of background from stable homotopy theory used by Freed–Hopkins in their work on invertible extended topological field theories. We provide a working guide to the stable homotopy category, to the Steenrod algebra and to computations using the Adams spectral sequence. Many examples are worked out in detail to illustrate the techniques.

Contents

1. Introduction and organization

1.1. Introduction. The main theorem of [**FH16**] states that deformation classes of reflection positive invertible n-dimensional extended topological field theories with symmetry group H_n are classified by the torsion in

$$[MTH, \Sigma^{n+1} I_{\mathbb{Z}}].$$

Here, MTH is the Madsen–Tillmann spectrum associated to a group H which is a stabilization of H_n, $I_{\mathbb{Z}}$ is the Anderson dual of the sphere spectrum, and $[-,-]$ denotes the *stable* homotopy classes of maps. These concepts will be discussed in Section 2.

In order to complete the classification problem, it is necessary to be able to compute stable homotopy classes of maps from a spectrum X to $I_{\mathbb{Z}}$. This problem can be reduced to the computation of the stable homotopy groups of X itself as will be described in Section 2.8.

2010 *Mathematics Subject Classification.* Primary 55Q10, 55T15, 55R40, 55S10.

This material is based upon work supported by the National Science Foundation under Grant No. DMS-1725563.

In general, it is notoriously difficult, if not impossible, to completely compute the homotopy groups of a spectrum X. However, homotopy theorists are very good at doing these computations in small ranges and the problems motivated by physics only require information in small dimensions, making us a perfect match.

The main tool used to compute low-dimensional homotopy groups of spectra is the Adams spectral sequence. Adams initially introduced this spectral sequence in order to resolve the Hopf invariant one problem [**Ada58, Ada60**]. It has been a standard tool in homotopy theory since then. In brief, the Adams spectral sequence takes in information about the cohomology of a space or spectrum and outputs information about its stable homotopy groups.

The Steenrod algebra, which we denote by $\mathcal{A}$, is one of the classical structures in homotopy theory. The mod-2 cohomology $H^*(X;\mathbb{Z}/2)$ of any space or spectrum X is a module over $\mathcal{A}$. This module is the input to the Adams spectral sequence. Although it can be difficult to compute the $\mathcal{A}$-module structure of the cohomology of an arbitrary space or spectrum X, we work under the favorable circumstance that the examples we consider are related to the classifying spaces of various Lie groups. With the $\mathcal{A}$-module structure of $H^*(X;\mathbb{Z}/2)$ in hand, and some knowledge of homological algebra over the Steenrod algebra, the E_2-page of the Adams spectral sequence can be computed. In the low-dimensional range, we are lucky, and every example we consider is fully computable by hand.

The aim of this paper is to introduce the reader to enough of the machinery of spectra, the Steenrod algebra and the Adams spectral sequence to understand the computation of the homotopy groups π_*MTH. To illustrate how one applies the theory, we do the computations for a few examples. In particular, we go over the cases when H is Spin^c and Pin^c in detail, an exercise which was left to the reader in Section 10 of [**FH16**] and was not covered in [**Cam17**].

1.2. Organization. In order to fully explicate the computations for readers unfamiliar with stable homotopy theory, we include an introduction to spectra in Section 2. Among other topics, we discuss the category of spectra and its homotopy category (Section 2.1 and Section 2.3), the homotopy groups of spectra (Section 2.2), the Anderson dual (Section 2.8) and the construction of Thom spectra (Section 2.9). These latter are integral to the Freed–Hopkins classification since it is Thom spectra that are tightly linked with cobordism groups and the cobordism hypothesis.

In Section 3 we discuss the Steenrod algebra, $\mathcal{A}$, which is a non-commutative, infinitely generated algebra that acts on the cohomology of all spaces and spectra. In Section 3.2, we introduce $\mathcal{A}_1$, an eight dimensional sub-algebra of $\mathcal{A}$ that will play a crucial role in the computations. In Section 3.4 we compute the $\mathcal{A}_1$-module structure for some examples of cobordism spectra. This computation depends on knowing how to determine the $\mathcal{A}_1$-module structure of the cohomology of classifying spaces, along with the Thom isomorphism and the Wu formula. These things are discussed in Section 3.3.

The Adams spectral sequence is introduced in Section 4. The primary tool for computation with the Adams spectral sequence is homological algebra over $\mathcal{A}$ and, in our examples, over $\mathcal{A}_1$. This section includes a discussion of resolutions (Section 4.4) and computations of $\mathrm{Ext}_{\mathcal{A}_1}$ for a menagerie of $\mathcal{A}_1$-modules. It includes explanations of Adams charts (Section 4.3), of certain mulitplicative structures on

Ext (Section 4.2) and a variety of useful tricks. In Section 4.7, we formally construct the spectral sequence and in Section 4.8, we provide a "user's manual".

In Section 5, we come to the main event. In the range $0 \leq n \leq 4$ we compute $\pi_n MTH(s)$ and $\pi_n MTH^c(s)$ in all of the cases that were not explained in further detail in [**Cam17**]. The computations rely on the Adams spectral sequence, and we use all of the material developed in Section 3 and Section 4 to compute the E_2-pages. In such a small range, and in these cases, the spectral sequences collapse and the homotopy groups can be read off of the Adams charts.

2. A working guide to spectra

In this section, we give an introduction to spectra. If one is interested in computing homotopy groups, then one can often get away with an understanding of the properties of the homotopy category of spectra (see Section 2.3). Some of the information we include is not strictly necessary for this understanding, but we tried to strike a balance between too little and too much information.

For a more in-depth introduction to spectra, a starting point would be Section 1.4 of Lurie [**Lur**] and the introduction by Elmendorff–Kriz–Mandell–May to Chapter 6 of [**Jam95**] (a book that contains other hidden gems). One could then move on to Part III of Adams [**Ada95**], Chapter 10.9 of Weibel [**Wei94**] and Chapter 12 of Aguilar–Gitler–Carlos [**AGP02**]. For serious treatments of different modern models of the category of spectra together with all of its structure, see the first parts of Schwede [**Sch**], Mandell–May–Schwede–Shipley [**MMSS01**], Elmendorff–Kriz–Mandell–May [**EKMM97**] or Lurie [**Lur**]. For the equivariant treatment, see Lewis–May–Steinberger [**LMSM86**].

NOTATION 2.0.1. We let Ab be the category of graded abelian groups. We let Top_* be a category of suitably nice based topological spaces with continuous maps that preserve the base points.

Motivation for the category of spectra comes from at least two directions. First, there is Brown's representability theorem that states that a cohomology theory $E^*\colon \mathrm{Top}_*^{\mathrm{op}} \to \mathrm{Ab}$ has a sequence of representing spaces E_n. That is, $E^n(X) \cong [X, E_n]$. We will let $\Sigma(-)$ be the reduced suspension and $\Omega(-)$ be the based loops functor. The isomorphism $[\Sigma X, E_n] \cong [X, \Omega E_n]$ together with the suspension isomorphism $E^n(X) \cong E^{n+1}(\Sigma X)$ give rise to an isomorphism

$$[X, E_n] \xrightarrow{\cong} [X, \Omega E_{n+1}]$$

which is natural in X. By the Yoneda Lemma, this corresponds to a weak equivalence $\omega_n\colon E_n \xrightarrow{\simeq} \Omega E_{n+1}$. Further, to discuss natural transformations between cohomology theories, one is led to discuss maps between these sequences of spaces. It thus behooves us to construct a category which consists of sequences of spaces.

Another motivation is via Freudenthal's suspension theorem. Let X be a k-connected topological space. Freudenthal's suspension theorem states that the map $\pi_n(X) \to \pi_{n+1}(\Sigma X)$ is an isomorphism if $n \leq 2k$. For a fixed n and connected X, this implies that $\pi_{n+k}(\Sigma^k X)$ stabilizes as k goes to infinity. This motivates the definition of the *nth stable homotopy group*

$$\pi_n^s X = \mathrm{colim}_k\, \pi_{n+k}\Sigma^k X \cong \pi_{n+m}(\Sigma^m X) \quad m \gg 0.$$

An amazingly useful fact is that $\pi_*^s\colon \mathrm{Top}_* \to \mathrm{Ab}$ is a homology theory, making the stable homotopy groups often (slightly) more computable than the usual, *unstable*, homotopy groups. It is useful to consider the sequences of spaces $\{\Sigma^n X\}$ as the fundamental objects, and we come again to a point where it is necessary to define some category of sequences of spaces.

We will define the following five categories in the next few sections:

(1) The category of prespectra, denoted PreSp. See Definition 2.1.1.
(2) The category of spectra, denoted Sp. See Definition 2.1.3.
(3) The category of CW-prespectra, denoted CWPreSp. See Definition 2.1.4.
(4) The category of CW-spectra, denoted CWSp. See Definition 2.1.12.
(5) The homotopy category of spectra, denoted hSp. See Section 2.3.

The first four are a means to the fifth. We justify this complication by the following analogy inspired from Chapter 10 of [**Wei94**] and, in particular, Analogy 10.9.7. The reader can skip this analogy now and come back to it at the end of Section 2.3.

ANALOGY 2.0.2. To justify having both spectra and prespectra, we make an analogy with the categories of sheaves and presheaves. Although we do homological algebra in the category of sheaves, some constructions are easier to make in the category of presheaves. The forgetful functor from sheaves to presheaves has a left adjoint, the sheafification functor. This allows one to transport constructions from presheaves to sheaves.

In this part of the analogy, spectra are the sheaves and prespectra are the presheaves. The analogue of the sheafification functor is called spectrification and is denoted $L\colon \mathrm{PreSp} \to \mathrm{Sp}$. It is the left adjoint to a forgetful functor from Sp to PreSp. See Remark 2.1.9.

Now, switching gears, we think of the category $\mathcal{C}$ of bounded below chain complexes of R-modules. There are two important kinds of equivalences in this category, the chain homotopy equivalences and the quasi-isomorphisms. The derived category $D(\mathcal{C})$ is characterized as the initial category which receives a functor $\mathcal{C} \to D(\mathcal{C})$ such that the quasi-isomorphisms are mapped to isomorphisms in $D(\mathcal{C})$. Chain homotopy equivalence is an equivalence relation, but the property of being quasi-isomorphic is not. In theory, it takes more work to invert the quasi-isomorphisms than it does to invert the chain homotopy equivalences. However, a quasi-isomorphism between bounded below projective chain complexes is a chain homotopy equivalence and, further, any chain complex is quasi-isomorphic to a projective one. Therefore, a model for $D(\mathcal{C})$ is the category whose objects are projective chain complexes and morphisms are chain homotopy equivalences of maps.

In this part of the analogy, the topological spaces are the R-modules and the category of spectra is the analogue of $\mathcal{C}$. The chain homotopy equivalences correspond to the homotopy equivalences and the quasi-isomorphisms to the weak homotopy equivalences. The homotopy category of spectra is analogous to $D(\mathcal{C})$. The projective chain complexes are the analogues to CW-spectra, and a model for the homotopy category of spectra is the category of CW-spectra together with homotopy classes of maps between them.

We have not mentioned CW-prespectra and use it to tie the knot between the two analogies: CW-prespectra are easy to define in prespectra, and the spectrification functor is used to transfer the definition to spectra.

2.1. The categories.

DEFINITION 2.1.1 (Prespectra). A *prespectrum* X is a sequence of spaces $X_n \in \mathrm{Top}_*$ for $n \geq 0$ and continuous maps $\sigma_n \colon \Sigma X_n \to X_{n+1}$. We let $\omega_n \colon X_n \to \Omega X_{n+1}$ be the adjoint of σ_n and note that giving the structure maps σ_n of a prespectrum is equivalent to specifying the maps ω_n. A map of prespectra $f \colon X \to Y$ of degree r is a sequence of continuous, based maps $f_n \colon X_n \to Y_{n-r}$ such that the following diagram commutes:

$$\begin{array}{ccc} \Sigma X_n & \xrightarrow{f_n} & \Sigma Y_{n-r} \\ \downarrow{\scriptstyle \sigma_n} & & \downarrow{\scriptstyle \sigma_{n-r}} \\ X_{n+1} & \xrightarrow[f_{n+1}]{} & Y_{n+1-r}. \end{array}$$

We let PreSp denote the category of prespectra. (The plural of prespectrum is *prespectra.*)

REMARK 2.1.2. If the maps ω_n are weak homotopy equivalences, then X is often called an *Ω-prespectrum.*

DEFINITION 2.1.3 (Spectra). A prespectrum is called a *spectrum* if the maps ω_n are homeomorphisms. We let Sp denote the full subcategory of prespectra generated by the objects which are spectra.

DEFINITION 2.1.4 (CW-prespectra). We call a prespectrum a *CW-prespectrum* if the spaces X_n are CW-complexes and the maps $\Sigma X_n \to X_{n+1}$ are cellular inclusions. We let CWPreSp denote the full subcategory of prespectra generated by the objects which are CW-prespectra.

EXAMPLE 2.1.5. The standard example is the suspension prespectrum $\Sigma^\infty A$ of a based topological space A. Its nth space is given by $\Sigma^n A$ and the structure maps are identities $\Sigma\Sigma^n A \cong \Sigma^{n+1} A \to \Sigma^{n+1} A$. In fact, this extends to a functor $\Sigma^\infty \colon \mathrm{Top}_* \to \mathrm{PreSp}$ which sends a space A to $\Sigma^\infty A$. The functor Σ^∞ is left adjoint to the functor $\Omega^\infty \colon \mathrm{PreSp} \to \mathrm{Top}_*$ which sends a prespectrum to its zeroth space.

EXAMPLE 2.1.6. The Eilenberg-MacLane prespectrum HG, where G is an abelian group, has nth space $K(G,n)$. The structure maps of HG are the adjoints to the homotopy equivalences $\omega_n \colon K(G,n) \to \Omega K(G, n+1)$. A homomorphism of abelian groups $G_1 \to G_2$ give rise to a map of prespectra $HG_1 \to HG_2$.

EXAMPLE 2.1.7. Another example is given by K-theory. The odd spaces of K are the infinite unitary group U and the even spaces are $\mathbb{Z} \times BU$, where BU is the classifying space of U. The structure maps ω_n are the equivalences given by Bott Periodicity. Similarly, real K-theory is denoted by KO. Its spaces repeat with period eight starting with $\mathbb{Z} \times BO$, where BO is the classifying space of the infinite orthogonal group O.

EXAMPLE 2.1.8. If $X = X_0$ is an infinite loop space so that there exists spaces X_k so that $X \simeq \Omega^k X_k$ for all $k \geq 0$, then the X_k assemble into a prespectrum.

DEFINITION 2.1.9. The *spectrification functor* $L \colon \mathrm{PreSp} \to \mathrm{Sp}$ is the left adjoint to the forgetful functor $U \colon \mathrm{Sp} \to \mathrm{PreSp}$.

REMARK 2.1.10. The functor L exists by Freyd's adjoint functor theorem. It can be constructed easily if the maps ω_n are inclusions (for example, if X is a CW-prespectrum). In this case, LX is the spectrum whose kth space is

$$LX_k = \operatorname{colim}_n \Omega^{n+k} X_n,$$

where the colimit is taken over $\Omega^{n+k}(\omega_n)\colon \Omega^{n+k}X_n \to \Omega^{n+k+1}X_{n+1}$. If X is already a spectrum, then $LX \cong X$ as these maps are all homeomorphisms. For a general definition, we refer the reader to Appendix A.1 of [**LMSM86**].

WARNING 2.1.11. We abuse notation and write ULX simply as LX. Further, we often omit the L if we are not emphasizing the replacement. For example, we write $\Sigma^\infty A = L(\Sigma^\infty A)$, $HG = L(HG)$, etc..

DEFINITION 2.1.12 (CW-Spectra). The category of *CW-spectra*, denoted CWSp, is the full subcategory of spectra generated by the image of the restriction of L to CW-prespectra. That is, $X \in \mathrm{CWSp}$ if it is of the form LY for some $Y \in \mathrm{CWPreSp}$.

We summarize the discussion by the following diagram of adjunctions, where $\Omega^\infty\colon \mathrm{Sp} \to \mathrm{Top}_*$ is also the zeroth space functor:

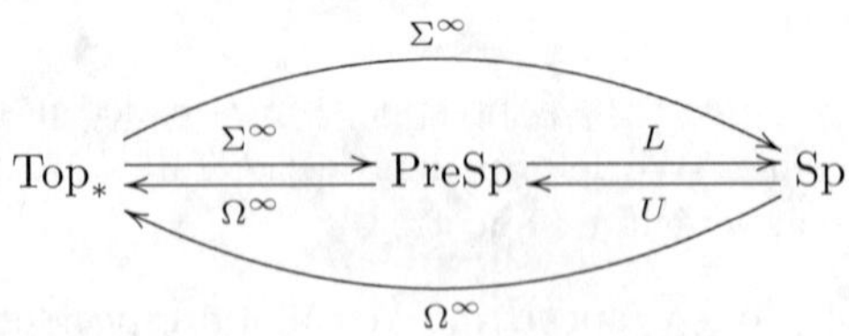

The coproduct in Top_* is the wedge $A \vee B$. The category Top_* is a closed symmetric monoidal category, where the hom objects are the spaces of continuous based maps $\mathrm{Maps}(A,B)$ and the symmetric monoidal product is the smash product $A \wedge B$. There is an associated homeomorphism

$$\mathrm{Maps}(A \wedge B, C) \cong \mathrm{Maps}(A, \mathrm{Maps}(B, C)).$$

We briefly discuss related constructions in (pre)spectra.

For X a prespectrum and A a based topological space, we let $X \wedge A$ be the prespectrum whose spaces are given by $X_n \wedge A$ and structure maps by $\sigma_n \wedge \mathrm{id}_A$. We define $\Sigma^r X = X \wedge S^r$ with $\Sigma = \Sigma^1$.

Similarly, we let $F(A,X)$ be the prespectrum whose nth space is $\mathrm{Maps}(A, X_n)$ and whose structure maps are given $f \mapsto \omega_n \circ f$, using the identification

$$\Omega\,\mathrm{Maps}(A, X_{n+1}) \cong \mathrm{Maps}(A, \Omega X_{n+1}).$$

We let $\Omega(X) = F(S^1, X)$. In the homotopy category (defined in Section 2.3), the functors $\Omega(-)$ and $\Sigma(-)$ become inverses, so we let $\Sigma^{-1}(-) = \Omega(-)$.

In prespectra, the coproduct is also a wedge construction. The spaces of $X \vee Y$ are $X_n \vee Y_n$ with structure maps $\sigma_n \vee \sigma_n$, using the fact that $\Sigma(X_n \vee Y_n) \cong \Sigma X_n \vee \Sigma Y_n$.

These constructions transfer to spectra via the spectrification functor L, and we abuse notation by dropping the L from the notation. For example, we write $\Sigma X = L(\Sigma X)$.

REMARK 2.1.13. Smash products of spectra and function spectra are harder to construct, and we will not do this here. We do note however that there are versions of the category of spectra which are closed symmetric monoidal with respect to an appropriate smash product. The first such construction is due to Elmendorf–Kriz–Mandell–May [**EKMM97**]. However, up to homotopy (see the definition of homotopy in these categories below), the smash product $X \wedge Y$ was constructed directly two decades prior. This is called Boardman's *handicrafted* smash product and the construction is described in [**Ada95**]. One can also construct a function spectrum $F(X,Y)$ so that $F(X,F(Y,Z)) \simeq F(X \wedge Y, Z)$. We will only use these constructions up to homotopy and we take them for granted.

2.2. Homotopies and homotopy groups. Let I_+ be the unit interval $[0,1]$ with a disjoint basepoint. Then the prespectrum $X \wedge I_+$ admits a map

$$X \vee X \xrightarrow{i_0 \vee i_1} X \wedge I_+$$

defined levelwise on each factor by the inclusions at 0 and 1 respectively. As in Top_*, we can use the prespectrum $X \wedge I_+$ to define homotopies between maps.

Two maps of prespectra $f, g\colon X \to Y$ are *homotopic*, denoted $f \simeq g$, if there is a map $H\colon X \wedge I_+ \to Y$ which restricts to $f \vee g$ along the inclusion

$$X \vee X \xrightarrow{i_0 \vee i_1} X \wedge I_+ \xrightarrow{H} Y.$$

Maps of spectra are homotopic if they are homotopic as maps of prespectra. We will let the set of homotopy classes of maps between two (pre)spectra X and Y be denoted by $\{X, Y\}$. If Y is an Ω-prespectrum, this is in fact an abelian group. Similarly, homotopy classes of maps of degree r are denoted by $\{X, Y\}_r$. Two (pre)spectra X and Y are *homotopy equivalent* if there are maps $f\colon X \to Y$ and $g\colon Y \to X$ such that $f \circ g \simeq \mathrm{id}_Y$ and $g \circ f \simeq \mathrm{id}_X$.

DEFINITION 2.2.1. Let X be a (pre)spectrum and $n \in \mathbb{Z}$. The *nth homotopy group* of X is

$$\pi_n X = \mathrm{colim}_k \, \pi_{n+k} X_k$$

where the maps in the colimit take an element $S^{n+k} \to X_k$ to the composite $S^{n+k+1} \to \Sigma X_k \xrightarrow{\sigma_k} X_{k+1}$. A map of (pre)spectra is a *weak homotopy equivalence* if it induces an isomorphism on homotopy groups.

REMARK 2.2.2. The unit of the adjunction $X \to LX$ is a functorial replacement of X by the weakly homotopy equivalent spectrum LX.

REMARK 2.2.3 (Whitehead's theorem). A map of CW-spectra which is a weak homotopy equivalence is also a homotopy equivalence.

2.3. The homotopy category of spectra and its triangulation. First, we recall the analogous object for Top_*. The homotopy category of based topological spaces hTop_* is the initial category receiving a functor from Top_* which sends weak homotopy equivalences to isomorphisms. Using Whitehead's theorem and CW-approximation, one model for hTop_* has objects the pointed CW-complexes and morphisms the based homotopy classes of maps between them. The map $\mathrm{Top}_* \to \mathrm{hTop}_*$ sends A to a CW-approximation ΓA, which is functorial up to homotopy, and a map f to the homotopy equivalence class of Γf.

There are many constructions of the homotopy category of spectra, which we denote by hSp, including through the theory of ∞-categories. These all give equivalent categories and hSp is one of the modern settings for homotopy theory. In this section, we give some of the standard tools to work in hSp.

The homotopy category hSp is initial among categories that admit a functor out of (Pre)Sp which sends the weak homotopy equivalences to isomorphisms. In particular, any functor (Pre)Sp $\to \mathcal{D}$ with this property factors through the functor (Pre)Sp $\to$ hSp:

$$\begin{array}{ccccc} \mathrm{PreSp} & \xrightarrow{L} & \mathrm{Sp} & \longrightarrow & \mathrm{hSp} \\ & \searrow & \downarrow & \swarrow & \\ & & \mathcal{D} & & \end{array}$$

The objects of hSp are simply called *spectra*. The category hSp is a triangulated category with shift operator given by the suspension $\Sigma(-)$, which in hSp becomes inverse to $\Sigma^{-1}(-) = \Omega(-)$. We define

$$[X, Y]_r := \mathrm{hSp}_r(X, Y) = \mathrm{hSp}(\Sigma^r X, Y)$$

and let $[X, Y] = [X, Y]_0$. These are abelian groups for all X, Y and r. The isomorphisms in hSp are denoted by $\simeq$ because of their relationship to the weak homotopy equivalences.

REMARK 2.3.1. The morphisms in hSp must be computed with care, and we remind the reader of Analogy 2.0.2. With this analogy in mind, note that $[X, Y]$ is not in general isomorphic to $\{X, Y\}$. Here, $\{X, Y\}$ denotes the homotopy classes of maps as defined in Section 2.2. This is the essence of the "cells now — maps later" discussion on p.142 of [**Ada95**].

REMARK 2.3.2 (CW-approximation). For any prespectrum X, there is a CW-spectrum ΓX connected to X by a zig-zag of weak homotopy equivalences. The construction is functorial up to homotopy.

REMARK 2.3.3. We use CW-approximation to describe models for hSp. The first has objects CW-spectra and morphisms homotopy classes of maps between them. In particular, if X and Y are CW-spectra, then $[X, Y] \cong \{X, Y\}$. The functor (Pre)Sp $\to$ hSp sends X to ΓX and a map f to the homotopy equivalence class of Γf. A slightly larger model is to let the objects be CW-prespectra and morphisms $[X, Y] \cong \{LX, LY\} \cong \{X, LY\}$. One can also take objects to be all prespectra and morphisms to be $[X, Y] \cong \{\Gamma X, \Gamma Y\}$.

The point we want to stress here is that, for any two X and Y, whether they be prespectra, spectra, CW-prespectra or CW-spectra, it makes sense to write down $[X, Y]$. Every point of view yields isomorphic abelian groups. In hSp, we forget the distinctions: All objects have equal dignity and are called *spectra*.

We extend Σ^∞ to a functor $\Sigma^\infty \colon \mathrm{Top}_* \to \mathrm{hSp}$ by sending A to the image of $\Sigma^\infty A \in \mathrm{hSp}$. We often simply write A to denote $\Sigma^\infty A \in \mathrm{hSp}$. For example, S^t as a spectrum is

$$S^t \simeq \Sigma^\infty S^t \simeq \Sigma^t \Sigma^\infty S^0 \simeq \Sigma^t S^0.$$

The *sphere spectrum* is the spectrum S^0. On the other hand, Ω^∞ induces a functor $\Omega^\infty \colon \mathrm{hSp} \to \mathrm{hTop}_*$.

We let $F(X, Y)$ and $X \wedge Y$ be the function spectrum and smash product in hSp. See Remark 2.1.13. The category hSp is a closed symmetric monoidal category so that

$$F(X \wedge Y, Z) \simeq F(X, F(Y, Z)). \tag{2.3.1}$$

The sphere spectrum S^0 is the unit for the symmetric monoidal structure and

$$S^0 \wedge X \simeq X, \qquad\qquad F(S^0, X) \simeq X.$$

If X is a spectrum and A is a based topological space, for the constructions described in Section 2.1, we have $A \wedge X \simeq (\Sigma^\infty A) \wedge X$ and $F(A, X) \simeq F(\Sigma^\infty A, X)$.

There is an identity

$$[X, Y]_t = \pi_t F(X, Y).$$

In particular, if $\pi_* X$ denotes the homotopy groups of X,

$$\pi_t X \cong [S^t, X] \cong \pi_0 F(S^t, X) \cong \pi_t F(S^0, X).$$

The category hSp has arbitary products and coproducts. Further, for a collection of objects X_α, $\alpha \in I$ with the property that, for every $k \in \mathbb{Z}$, $\pi_k X_\alpha = 0$ for all but finitely many $\alpha \in I$, the map

$$\bigvee_{\alpha\in I} X_\alpha \xrightarrow{\simeq} \prod_{\alpha\in I} X_\alpha \tag{2.3.2}$$

is an isomorphism.

Pushout and pullback diagrams also coincide in hSp. The exact triangles

$$X \to Y \to Z \to \Sigma X$$

are equivalently called *cofiber* and *fiber* sequences. The spectrum Z is called the *cofiber* of $X \to Y$, while X is called the *fiber* of $Y \to Z$. A map $X \to Y$ is null homotopic if and only if $Z \simeq Y \vee \Sigma X$.

A standard example of an exact triangle in hSp is constructed by killing an element in homotopy. For example, if $\alpha\colon S^n \to S^m$ is an element of $\pi_n S^m$, then $C(\alpha)$ is defined by the exact triangle

$$S^n \xrightarrow{\alpha} S^m \to C(\alpha) \to S^{n+1}.$$

If $X \to Y \to Z \to \Sigma X$ is an exact triangle, then so are the four term sequences obtained by applying $W \wedge (-)$, $F(W, -)$ or $F(-, W)$. Further, applying either of $[-, X]$ and $[X, -]$ to an exact triangle gives rise to a long exact sequence of abelian groups. In particular, there are long exact sequences on homotopy groups $\pi_*(-)$.

A useful fact about the functor Σ^∞ is that it commutes with $\wedge$ and with $\vee$. Also, applying Σ^∞ to a homotopy cofiber sequence $A \to B \to C$ of spaces gives an exact triangle in hSp. In particular, the cofiber sequence $A \vee B \to A \times B \to A \wedge B$ gives rise to a split cofiber sequence of spectra so that

$$\Sigma^\infty(A \times B) \simeq \Sigma^\infty A \vee \Sigma^\infty B \vee \Sigma^\infty(A \wedge B).$$

WARNING 2.3.4. From this point onwards, when we say "spectrum", we mean an element of hSp unless otherwise specified.

2.4. Cohomology and homology theories. A generalized homology theory is a collection of functors $E_n\colon \mathrm{hSp} \to \mathrm{Ab}$ indexed by $\mathbb{Z}$, together with natural isomorphisms $E_{n+1}(\Sigma -) \xrightarrow{\cong} E_n(-)$ such that E_n takes arbitrary coproducts to direct sums and exact triangles to exact sequences. A generalized cohomology theory is a collection of contravariant functors $E^n\colon \mathrm{hSp}^{\mathrm{op}} \to \mathrm{Ab}$ indexed by $\mathbb{Z}$ and natural isomorphisms $E^n(-) \xrightarrow{\cong} E^{n+1}(\Sigma -)$ such that E^n that takes arbitrary coproducts to direct products and exact triangles to exact sequences. We refer the reader to Whitehead [**Whi62**, Section 5] for more on generalized homology and cohomology theories.

Any spectrum in hSp gives rise to generalized homology and cohomology theories $E_*\colon \mathrm{hSp} \to \mathrm{Ab}$ and $E^*\colon \mathrm{hSp}^{\mathrm{op}} \to \mathrm{Ab}$. Further, by precomposing with $\Sigma^\infty\colon \mathrm{Top}_* \to \mathrm{hSp}$, we obtain (reduced) theories defined on topological spaces. If $E \in \mathrm{hSp}$,

$$E^n(X) = [X, E]_{-n} \cong \pi_{-n}F(X, E) \cong [X, \Sigma^n E]$$

and

$$E_n(X) = \pi_n(E \wedge X).$$

Conversely, the Brown representability theorem implies that any homology or cohomology theory is represented by a spectrum $E = \{E_n\}$ so that $E^n(X) = [X, E_n]$.

REMARK 2.4.1. If $E \in \mathrm{PreSp}$ is a prespectrum, and A is a topological space,

$$E^n(A) \cong [A, (LE)_n]$$

where the right hand side denotes homotopy classes of maps in Top_*. In particular, if $E \in \mathrm{Sp}$, then $E^n(A) \cong [A, E_n]$. In fact, for this to hold, it is enough that the structure maps ω_n be weak homotopy equivalences (i.e., that E be an Ω-prespectrum).

If $E \in \mathrm{CWPreSp}$ is such that E_n is $n-1$-connected, then

$$E_n(A) \cong \pi_n(E \wedge A) \cong \mathrm{colim}_k\, \pi_{n+k}(E_k \wedge A).$$

EXAMPLE 2.4.2. If $E = HG$, the Eilenberg–MacLane spectrum for an abelian group G and A is a based space, or B_+ is an unbased space with a disjoint base point, then

$$HG^*(A) = \widetilde{H}^*(A; G) \qquad\qquad HG^*(B_+) = H^*(B; G).$$

Further, by definition, $\widetilde{H}^*(A; G) \cong HG^*(\Sigma^\infty A)$.

2.5. Connective spectra. Let $A \in \mathrm{Top}_*$ be a connected CW-complex. For every $m \geq 0$, there is a space $A_{\tau \geq m}$ with the property that $\pi_n A_{\tau\geq m} = 0$ if $n < m$, together with a map $A_{\tau\geq m} \to A$ which is an isomorphism on π_n if $n \geq m$. The space $A_{\tau\geq m}$ is called the *mth connective cover of* A, and is obtained as the mth stage of the Whitehead tower. This can be done functorially and the spaces $A_{\tau\geq m}$ are unique up to canonical isomorphism in hTop_*.

Note that the homotopy groups of spectra are defined for any integer $n \in \mathbb{Z}$. In particular, some spectra have negative homotopy groups. If $X \in \mathrm{hSp}$ and $m \in \mathbb{Z}$, the *mth connective cover* of X is a spectrum $X_{\tau\geq m}$ with the property that $\pi_n X_{\tau\geq m} = 0$ for $n < m$, together with a map $X_{\tau \geq m} \to X$ which is an isomorphism on π_n if $n \geq m$. If $X \in \mathrm{hSp}$ is represented by a prespectrum with spaces X_n, then $X_{\tau\geq m}$ is represented by a prespectrum whose spaces are $(X_{\tau\geq m})_n = (X_n)_{\tau \geq m+n}$ and whose structure maps are obtained from those of X using the functoriality and uniqueness. The spectrum $X_{\tau\geq 0}$ is called the *connective cover* of X.

NOTATION 2.5.1. The spectrum ku denotes the connective cover of the K-theory spectrum K. The spectrum ko denotes the connective cover of the real K-theory spectrum KO.

2.6. Multiplicative homology theories. One of the main reasons for introducing a symmetric monoidal products on the category of spectra Sp or on its homotopy category hSp is the discussion of *ring spectra.* The cohomology theory that one first encounters, singular cohomology, has the structure of a graded ring. By the Brown representability theorem, this gives rise to maps $H\mathbb{Z} \wedge H\mathbb{Z} \to H\mathbb{Z}$ for the Eilenberg–MacLane spectrum $H\mathbb{Z}$. Many cohomology theories come equipped with this structure; for example, K and KO-theory and nearly all cobordism theories.

We give some definitions in hSp. A *ring spectrum* is a spectrum $R \in$ hSp together with a multiplication map $\mu\colon R \wedge R \to R$ and a unit map $\eta\colon S^0 \to R$ such that the diagram

$$\begin{array}{ccccc} S^0 \wedge R & \xrightarrow{\eta \wedge \mathrm{id}_R} & R \wedge R & \xleftarrow{\mathrm{id}_R \wedge \eta} & R \wedge S \\ & \searrow^{\simeq} & \downarrow{\scriptstyle \mu} & \swarrow^{\simeq} & \\ & & R & & \end{array}$$

commutes (in hSp). Granted a notion of ring spectrum, we can define commutative ring spectra, and module spectra. A commutative ring spectrum is one such that the diagram

$$\begin{array}{ccc} R \wedge R & \xrightarrow{\mathrm{tw}} & R \wedge R \\ & \searrow^{\mu} \quad \swarrow^{\mu} & \\ & R & \end{array}$$

commutes, where tw is the map that exchanges the two copies of R. For R a ring spectrum, an R-module spectrum is a spectrum M together with a map $R \wedge M \to M$ which fits into the commutative diagrams that categorify the notion of a module over a ring.

Much of the intuition from homological algebra can be carried over to the context of ring spectra and module spectra. For example, one can define resolutions in this context. The homotopy groups of a resolution will reflect properties of the homotopy groups of the spectrum it resolves. See, for example, [**Mil81**]. This is one of the ideas in the construction of the Adams spectral sequence. See Section 4.7.

A construction from algebra that requires more care with the smash product when being adapted to spectra is the notion of quotient modules. This is solved in the modern categories of spectra, but is not needed here.

2.7. Spanier–Whitehead duality. The functional dual of a spectrum X is the function spectrum $F(X, S^0)$. This is often denoted by DX in analogy with Spanier–Whitehead duality. If $X \simeq \Sigma^\infty A$ for a finite CW-complex A, then DX is the classical Spanier–Whitehead dual of A.

The enriched adjunction (2.3.1) gives rise to certain important maps. First, there are the units and the counits which are "coevaluations" and "evaluations" respectively:

$$coev\colon Y \to F(X, X \wedge Y) \qquad\qquad ev\colon X \wedge F(X,Y) \to Y$$

Using the adjunction (2.3.1) and ev, for any spectra X, Y and Z, the adjoint to the $X \wedge F(X,Y) \wedge Z \xrightarrow{ev \wedge Z} Y \wedge Z$ gives a map

$$F(X,Y) \wedge Z \to F(X, Y \wedge Z) \tag{2.7.1}$$

which may or may not be an isomorphism in hSp.

The spectrum Z is called *dualizable* if this is an isomorphism in hSp for all spectra X and Y. Examples of dualizable spectra are the spheres $S^t = \Sigma^t S^0$ and, more generally, the suspension spectrum $\Sigma^\infty A$ of any finite CW-complex A. Finally, to verify that Z is dualizable, it is enough to check that

$$DZ \wedge Z \simeq F(Z, S^0) \wedge Z \to F(Z, Z)$$

is a weak equivalence.

2.8. Brown-Comenetz and Anderson duality. For any injective abelian group A, the functor from Top_* to abelian groups given by

$$I_A^n(X) = \mathrm{Hom}_{\mathbb{Z}}(\pi_n(\Sigma^\infty X), A)$$

defines a cohomology theory, which is represented by a spectrum denoted I_A. For example, if $A = \mathbb{Q}$, then

$$I_{\mathbb{Q}}^n(X) \cong \widetilde{H}^n(X; \mathbb{Q}),$$

and $I_{\mathbb{Q}}$ is equivalent to $H\mathbb{Q}$.

Since $\mathbb{Q}/\mathbb{Z}$ is an injective abelian group, we also obtain a spectrum $I_{\mathbb{Q}/\mathbb{Z}}$, which is often called the *Brown-Comenetz spectrum.* The natural map $\mathbb{Q} \to \mathbb{Q}/\mathbb{Z}$ together with the Yoneda Lemma gives rise to a map of spectra $I_{\mathbb{Q}} \to I_{\mathbb{Q}/\mathbb{Z}}$. Then $I_{\mathbb{Z}}$ is defined by the exact triangle in hSp

$$I_{\mathbb{Z}} \to I_{\mathbb{Q}} \to I_{\mathbb{Q}/\mathbb{Z}} \to \Sigma I_{\mathbb{Z}}. \tag{2.8.1}$$

The spectrum $I_{\mathbb{Z}}$ is called the *Anderson dual spectrum.*

Associated to (2.8.1) is a long exact sequence on cohomology

$$\ldots \to I_{\mathbb{Q}}{}^{*-1}(X) \to I_{\mathbb{Q}/\mathbb{Z}}^{*-1}(X) \to I_{\mathbb{Z}}^*(X) \to I_{\mathbb{Q}}{}^*(X) \to I_{\mathbb{Q}/\mathbb{Z}}^*(X) \to \ldots$$

If the homotopy groups $\pi_*(\Sigma^\infty X)$ are finitely generated abelian groups in each degree, one can deduce from this long exact sequence that there is an isomorphism

$$I_{\mathbb{Z}}^*(X) \cong \mathrm{Torsion}(\pi_{*-1}(\Sigma^\infty X)) \oplus \mathrm{Free}(\pi_* X).$$

So, computing $I_{\mathbb{Z}}^*(X) \cong [X, \Sigma^* I_{\mathbb{Z}}]$ is equivalent to computing the (stable) homotopy groups of X.

2.9. Thom spectra. Let B be a topological space and $\nu\colon E \to B$ be a n-dimensional real vector bundle on B. Then $\mathrm{Sph}(\nu)\colon \mathrm{Sph}(E) \to B$ is the n-sphere bundle whose fibers are the one-point compactification of the fibers of ν. The bundle $\mathrm{Sph}(E)$ has a section $s\colon B \to \mathrm{Sph}(E)$ which sends b to the point at infinity in the fiber $\mathrm{Sph}(E)_b$. Then the Thom space of ν is defined as

$$B^\nu = \mathrm{Sph}(E)/s(B).$$

The *Thom spectrum,* also denoted by B^ν, is the suspension spectrum of the Thom space. The composite

$$\mathrm{Sph}(E) \to \mathrm{Sph}(E) \times \mathrm{Sph}(E) \to B \times B^\nu,$$

which is the diagonal map followed by the product of $\mathrm{Sph}(\nu)$ and the quotient map, induces a map $B^\nu \to B_+ \wedge B^\nu$ called the *Thom diagonal.*

If $\nu = \alpha \oplus \mathbf{n}$ where $\mathbf{n}$ is the trivial n-dimensional bundle, then

$$B^\nu \simeq \Sigma^n B^\alpha. \tag{2.9.1}$$

In particular, if $\mathbf{0}$ is the zero bundle, then $B^{\mathbf{0}} = \Sigma^\infty B_+$.

The identity (2.9.1) motivates the definition of Thom spectra for virtual bundles. We give the definition for based spaces B which are CW-complexes with finitely many cells in each dimension. Recall that a virtual bundle ν over B is the formal difference $\nu = \alpha - \beta$ of vector bundles α and β over B. If α is an n-dimensional bundle and β is an m-dimensional bundle, we say that ν has dimension $n - m$.

If B is compact, we can choose a bundle $\beta^{\perp}$ and an integer k so that $\beta \oplus \beta^{\perp} \cong \mathbf{k}$. In this case, we define

$$B^{\nu} := \Sigma^{-k} B^{\alpha \oplus \beta^{\perp}}.$$

This is independent of the choice of complement $\beta^{\perp}$. Now, let B_q be the q-skeleton of B. By our assumption on B, the space B_q is compact. The bundle ν pulls back to virtual bundles ν_q over B_q for each q. There are induced maps of Thom spectra $B_q^{-\nu_q} \to B_{q+1}^{-\nu_{q+1}}$, and

$$B^{-\nu} := \operatorname{colim}_q B_q^{-\nu_q}.$$

EXAMPLE 2.9.1. Let O_n be the nth orthogonal group and BO_n its classifying space. A model for BO_n is given by the Grassmanian $G_n = \varinjlim_k \mathrm{Gr}_n(\mathbb{R}^k)$, where $\mathrm{Gr}_n(\mathbb{R}^k)$ is the space of n-dimensional subspaces of $\mathbb{R}^k$ and the maps in the colimit are induced by the inclusions $\mathbb{R}^k \subseteq \mathbb{R}^{k+1}$ into the first k-coordinates. This has the homotopy type of a CW-complex with finitely many cells in each dimension. Consider the subspace of $G_n \times \mathbb{R}^{\infty}$ given by

$$E_n = \{(P, v) \in G_n \times \mathbb{R}^{\infty} : P \in G_n, v \in P\}.$$

The map

$$\gamma_n \colon E_n \to G_n$$

which sends (P, v) to P is an n-dimensional vector bundle. This is often called the *universal bundle* over BO_n. The associated Thom space is denoted by MO_n, which is also used to denote the associated Thom spectrum.

If $H_n \to O_n$ is a group homomorphism, then the universal bundle γ_n pulls back to a bundle over BH_n that we will also denote by γ_n. The associated Thom space/spectrum is denoted by MH_n.

Finally, in these examples, the Thom spectrum of the virtual bundle $-\gamma_n$ is denoted by MTH_n and is called the *Madsen–Tillmann spectrum.*

REMARK 2.9.2. Thom spectra are related to Spanier–Whitehead duality via the *Atiyah duality* isomorphism. Let M be an n-manifold and TM be the tangent space of M, then Atiyah duality is the equivalence $M^{-TM} \simeq D(\Sigma^{\infty} M_+)$.

The cohomology of the Thom space is related to the cohomology of the base space. We treat the case $H^*(-;\mathbb{Z}/2)$ as it comes free of orientability conditions. Given any virtual n-bundle ν, there is an isomorphism

$$\mathrm{Th} \colon H^*(B;\mathbb{Z}/2) \cong \widetilde{H}^*(B^{\mathbf{0}};\mathbb{Z}/2) \to \widetilde{H}^{*+n}(B^{\nu};\mathbb{Z}/2).$$

called the *Thom isomorphism.* The isomorphism is given by an external cup product with a class

$$U = U(\nu) \in \widetilde{H}^n(B^{\nu};\mathbb{Z}/2)$$

called the *Thom class.*

3. The Steenrod algebra

In this section, we review some basic facts about the Steenrod algebra $\mathcal{A}$ at the prime $p = 2$. A very good reference for this material is Mosher–Tangora [**MT68**] and the interested reader should consult it for a more thorough presentation.

We focus on the prime $p = 2$, although much of this story has an analogue at odd primes. We will let

$$H^*(X) = \widetilde{H}^*(X; \mathbb{Z}/2)$$

denote the reduced mod 2 cohomology of X if it is a space, or simply the mod 2 cohomology of X if it is a (pre)spectrum. If $X \in \mathrm{Top}$ and we want to refer to the unreduced cohomology, we will use the notation $H^*(X; \mathbb{Z}/2)$.

3.1. Cohomology operations and the Steenrod algebra. Let $\mathrm{Vect}(\mathbb{Z}/2)$ denote the category of $\mathbb{Z}$-graded $\mathbb{Z}/2$ vector spaces, so that mod 2 cohomology is a functor

$$H^*(-; \mathbb{Z}/2) \colon \mathrm{Top} \to \mathrm{Vect}(\mathbb{Z}/2).$$

A cohomology operation of degree k is a natural transformation

$$\gamma \colon H^*(-; \mathbb{Z}/2) \to H^{*+k}(-; \mathbb{Z}/2).$$

The operation γ is said to be *stable* if it commutes with the suspension isomorphism

$$\Sigma \colon H^*(-) \xrightarrow{\cong} H^{*+1}(\Sigma(-)).$$

EXAMPLE 3.1.1. The short exact sequence

$$0 \to \mathbb{Z}/2 \to \mathbb{Z}/4 \to \mathbb{Z}/2 \to 0$$

induces a long exact sequence on cohomology

$$\cdots \longrightarrow H^*(-; \mathbb{Z}/4) \longrightarrow H^*(-; \mathbb{Z}/2) \longrightarrow H^{*+1}(-; \mathbb{Z}/2) \longrightarrow \cdots$$

The connecting homomorphism $H^*(-; \mathbb{Z}/2) \to H^{*+1}(-; \mathbb{Z}/2)$ is natural and commutes with the suspension isomorphism, so it is a stable cohomology operation of degree one. We call this operation Sq^1; it is also known as the Bockstein homomorphism.

EXAMPLE 3.1.2. Consider the real projective plane $\mathbb{R}P^2$. Then

$$H^*(\mathbb{R}P^2; \mathbb{Z}/2) \cong \mathbb{Z}/2[w_1]/w_1^3$$

for a class w_1 in degree 1. (The name w_1 will reappear in Section 3.3 and is used consistently here.) Then $Sq^1(w_1) = w_1^2$. In fact, $\mathbb{R}P^2$ can be constructed from the circle S^1 via the following pushout diagram:

$$\begin{array}{ccc} S^1 & \longrightarrow & D^2 \\ {\scriptstyle 2}\downarrow & & \downarrow \\ S^1 & \longrightarrow & \mathbb{R}P^2 \end{array}$$

The element w_1 is dual to the homology class represented by the 1-cell and the element w_1^2 is dual to that represented by the 2-cell. The cohomology operation $Sq^1(w_1) = w_1^2$ is recording the fact that the 2-cell of $\mathbb{R}P^2$ is attached to the 1-cell via the multiplication by 2 map. See Figure 2.

DEFINITION 3.1.3. The Steenrod algebra $\mathcal{A}$ is the graded non-commmutative $\mathbb{Z}/2$-algebra generated in degree k by the stable cohomology operations of that degree and with multiplication given by composition of operations.

REMARK 3.1.4. Let $H\mathbb{Z}/2$ be the mod-2 Eilenberg–MacLane spectrum whose nth space is given by $K(\mathbb{Z}/2, n)$. Since

$$H^t(-;\mathbb{Z}/2) \cong [(-)_+, K(\mathbb{Z}/2,n)] \cong [\Sigma^\infty(-)_+, \Sigma^t H\mathbb{Z}/2]$$

it follows from the Yoneda Lemma that degree t cohomology operations are in one to one correspondence with maps $[H\mathbb{Z}/2, \Sigma^t H\mathbb{Z}/2]$. Therefore,

$$\mathcal{A} \cong H\mathbb{Z}/2^*(H\mathbb{Z}/2).$$

Constructing all cohomology operations is rather difficult and a good reference is given by [**MT68**]. However, $\mathcal{A}$ can be described axiomatically and this is the approach we take here.

THEOREM 3.1.5. *For each $k \geq 0$, there exists a stable cohomology operation of degree k*

$$Sq^k \colon H^*(-;\mathbb{Z}/2) \to H^{*+k}(-;\mathbb{Z}/2)$$

called the kth Steenrod square. *For X a topological space, the Steenrod squares satisfy the following properties:*

(a) $Sq^0 = 1$
(b) *For $x \in H^k(X;\mathbb{Z}/2)$, $Sq^k(x) = x^2$.*
(c) *If $x \in H^i(X;\mathbb{Z}/2)$ and $i < k$, then $Sq^k(x) = 0$.*
(d) *(Cartan Formula) $Sq^k(xy) = \sum_{i+j=k} Sq^i(x)Sq^j(y)$, where the multiplication on $H^*(X;\mathbb{Z}/2)$ is given by the cup product.*

REMARK 3.1.6. In Theorem 3.1.5, the Cartan Formula is only expressed for the cup product of elements in $H^*(X;\mathbb{Z}/2)$. However, it also holds for the cross product. That is, if $x \in H^*(X;\mathbb{Z}/2)$ and $y \in H^*(Y;\mathbb{Z}/2)$, then for

$$x \otimes y \in H^*(X \times Y;\mathbb{Z}/2) \cong H^*(X;\mathbb{Z}/2) \otimes_{\mathbb{Z}/2} H^*(Y;\mathbb{Z}/2),$$

then

$$Sq^k(x \otimes y) = \sum_{i+j=k} Sq^i(x) \otimes Sq^j(y).$$

If one is working with the reduced cohomology groups, then the same formula holds for $H^*(X \wedge Y) \cong H^*(X) \otimes_{\mathbb{Z}/2} H^*(Y)$.

Finally, if there is a continuous map $Y \to X \times Y$, so that $H^*(Y;\mathbb{Z}/2)$ becomes a module over $H^*(X;\mathbb{Z}/2)$, then the Cartan Formula implies that

$$Sq^k(x \cdot y) = \sum_{i+j=k} Sq^i(x) \cdot Sq^j(y)$$

where $\cdot$ denotes the action of $H^*(X;\mathbb{Z}/2)$ on $H^*(Y;\mathbb{Z}/2)$.

THEOREM 3.1.7. *The Steenrod algebra $\mathcal{A}$ is the tensor algebra over $\mathbb{Z}/2$ generated by the Sq^i subject to the following relations:*

(1) $Sq^0 = 1$
(2) *The Adem relations: For $0 < a < 2b$,*

$$Sq^a Sq^b = \sum_{c=0}^{[a/2]} \binom{b-c-1}{a-2c} Sq^{a+b-c} Sq^c.$$

REMARK 3.1.8. The Steenrod algebra $\mathcal{A}$ is a graded, non-commutative, augmented algebra. In fact, it is a cocommutative Hopf algebra over $\mathbb{Z}/2$[1] whose coproduct $\psi\colon \mathcal{A} \to \mathcal{A} \otimes \mathcal{A}$ is determined by

$$\psi(Sq^k) = \sum_{i+j=k} Sq^i \otimes Sq^j .$$

The antipode $\chi\colon \mathcal{A} \to \mathcal{A}$ is defined inductively by the identities

$$\chi(Sq^0) = Sq^0, \qquad \sum_{i=0}^{k} Sq^i \chi(Sq^{k-i}) = 0, \;\; k > 0.$$

We note that $\mathcal{A}_0 = \mathbb{Z}/2$ and let $I(\mathcal{A})$ be the kernel of the augmentation $\varepsilon\colon \mathcal{A} \to \mathbb{Z}/2$.

REMARK 3.1.9. We let $\mathrm{Mod}_{\mathcal{A}}$ be the category of graded left modules over $\mathcal{A}$. These are $\mathbb{Z}$-graded $\mathbb{Z}/2$-vector spaces together with a left action of $\mathcal{A}$. Given M and N in $\mathrm{Mod}_{\mathcal{A}}$, we let $M \otimes_{\mathbb{Z}/2} N$ be the module whose structure is given by $a(m \otimes n) = \sum a_i m \otimes a_j n$, where $a \in \mathcal{A}$ and $\psi(a) = \sum a_i \otimes a_j$.

Modules which satisfy the conditions of Theorem 3.1.5 are called unstable modules. The cohomology of a spectrum need not be an unstable module in general.

To specify an $\mathcal{A}$-module structure on a graded $\mathbb{Z}/2$-vector space M, one must describe the action of the Steenrod squares on M. We record this information in a picture we call an *cell diagram.* See Figure 1. The following result implies that specifying the action of Sq^{2^n} for $n \geq 0$ is enough to describe an $\mathcal{A}$-module.

THEOREM 3.1.10. *$\mathcal{A}$ is generated as an algebra by Sq^{2^n} for $n \geq 0$.*

EXAMPLE 3.1.11. Consider the complex projective plane $\mathbb{C}P^2$. Then

$$H^*(\mathbb{C}P^2; \mathbb{Z}/2) \cong \mathbb{Z}/2[w_2]/w_2^3$$

for a class w_2 in degree 2. (The name w_2 reappears in Section 3.3 and is used consistently here.) It follows from the properties of the squares that $Sq^2(w_2) = w_2^2$. In fact, $\mathbb{C}P^2$ can be constructed from the sphere S^2 via the following pushout diagram:

$$\begin{array}{ccc} S^3 & \longrightarrow & D^4 \\ {\scriptstyle \eta}\big\downarrow & & \big\downarrow \\ S^2 & \longrightarrow & \mathbb{C}P^2 \end{array}$$

where $\eta\colon S^3 \to S^2$ is the Hopf fibration. The element w_2 is dual to the homology class represented by the 2-cell and the element w_2^2 is dual to that represented by the 4-cell. The cohomology operation $Sq^2(w_2) = w_2^2$ is recording the fact that the 4-cell of $\mathbb{C}P^2$ is attached to the 2-cell via the map η. See Figure 2.

EXAMPLE 3.1.12. The Steenrod operations for the cohomology of $\mathbb{R}P^\infty \simeq BO_1$ are completely explicit. Writing $H^*(\mathbb{R}P^\infty; \mathbb{Z}/2) \cong \mathbb{Z}/2[w_1]$ for w_1 in degree 1, we have

$$Sq^n(w_1^m) = \binom{m}{n} w_1^{m+n}.$$

---

[1]The authors have heard the following anecdote from Doug Ravenel: During a lecture of Milnor on Hopf algebras at Princeton many years ago, Steenrod asked if there were any interesting examples.

Using the naturality of the squares, this example often comes in handy in computing operations in the cohomology of other spaces. See Figure 4

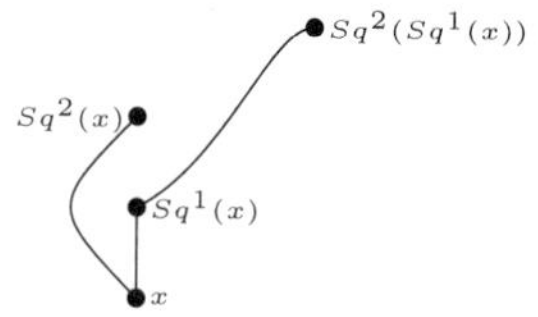

FIGURE 1. A *cell diagram*, used to depict the Steenrod operations on the cohomology of a space or spectrum. Each • denotes a generator of $\mathbb{Z}/2$. The difference in cohomological degree of the generators is represented vertically. Straight lines denote the action of Sq^1 and curved lines denote the action of Sq^2.

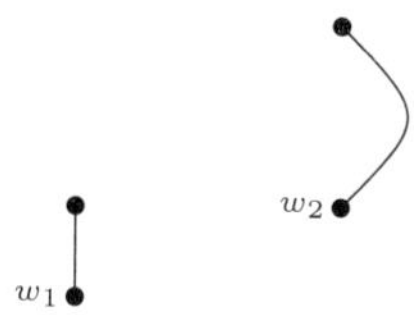

FIGURE 2. The structure of $H^*(\mathbb{R}P^2)$ (left), $H^*(\mathbb{C}P^2)$ (right) as modules over $\mathcal{A}$. The class w_1 is in $H^1(\mathbb{R}P^2)$. The class w_2 is in $H^2(\mathbb{C}P^2)$.

3.2. The subalgebras $\mathcal{A}_n$. The Steenrod algebra is an infinitely generated non-commutative algebra. However, it is finitely generated in each degree. In fact, it is filtered by the finite sub-Hopf algebras generated by $Sq^1, \ldots, Sq^{2^n}$, which are denoted $\mathcal{A}_n$. Further, each algebra $\mathcal{A}_n$ contains a commutative subalgebra generated by elements $Q_0, \ldots, Q_n$ which are defined inductively by

$$Q_0 = Sq^1,$$
$$Q_i = Sq^{2^i} Q_{i-1} + Q_{i-1} Sq^{2^i}.$$

In fact, the Q_i's generate an exterior algebra and we let $\mathcal{E}_n = E(Q_0, \ldots, Q_n)$.

For example, the algebra $\mathcal{A}_1$ is the subalgebra of $\mathcal{A}$ generated by Sq^1 and Sq^2. As a module over itself, $\mathcal{A}_1$ admits the cell diagram depicted in Figure 3.

DEFINITION 3.2.1. Let $\mathcal{B}$ be a subalgebra of a $\mathbb{Z}/2$-algebra $\mathcal{C}$. Then

$$\mathcal{C}/\!/\mathcal{B} := \mathcal{C} \otimes_{\mathcal{B}} \mathbb{Z}/2$$

where $\mathbb{Z}/2$ denotes the trivial $\mathcal{B}$ module concentrated in degree zero.

REMARK 3.2.2. The subalgebras $\mathcal{A}_1$ and $\mathcal{E}_1$ appear naturally in classical computations as they are related to K-theory. Let ku be the connective K-theory spectrum and ko its real version. See Section 2.5. Then there are isomorphisms of $\mathcal{A}$-modules

$$H^*(\mathrm{ku}) \cong \mathcal{A}/\!/\mathcal{E}_1, \qquad H^*(\mathrm{ko}) \cong \mathcal{A}/\!/\mathcal{A}_1.$$

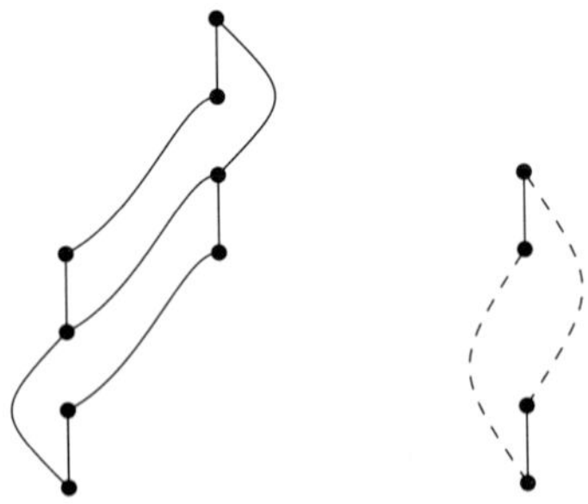

FIGURE 3. $\mathcal{A}_1$ (left) and its subalgebra $\mathcal{E}_1$ (right). The dashed lines represent the action of $Q_1 = Sq^1Sq^2 + Sq^2Sq^1$.

Similarly, if tmf is the connective spectrum of topological modular forms and $BP\langle 2\rangle$ is a spectrum obtained from the Brown-Peterson spectrum BP by killing a choice of generators v_k for $k \geq 3$, then

$$H^*(BP\langle 2\rangle) \cong \mathcal{A}/\!/\mathcal{E}_2, \qquad H^*(\mathrm{tmf}) \cong \mathcal{A}/\!/\mathcal{A}_2.$$

These spectra are the chromatic height 2 analogues of ku and ko respectively.

3.3. Thom spectra and Stiefel–Whitney classes. Given an n-dimensional vector bundle $\nu\colon E \to B$, we recalled the definition of the Thom space B^ν of ν in Section 2.9. Further, we recalled the Thom isomorphism

$$\mathrm{Th}\colon H^*(B;\mathbb{Z}/2) \to \widetilde{H}^{*+n}(B^\nu;\mathbb{Z}/2)$$

which was given by the cup product with a Thom class $U \in \widetilde{H}^n(B^\nu;\mathbb{Z}/2)$. We note that $\mathrm{Th}(1) = U$ and write

$$\mathrm{Th}(x) = xU.$$

WARNING 3.3.1. The Steenrod operations do not commute with the Thom isomorphism. This fact is crucial for Definition 3.3.2 below.

The Thom isomorphism is used to define classical invariants of a bundle ν called the *Stiefel–Whitney classes.*

DEFINITION 3.3.2. The *ith Stiefel–Whitney class* $w_i = w_i(\nu)$ of a vector bundle ν is defined by

$$w_i = \mathrm{Th}^{-1}(Sq^i(U)) \in H^i(B;\mathbb{Z}/2).$$

In particular, they satisfy the identity

$$w_iU = Sq^i(U).$$

The *total Stiefel–Whitney class* is the formal sum

$$w = w(\nu) = 1 + w_1 + w_2 + \dots$$

REMARK 3.3.3. If ν is a trivial bundle, the Stiefel–Whitney classes are trivial except for $w_0 = 1$. Given two vector bundles ν and η, one can show that

$$w(\nu \oplus \eta) = w(\nu)w(\eta).$$

If follows that

$$w(\nu \oplus \nu^\perp) = 1$$

for any orthogonal complement of an embedding of ν into a trivial bundle $\mathbf{m}$. This identity allows us to determine the Stiefel–Whitney classes of $\nu^{\perp}$ given those of ν. It also allows us to define the Stiefel–Whitney classes of a virtual bundle. In particular,

$$w(-\nu) = w(\nu)^{-1}.$$

The effect of the Steenrod squares on the Stiefel–Whitney classes is given by the Wu formula.

THEOREM 3.3.4 (Wu Formula). *Let $\nu\colon E \to B$ be a vector bundle over B. Then*

$$Sq^i(w_j) = \sum_{k=0}^{i} \binom{(j-i)+(k-1)}{k} w_{i-k}w_{j+k}.$$

REMARK 3.3.5. Applying $\mathrm{Th}(-)$ to both sides of the display in Definition 3.3.2, one deduces that

$$Sq^i(U) = w_iU \in H^{i+n}(B^{\nu}).$$

Further, the Thom diagonal gives $H^*(B^{\nu})$ the structure of an $H^*(B;\mathbb{Z}/2)$-module. So using Remark 3.1.6, for $x \in H^*(B;\mathbb{Z}/2)$ we have

$$Sq^k(xU) = \sum_{i+j=k} Sq^i(x)Sq^j(U) = \sum_{i+j=k} Sq^i(x)w_jU.$$

This determines the structure of $H^*(B^{\nu})$ as an $\mathcal{A}$-module based on that of $H^*(B)$.

3.4. Examples of computations of Steenrod operations. In this section, we go through a few selected computations of Steenrod operations. Most of the examples play a role in Section 10 of [**FH16**]. Further, the computations illustrate many of the concepts and techniques mentioned above. We do not do all the computations in detail but try to give enough information for the reader to learn the techniques and be able to reproduce them on their own.

EXAMPLE 3.4.1. The classifying space BO_n carries the universal n-plane bundle γ_n, and its Thom space is denoted MO_n. The cohomology of BO_n is

$$H^*(BO_n;\mathbb{Z}/2) = \mathbb{Z}/2[w_1,\dots,w_n], \qquad H^*(MO_n) = Z/2[w_1,\dots,w_n]\{U\}.$$

Similarly, BSO_n carries the universal oriented n-plane bundle and its Thom space is denoted by MSO_n. A bundle is oriented if and only if $w_1 = 0$, and so

$$H^*(BSO_n;\mathbb{Z}/2) = \mathbb{Z}/2[w_2,\dots,w_n], \qquad H^*(MSO_n) = \mathbb{Z}/2[w_2,\dots,w_n]\{U\}.$$

EXAMPLE 3.4.2. As special cases of Example 3.4.1 we have

$$H^*(\mathbb{R}P^{\infty};\mathbb{Z}/2) \cong H^*(BO_1;\mathbb{Z}/2) \cong \mathbb{Z}/2[w_1], \qquad H^*(MO_1) \cong \mathbb{Z}/2[w_1]\{U\}$$

Further, $Sq^1(w_1^kU) = w_1^{k+1}U$ if k is even and zero if k is odd. Using the Cartan Formula as in Remark 3.1.6, one deduces that $Sq^2(w_1^kU) = \binom{k-1}{2}w_1^{k+2}U$. In fact, $MO_1 \simeq \mathbb{R}P^{\infty}$.

Similarly,

$$H^*(\mathbb{C}P^{\infty};\mathbb{Z}/2) \cong H^*(BSO_2;\mathbb{Z}/2) \cong \mathbb{Z}/2[w_2], \qquad H^*(MU_1) \cong \mathbb{Z}/2[w_2]\{U\},$$

and $Sq^2(w_2^kU) = w_2^{k+1}U$ if k is even and zero if k is odd. All of the Sq^1s are zero. In fact, $MU_1 \simeq \mathbb{C}P^{\infty}$.

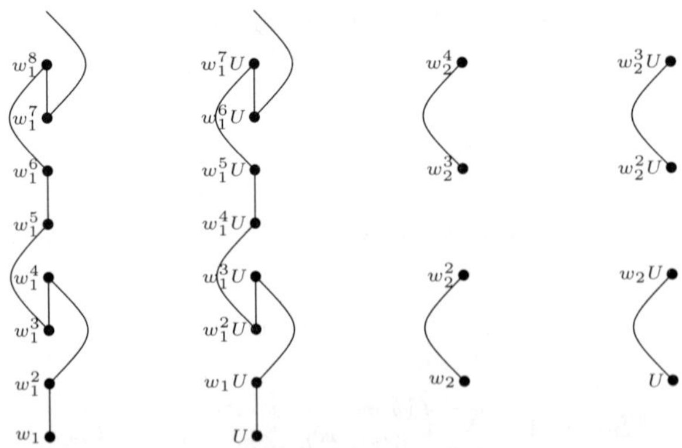

FIGURE 4. From the left, the structures of $H^*(BO_1)$, $H^*(MO_1)$, $H^*(BU_1)$ and $H^*(MU_1)$ as $\mathcal{A}_1$-modules.

EXAMPLE 3.4.3. As an exercise that will be relevant in Section 5, we consider the structure of $H^*(MU_1 \wedge MO_1)$ as modules over $\mathcal{A}_1$. By the Künneth isomorphism, we have

$$H^*(MU_1 \wedge MO_1) \cong H^*(MU_1) \otimes_{\mathbb{Z}/2} H^*(MO_1).$$

We use the Cartan formula as discussed in Remark 3.1.6. Since all of the Sq^1s vanish in $H^*(MU_1)$, we deduce from the Cartan formula that for any $a \in H^*(MU_1)$ and $b \in H^*(MO_1)$,

$$Sq^1(a \otimes b) = Sq^1(a) \otimes b + a \otimes Sq^1(b) = a \otimes Sq^1(b),$$
$$Sq^2(a \otimes b) = Sq^2(a) \otimes b + Sq^1(a) \otimes Sq^1(b) + a \otimes Sq^2(b) = Sq^2(a) \otimes b + a \otimes Sq^2(b).$$

The $\mathcal{A}_1$-module structure is illustrated in a small range in Figure 5.

EXAMPLE 3.4.4. In this example, we compute part of the structure of $H^*(BO_3)$ as a module over $\mathcal{A}_1$. We recall from Section 3.3 that

$$H^*(BO_3; \mathbb{Z}/2) \cong \mathbb{Z}/2[w_1, w_2, w_3]$$

and using the Wu formula, we compute that

$$\begin{aligned} &Sq^1(w_1) = w_1^2 && Sq^1(w_2) = w_1w_2 + w_3 && Sq^1(w_3) = w_1w_3 \\ &Sq^2(w_1) = 0 && Sq^2(w_2) = w_2^2 && Sq^2(w_3) = w_2w_3. \end{aligned}$$

With the Cartan formula, this determines all of the operations for $\mathcal{A}_1$ on $H^*(BO_3)$. For example,

$$\begin{aligned} Sq^2(Sq^1(w_2)) &= Sq^2(w_1w_2) + Sq^2(w_3) \\ &= Sq^2(w_1)Sq^0(w_2) + Sq^1(w_1)Sq^1(w_2) + Sq^0(w_1)Sq^2(w_2) + Sq^2(w_3) \\ &= w_1^2Sq^1(w_2) + w_1w_2^2 + w_2w_3 \\ &= (w_1^2 + w_2)Sq^1(w_2). \end{aligned}$$

We let

$$x = Sq^1(w_2) = w_1w_2 + w_3, \qquad \overline{w}_2 = w_1^2 + w_2. \tag{3.4.1}$$

A part of the cell diagram for $H^*(BO_3)$ is depicted in Figure 6.

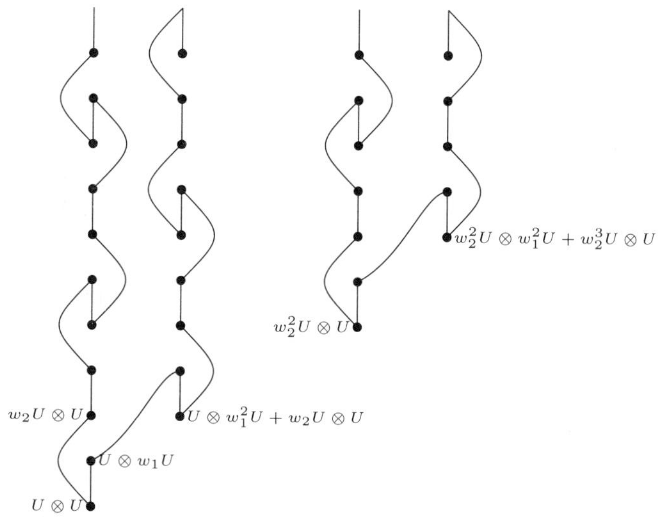

FIGURE 5. The $\mathcal{A}_1$-submodule of $H^*(MU_1 \wedge MO_1)$ generated by $U \otimes U$, $U \otimes w_1^2U + w_2U \otimes U$, $w_2^2U \otimes U$ and $w_2^2U \otimes w_1^2U + w_2^3U \otimes U$. The class $U \otimes U \in H^3(MU_1 \wedge MO_1)$. All classes of of degree $* \leq 5$ in $H^*(\Sigma^{-3}MU_1 \wedge MO_1)$ are contained in this submodule.

EXAMPLE 3.4.5. To compute the structure of $H^*(MO_3)$ as a module over $\mathcal{A}_1$, we use the Thom isomorphism and Remark 3.1.6. The former gives the identification

$$H^*(MO_3) \cong \mathbb{Z}/2[w_1, w_2, w_3]\{U\}$$

where the Thom class U is in $H^3(MO_3)$. Remark 3.1.6 allows us to compute the action of $\mathcal{A}_1$ on $H^*(MO_3)$ and the result is illustrated in Figure 7. For example,

$$\begin{aligned} Sq^2(w_2U) &= Sq^2(w_2)U + Sq^1(w_2)Sq^1(U) + w_2Sq^2(U) \\ &= w_2^2U + xw_1U + w_2^2U = w_1xU. \end{aligned}$$

A few other relations are given by

$$\begin{array}{llll} Sq^1(U) = w_1U & Sq^1(w_1U) = 0 & Sq^1(w_2U) = w_3U & Sq^1(w_3U) = 0 \\ Sq^2(U) = w_2U & Sq^2(w_1U) = w_1\overline{w}_2U & Sq^2(w_2U) = w_1xU & Sq^2(w_3U) = w_1^2w_3U. \end{array}$$

EXAMPLE 3.4.6. We turn to the computation of part of the structure of the cohomology of MTO_3 as a module over $\mathcal{A}_1$. Recall that MTO_3 is the Thom space for the virtual bundle $-\gamma_3$ over BO_3. Again, we have a Thom isomorphism

$$H^*(MTO_3) \cong \mathbb{Z}/2[w_1, w_2, w_3]\{\overline{U}\}$$

where the Thom class $\overline{U} = U(-\gamma_3)$ is in degree -3. However, here $w_i = w_i(\gamma_3)$, the Stiefel–Whitney classes of the universal bundle. Let $\overline{w}_i = w_i(-\gamma_3)$. To compute the Steenrod operations using the formula

$$Sq^i(\overline{U}) = \overline{w}_i\overline{U},$$

of Definition 3.3.2, we need a formula for the $\overline{w}_i$s in terms of the w_is. Letting $w = w(\gamma_3)$ and $\overline{w} = w(-\gamma_3)$ be the total Stiefel–Whitney classes, Remark 3.3.3

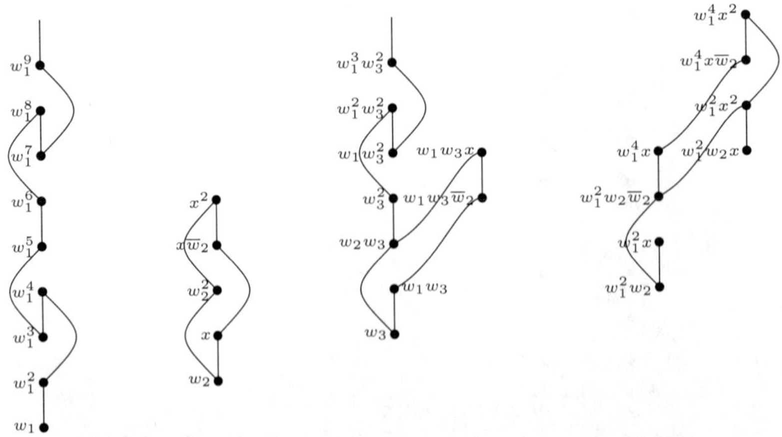

FIGURE 6. The $\mathcal{A}_1$-submodule of $H^*(BO_3)$ generated by w_1, w_2, w_3 and $w_1^2w_2$. The class $w_1 \in H^1(BO_3)$ and $w_2 \in H^2(BO_3)$.

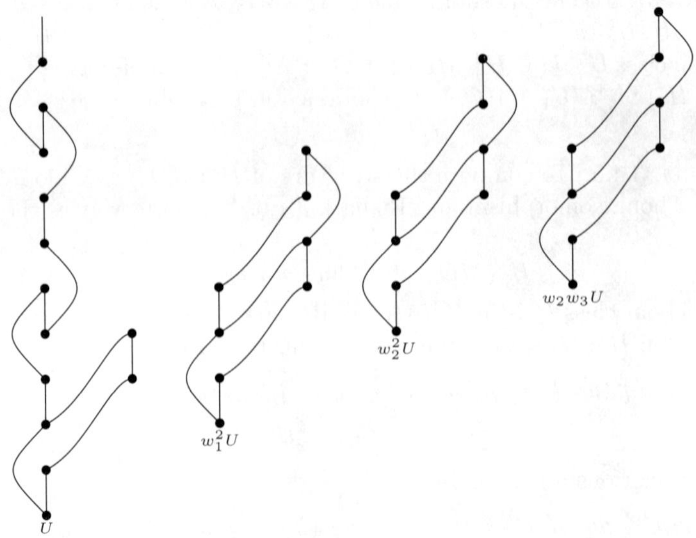

FIGURE 7. The $\mathcal{A}_1$-submodule of $H^*(MO_3)$ generated by the classes U, w_1^2U, w_2^2U and w_2w_3U. The class $U \in H^3(MO_3)$. This submodule contains all cohomology classes in $H^*(\Sigma^{-3}MO_3)$ of degree $* \leq 5$.

gives an identity

$$\overline{w} = w^{-1} = \frac{1}{1 + w_1 + w_2 + w_3} = \sum_{i \geq 0} (w_1 + w_2 + w_3)^i.$$

Collecting the terms of the same degree, we get that

$$\overline{w}_1 = w_1$$
$$\overline{w}_2 = w_1^2 + w_2.$$

Therefore, a few relations are given by

$$Sq^1(\overline{U})=w_1\overline{U}, \quad Sq^1(w_1\overline{U})=0 \quad Sq^1(w_2\overline{U})=w_3\overline{U} \quad Sq^1(w_3\overline{U})=0$$
$$Sq^2(\overline{U})=(w_1^2+w_2)\overline{U} \quad Sq^2(w_1\overline{U})=w_1w_2\overline{U} \quad Sq^2(w_2\overline{U})=w_1w_3\overline{U} \quad Sq^2(w_3\overline{U})=0.$$

A part of the cell diagram for $H^*(MTO_3)$ is depicted in Figure 8.

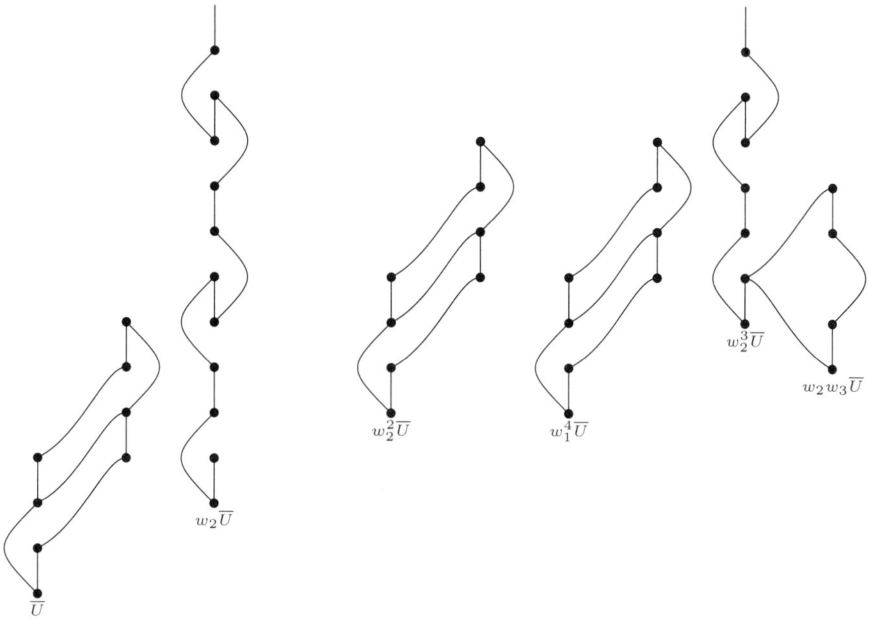

FIGURE 8. The $\mathcal{A}_1$-submodule of $H^*(MTO_3)$ generated by the classes $\overline{U}$, $w_2\overline{U}$, $w_2^2\overline{U}$, $w_1^4\overline{U}$, $w_2^3\overline{U}$ and $w_2w_3\overline{U}$. The class $\overline{U} \in H^{-3}(MTO_3)$. This submodule contains all cohomology classes in $H^*(\Sigma^3 MTO_3)$ of degree $* \leq 5$.

EXERCISE 3.4.7. Use the formulas of Example 3.4.5 and Remark 3.1.6 to compute that the $\mathcal{A}_1$-submodule of $H^*(MO_3)$ generated by U, w_1^2U and w_2^2U has the structure depicted in Figure 7. Do the same thing for Figure 8 using the results of Example 3.4.6.

EXAMPLE 3.4.8. In this example, we compute the structure of $H^*(MSO_3)$ as a module over $\mathcal{A}_1$. Let $\iota\colon MSO_3 \to MO_3$ be the map of Thom spectra induced by the inclusion of SO_3 into O_3. The induced map $\iota^*\colon H^*(MO_3) \to H^*(MSO_3)$ is given by moding out w_1. The Thom class of γ_3 maps to that of the universal bundle on BSO_3. We get an isomorphism

$$H^*(MSO_3) \cong \mathbb{Z}/2[w_2, w_3]\{U\}.$$

Further, the Steenrod operations are natural with maps of spaces or spectra, so $Sq^k\iota^* = \iota^* Sq^k$.

We use that $x \equiv w_3 \mod (w_1)$ for x as in (3.4.1). We get the following formulas from Example 3.4.5. First, in the cohomology of BSO_3, we have

$$Sq^1(w_2) = w_3 \qquad Sq^1(w_3) = 0 \qquad Sq^2(w_2) = w_2^2 \qquad Sq^2(w_3) = w_2w_3.$$

So, in the cohomology of MSO_3, we have

$$\begin{array}{lll} Sq^1(U) = 0 & Sq^1(w_2U) = w_3U & Sq^1(w_3U) = 0 \\ Sq^2(U) = w_2U & Sq^2(w_2U) = 0 & Sq^2(w_3U) = 0. \end{array}$$

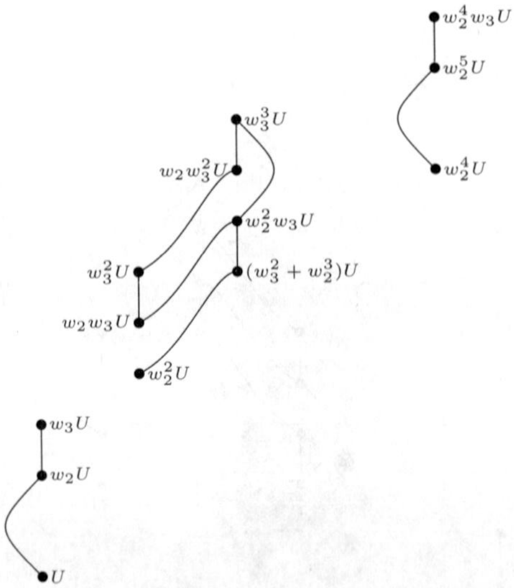

FIGURE 9. The $\mathcal{A}_1$-submodule of $H^*(MSO_3)$ generated by the classes U, w_2^2U, w_2w_3U and w_2^4U. The class $U \in H^3(MSO_3)$. This submodule contains all cohomology classes in $H^*(\Sigma^{-3}MSO_3)$ of degree $* \leq 5$.

4. The Adams spectral sequence

One of the most effective methods for computing stable homotopy groups is the Adams spectral sequence. The idea is roughly as follows. Take a space or a spectrum X and resolve it into pieces whose homotopy we understand. The Eilenberg–MacLane spectra are good candidates — they are constructed to have homotopy in a single degree. Then, reconstruct the stable homotopy groups of X from algebraic data associated to this resolution.

We will make this more precise and give a sketch of the construction of the Adams spectral sequence. In the cases of interest, it has the form

$$E_2^{s,t} = \mathrm{Ext}_{\mathcal{A}}^{s,t}(H^*(X), \mathbb{Z}/2) \Longrightarrow (\pi_{t-s}X)_2^{\wedge} \tag{4.0.1}$$

We will explain the terms in (4.0.1) throughout this section. We begin by defining $\mathrm{Ext}_{\mathcal{A}}$ and giving tools to compute it.

4.1. Computing Ext **over the Steenrod algebra.** Let $\mathcal{B}$ be a graded ring. For any $\mathcal{B}$-module M and $r \in \mathbb{Z}$, let $\Sigma^r M = M[r]$ be the graded $\mathcal{B}$-module given in degree t by

$$(\Sigma^r M)^t = (M[r])^t = M^{t-r}.$$

Let $\mathrm{Hom}_{\mathcal{B}}^*(M, N)$ be the graded abelian group given in degree t by

$$\mathrm{Hom}_{\mathcal{B}}^t(M, N) = \mathrm{Hom}_{\mathcal{B}}(M, \Sigma^t N).$$

The contravariant functor

$$\mathrm{Hom}^*_{\mathcal{B}}(-, N)\colon \mathcal{B}\text{-}\mathrm{Mod} \to \mathrm{Ab}$$

is left exact and has right derived functors $\mathrm{Ext}^s_{\mathcal{B}}(-, N)$. We let

$$\mathrm{Ext}^{s,t}_{\mathcal{B}}(-, N) = (\mathrm{Ext}^s_{\mathcal{B}}(-, N))^t$$

and treat $\mathrm{Ext}^{*,*}_{\mathcal{B}}(-, N)$ as a functor with values in bi-graded abelian groups. As always, the value of these functors on a $\mathcal{B}$-module M can be computed by choosing a resolution $P_\bullet$ of M by projective $\mathcal{B}$-modules and forming the cochain complex $\mathrm{Hom}^*_{\mathcal{B}}(P_\bullet, N)$. Then

$$\mathrm{Ext}^{s,t}_{\mathcal{B}}(M, N) = H^s(\mathrm{Hom}^t_{\mathcal{B}}(P_\bullet, N)).$$

A useful tool is the interpretation of elements in $\mathrm{Ext}^{s,t}_{\mathcal{B}}(M, N)$ as equivalence classes of extensions when $s \geq 1$. That is, an element of $\mathrm{Ext}^{s,t}_{\mathcal{B}}(M, N)$ is an exact complex, or *extension*,

$$0 \longrightarrow \Sigma^t N \longrightarrow P_1 \longrightarrow \dots \longrightarrow P_s \longrightarrow M \longrightarrow 0$$

where two extensions are equivalent if there exists a commutative diagram

$$\begin{array}{ccccccccccccc}
0 & \longrightarrow & \Sigma^t N & \longrightarrow & P_1 & \longrightarrow & \dots & \longrightarrow & P_s & \longrightarrow & M & \longrightarrow & 0 \\
 & & \Big\downarrow{\scriptstyle \mathrm{id}_N} & & \Big\downarrow & & & & \Big\downarrow & & \Big\downarrow{\scriptstyle \mathrm{id}_M} & & \\
0 & \longrightarrow & \Sigma^t N & \longrightarrow & P_1' & \longrightarrow & \dots & \longrightarrow & P_s' & \longrightarrow & M & \longrightarrow & 0.
\end{array}$$

EXAMPLE 4.1.1. The class in $\mathrm{Ext}^{1,1}_{\mathcal{A}}(\mathbb{Z}/2, \mathbb{Z}/2)$ represented by the extension

$$0 \to \Sigma\mathbb{Z}/2 \to \Sigma^{-1}H^*(\mathbb{R}P^2) \to \mathbb{Z}/2 \to 0,$$

which is depicted in Figure 10, is called h_0. The class in $\mathrm{Ext}^{1,2}_{\mathcal{A}}(\mathbb{Z}/2, \mathbb{Z}/2)$ represented by the extension

$$0 \to \Sigma^2\mathbb{Z}/2 \to \Sigma^{-2}H^*(\mathbb{C}P^2) \to \mathbb{Z}/2 \to 0,$$

which is depicted in Figure 11, is called h_1.

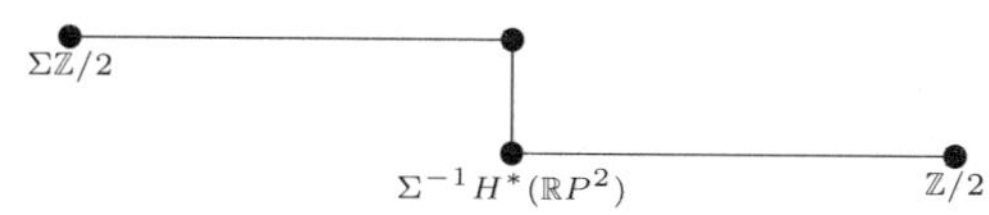

FIGURE 10. The extension representing h_0 in $\mathrm{Ext}^{1,1}_{\mathcal{A}}(\mathbb{Z}/2, \mathbb{Z}/2)$.

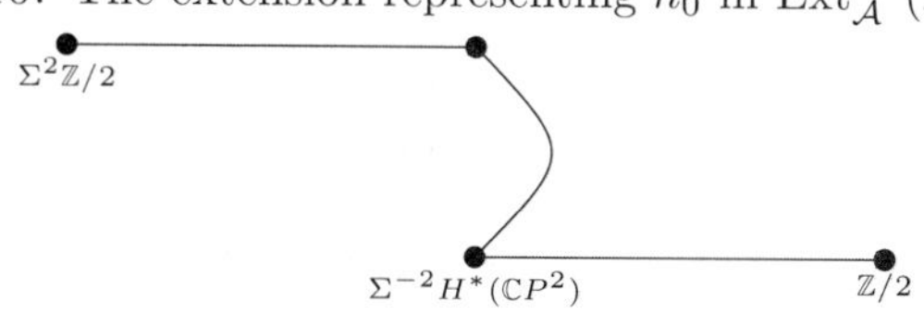

FIGURE 11. The extension representing h_1 in $\mathrm{Ext}^{1,2}_{\mathcal{A}}(\mathbb{Z}/2, \mathbb{Z}/2)$.

4.2. Module structure on Ext. Let $\mathcal{B}$ be a sub-Hopf algebra of the Steenrod algebra $\mathcal{A}$. Then for any $\mathcal{B}$-module M, there is a map

$$\mathrm{Ext}_{\mathcal{B}}^{s,t}(M,\mathbb{Z}/2)\otimes_{\mathbb{Z}/2}\mathrm{Ext}_{\mathcal{B}}^{s',t'}(\mathbb{Z}/2,\mathbb{Z}/2)\to\mathrm{Ext}_{\mathcal{B}}^{s+s',t+t'}(M,\mathbb{Z}/2).$$

This is called the *Yoneda product.* It is straightforward to describe the product in terms of extensions. Suppose that $s, s' \geq 1$. Given two extensions

$$0\longrightarrow\Sigma^t\mathbb{Z}/2\longrightarrow P_1\longrightarrow\dots\longrightarrow P_s\longrightarrow M\longrightarrow 0 \tag{4.2.1}$$

and

$$0\longrightarrow\Sigma^{t'}\mathbb{Z}/2\longrightarrow Q_1\longrightarrow\dots\xrightarrow{\varphi_{s'}} Q_{s'}\longrightarrow\mathbb{Z}/2\longrightarrow 0, \tag{4.2.2}$$

where (4.2.1) represents an element of $\mathrm{Ext}_{\mathcal{B}}^{s,t}(M,\mathbb{Z}/2)$ and (4.2.2) an element of $\mathrm{Ext}_{\mathcal{B}}^{s',t'}(\mathbb{Z}/2,\mathbb{Z}/2)$, we can splice the complexes to obtain an extension of length $s+s'$:

$$\begin{array}{ccccccccccccc}
0\to\Sigma^{t'+t}\mathbb{Z}/2\to\Sigma^tQ_1\to\dots\to\Sigma^tQ_{s'} & & \longrightarrow & & P_1\to\dots\to P_s\to M\to 0\\
& \searrow & & \nearrow & \\
& & \Sigma^t\mathbb{Z}/2 & &
\end{array}$$

which represents the product in $\mathrm{Ext}_{\mathcal{B}}^{s+s',t+t'}(M,\mathbb{Z}/2)$. This defines the module structure for elements of degree $s\geq 1$ in $\mathrm{Ext}_{\mathcal{B}}^{s,t}(M,\mathbb{Z}/2)$. If $s=0$, then given a homomorphism $M\to\Sigma^t\mathbb{Z}/2$ in

$$\mathrm{Ext}_{\mathcal{B}}^{0,t}(M,\mathbb{Z}/2)\cong\mathrm{Hom}_{\mathcal{B}}(M,\Sigma^t\mathbb{Z}/2)$$

and an element of $\mathrm{Ext}_{\mathcal{B}}^{s',t'}(\mathbb{Z}/2,\mathbb{Z}/2)$ represented by (4.2.2), we obtain an element in $\mathrm{Ext}_{\mathcal{B}}^{s',t+t'}(M,\mathbb{Z}/2)$ represented by

$$0\longrightarrow\Sigma^{t'}\mathbb{Z}/2\longrightarrow Q_1\longrightarrow\dots\longrightarrow\Sigma^tQ_{s'-1}\longrightarrow\Sigma^tQ_{s'}\times_{\Sigma^t\mathbb{Z}/2}M\longrightarrow M\longrightarrow 0$$

where $\Sigma^tQ_{s'}\times_{\Sigma^t\mathbb{Z}/2}M$ is the pull-back of $\mathcal{A}_1$-modules. There is a commutative diagram of exact sequences

$$\begin{array}{ccccccccc}
0 & \longrightarrow & \ker(\Sigma^t\varphi_{s'}) & \longrightarrow & \Sigma^tQ_{s'}\times_{\Sigma^t\mathbb{Z}/2}M & \longrightarrow & M & \longrightarrow & 0\\
 & & \downarrow & & \downarrow & & \downarrow & & \\
0 & \longrightarrow & \ker(\Sigma^t\varphi_{s'}) & \longrightarrow & \Sigma^tQ_{s'} & \longrightarrow & \Sigma^t\mathbb{Z}/2 & \longrightarrow & 0
\end{array}$$

so that we really do get an exact complex. An example when $\mathcal{B}=\mathcal{A}_1$ is given in Figure 12.

4.3. Adams charts. For $\mathcal{B}$ a sub-Hopf algebra of $\mathcal{A}$, we depict the information contained in $\mathrm{Ext}_{\mathcal{B}}^{s,t}(M,\mathbb{Z}/2)$ in a picture which we call an *Adams chart.* See Figure 13. An Adams chart is an illustration of $\mathrm{Ext}_{\mathcal{B}}^{s,t}(M,\mathbb{Z}/2)$ in the $(t-s,s)$-plane. A generator for a copy of $\mathbb{Z}/2$ in $\mathrm{Ext}_{\mathcal{B}}^{s,t}(M,\mathbb{Z}/2)$ is denoted by a $\bullet$. Multiplication by h_0 is recorded by drawing a vertical line between two classes and multiplication by h_1 by a line of slope $(1,1)$. An infinite string of classes connected by multiplications by h_0 is called an *h_0-tower.* Note that the Adams chart for Σ^rM is the same as that of M, but horizontally shifted to the right by r.

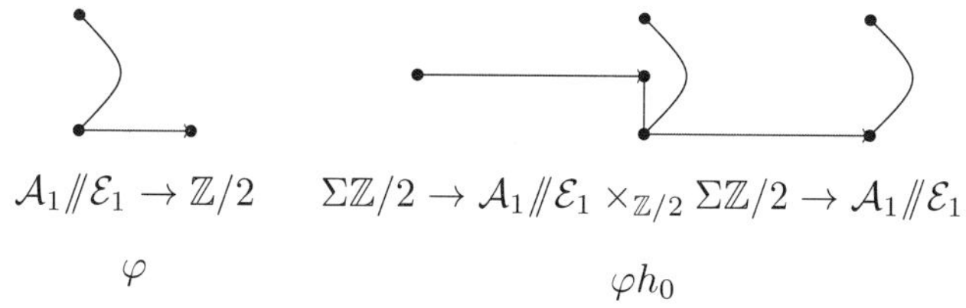

FIGURE 12. A representative extension for the element φh_0 in $\mathrm{Ext}^{1,1}_{\mathcal{A}_1}(\mathcal{A}_1/\!/\mathcal{E}_1, \mathbb{Z}/2)$ where the element $\varphi\colon \mathcal{A}_1/\!/\mathcal{E}_1 \to \mathbb{Z}/2$ of $\mathrm{Ext}^{0,0}_{\mathcal{A}_1}(\mathcal{A}_1/\!/\mathcal{E}_1, \mathbb{Z}/2)$ is the map which sends the element of degree two to zero.

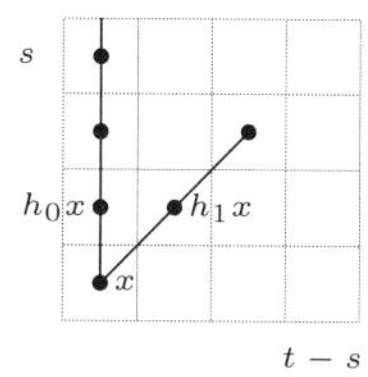

FIGURE 13. An example of an Adams chart for $\mathrm{Ext}^{s,t}_{\mathcal{B}}(M, \mathbb{Z}/2)$.

4.4. Minimal resolutions. Let $\mathcal{B}$ be a sub-Hopf algebra of $\mathcal{A}$. Recall that $\mathcal{A}$ is an augmented algebra with $\mathcal{A}_0 = \mathbb{Z}/2$. So this holds for any of its subalgebras. We let $I(\mathcal{B})$ be the kernel of the augmentation of $\mathcal{B}$. Note that for any $\mathcal{B}$-module P and $\mathbb{Z}/2$ the trivial $\mathcal{B}$-module, the map

$$\mathrm{Hom}^*_{\mathcal{B}}(P, \mathbb{Z}/2) \to \mathrm{Hom}^*_{\mathcal{B}}(I(\mathcal{B})P, \mathbb{Z}/2)$$

induced by the inclusion $I(\mathcal{B})P \hookrightarrow P$ is zero. So, if $P_\bullet$,

$$\ldots \to P_s \xrightarrow{f_s} P_{s-1} \to \ldots \to P_0 \to M$$

is a projective resolution of M which satisfies

$$f_s(P_s) \subseteq I(\mathcal{B})P_{s-1},$$

then the maps in the cochain complex $\mathrm{Hom}^*_{\mathcal{B}}(P_\bullet, \mathbb{Z}/2)$ are trivial and it follows that

$$\mathrm{Ext}^{s,t}_{\mathcal{B}}(M, \mathbb{Z}/2) \cong \mathrm{Hom}^t_{\mathcal{B}}(P_s, \mathbb{Z}/2).$$

Such a resolution is called a *minimal resolution.*

If M is a $\mathcal{B}$-module which is bounded below, then M has a minimal resolution by free $\mathcal{B}$-modules. In such a resolution $P_\bullet \to M$, the P_s are direct sums of suspensions of $\mathcal{B}$ and $\mathrm{Ext}^{s,t}_{\mathcal{B}}(M, \mathbb{Z}/2)$ is a product of $\mathbb{Z}/2$s indexed over the summands $\Sigma^t\mathcal{B} \subseteq P_s$. If there are finitely many of these, the product is isomorphic to a direct sum and each summand corresponds to a generator in $\mathrm{Ext}^{s,t}_{\mathcal{B}}(M, \mathbb{Z}/2)$.

If $\mathcal{B}$ and M are small, these are straightforward to construct and we do a few examples here in the case when $\mathcal{B} = \mathcal{A}_1$.

REMARK 4.4.1. Let $\mathcal{B}$ a sub-Hopf algebra of $\mathcal{A}$ and M a graded $\mathcal{B}$-module of finite type which is zero in degrees $t < n$. Using a free minimal resolution of M to compute $\mathrm{Ext}^{s,t}_{\mathcal{B}}(M, \mathbb{Z}/2)$, one deduces that $\mathrm{Ext}^{s,t}_{\mathcal{B}}(M, \mathbb{Z}/2) = 0$ for $t - s < n$.

EXAMPLE 4.4.2. We begin by constructing a resolution of the $\mathcal{A}_1$-module $M_0 = \mathcal{A}_1/\!/\mathcal{E}_0$, where $\mathcal{E}_0$ is the subalgebra generated by Sq^1. It is depicted below. This example is also treated by a different method in Example 4.5.5. The module M_0 has a periodic minimal resolution of the form

$$M_0 \longleftarrow \mathcal{A}_1 \longleftarrow \Sigma\mathcal{A}_1 \longleftarrow \Sigma^2\mathcal{A}_1 \longleftarrow \dots \tag{4.4.1}$$

See Figure 16. The horizontal (blue online) arrows indicate the maps in (4.4.1). The circled (in red online) classes are in the kernel. We have redrawn the kernels to the right (in red online) to make the next map easier to visualize. The duals of the boxed classes (in blue online) will form a basis of $\mathrm{Ext}^{*,*}_{\mathcal{A}_1}(M_0, \mathbb{Z}/2)$. Recall that h_0 was defined in Example 4.1.1. See also Figure 10. The class in $\mathrm{Ext}^{1,1}_{\mathcal{A}_1}(M_0, \mathbb{Z}/2)$ is the h_0 multiple of the class in $\mathrm{Ext}^{0,0}_{\mathcal{A}_1}(M_0, \mathbb{Z}/2)$. This is read off of the part of Figure 16 that has been framed (in gray).

The $\mathcal{A}_1$-module M_0 is not the restriction of any $\mathcal{A}$-module, but it has such a nice projective resolution that it is often used as a tool to compute resolutions for other modules. This will be explained below. There are larger versions of the module M_0 that we will denote by M_n obtained by stringing together copies of M_0, including the case $n = \infty$. For example, M_1 is drawn in Figure 17. These all have periodic minimal resolutions. For example,

$$M_1 \longleftarrow \mathcal{A}_1 \oplus \Sigma^4\mathcal{A}_1 \longleftarrow \Sigma(\mathcal{A}_1 \oplus \Sigma^4\mathcal{A}_1) \longleftarrow \Sigma^2(\mathcal{A}_1 \oplus \Sigma^4\mathcal{A}_1) \longleftarrow \dots$$

The Adams chart of M_n has h_0-towers starting in $(4k, 0)$ for $0 \le k \le n$. For example, the Adams chart for M_1 is depicted in Figure 17.

FIGURE 14. The $\mathcal{A}_1$ module M_0.

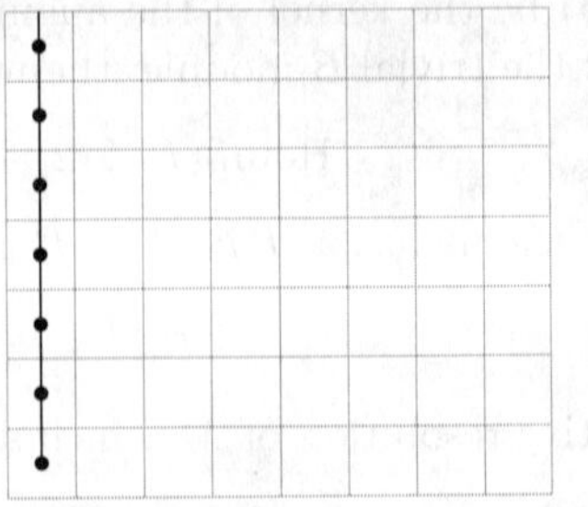

FIGURE 15. The Adams chart for $\mathrm{Ext}^{s,t}_{\mathcal{A}_1}(M_0, \mathbb{Z}/2)$.

EXAMPLE 4.4.3. The module $\mathbb{Z}/2$ has a rather complicated minimal resolution. It is an excellently annoying exercise to work it out. We have illustrated the first two terms of such a resolution in Figure 19. We will give a different approach in Example 4.6.5 to computing the Adams chart for $\mathrm{Ext}_{\mathcal{A}_1}(\mathbb{Z}/2, \mathbb{Z}/2)$ but we include it here in Figure 20.

4.5. Change-of-rings. Let $\mathcal{B}$ be a subalgebra of $\mathcal{A}$. We defined $\mathcal{A}/\!/\mathcal{B}$ in Definition 3.2.1.

LEMMA 4.5.1 (Shearing Isomorphism). *Let $\mathcal{B}$ be a sub Hopf-algebra of $\mathcal{A}$. Let M be an $\mathcal{A}$-module. Then there is an isomorphism of $\mathcal{A}$-modules*

$$\mathcal{A} \otimes_{\mathcal{B}} M \cong \mathcal{A}/\!/\mathcal{B} \otimes_{\mathbb{Z}/2} M$$

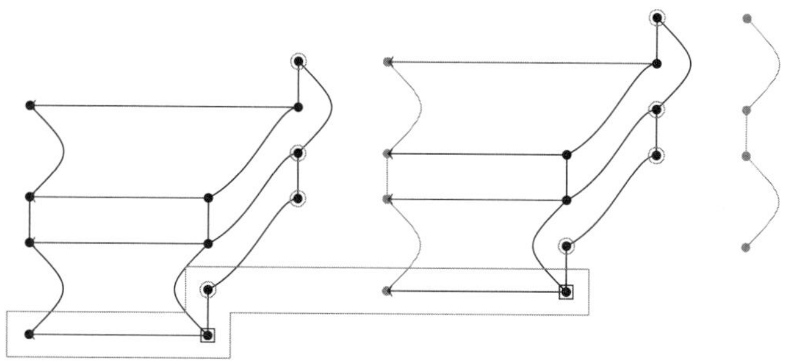

FIGURE 16. A minimal projective resolution for $M_0 = \mathcal{A}_1/\!/\mathcal{E}_0$. The horizontal (blue online) arrows indicate the maps in the resolution. The circled (in red online) classes are in the kernel. The kernels are redrawn to the right (in red online). The duals of the boxed classes (in blue online) form a basis of $\mathrm{Ext}^{*,*}_{\mathcal{A}_1}(M_0, \mathbb{Z}/2)$.

FIGURE 17. The $\mathcal{A}_1$-module M_1.

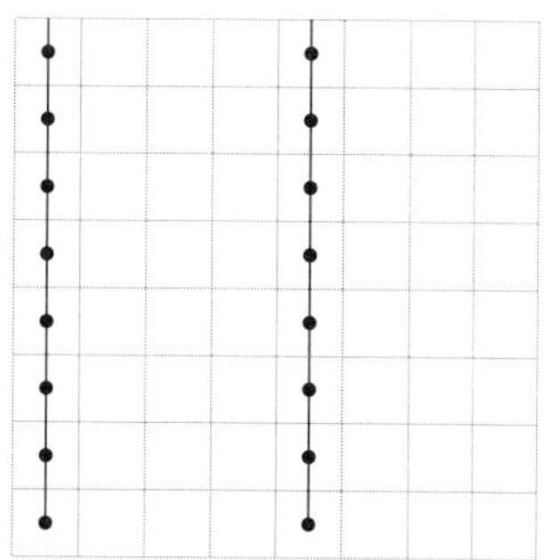

FIGURE 18. The Adams chart for $\mathrm{Ext}^{s,t}_{\mathcal{A}_1}(M_1, \mathbb{Z}/2)$.

where the action of $\mathcal{A}$ on $\mathcal{A} \otimes_{\mathcal{B}} M$ is via the left action of $\mathcal{A}$ on itself and the action of $\mathcal{A}$ on $\mathcal{A}/\!/\mathcal{B} \otimes_{\mathbb{Z}/2} M$ is the one described in Remark 3.1.9.

REMARK 4.5.2. If $\mathcal{B} = \mathbb{Z}/2$, the isomorphism of Lemma 4.5.1 is induced by the composite

$$\mathcal{A} \xrightarrow{\psi \otimes M} \mathcal{A} \otimes \mathcal{A} \otimes M \xrightarrow{\mathcal{A} \otimes f} \mathcal{A} \otimes M$$

where $f\colon \mathcal{A} \otimes M \to M$ is the structure map of the $\mathcal{A}$-module M. The maps ψ and χ below are as in Remark 3.1.8. The inverse is induced by the composite

$$\mathcal{A} \xrightarrow{\psi \otimes M} \mathcal{A} \otimes \mathcal{A} \otimes M \xrightarrow{\mathcal{A} \otimes \chi \otimes M} \mathcal{A} \otimes \mathcal{A} \otimes M \xrightarrow{\mathcal{A} \otimes f} \mathcal{A} \otimes M.$$

One verifies that these maps descend to the quotients for more general $\mathcal{B}$.

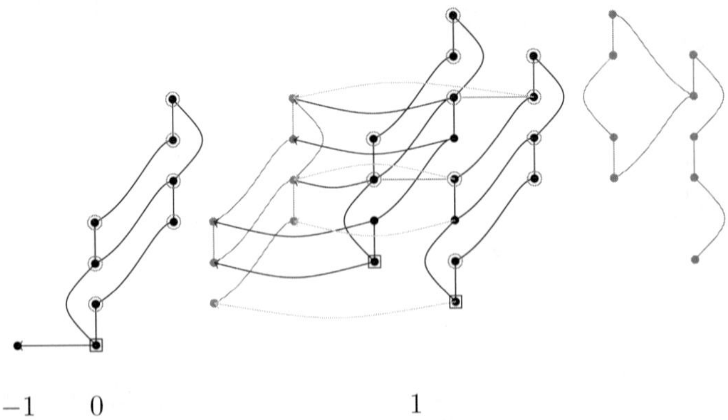

FIGURE 19. The beginning of a minimal resolution for the trivial $\mathcal{A}_1$-module $\mathbb{Z}/2$. Circled (red online) classes joined by a horizontal line indicate that the sum of the classes are in the kernel.

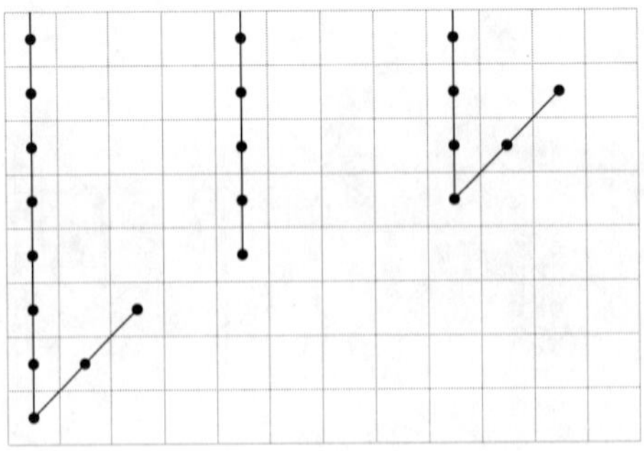

FIGURE 20. The Adams chart for $\mathrm{Ext}^{s,t}_{\mathcal{A}_1}(\mathbb{Z}/2, \mathbb{Z}/2)$.

From the shearing isomorphism and, from the adjunction

$$\mathrm{Hom}_{\mathcal{B}}(M, N) \cong \mathrm{Hom}_{\mathcal{A}}(\mathcal{A} \otimes_{\mathcal{B}} M, N)$$

one can prove that

$$\mathrm{Ext}^{*,*}_{\mathcal{A}}(\mathcal{A}/\!/\mathcal{B} \otimes_{\mathbb{Z}/2} M, N) \cong \mathrm{Ext}^{*,*}_{\mathcal{B}}(M, N)$$

for any $\mathcal{A}$-modules M and N. Therefore, in the case of extended modules, computations over $\mathcal{A}$ can be reduced to potentially easier computations over smaller sub-Hopf algebras $\mathcal{B}$. Some common examples are described below.

EXAMPLE 4.5.3. Many of the modules relevant in the computations of [**FH16**] are of the form $\mathcal{A}/\!/\mathcal{A}_1 \otimes_{\mathbb{Z}/2} M_0$ in some range. By the adjunction

$$\mathrm{Ext}^{*,*}_{\mathcal{A}}(\mathcal{A}/\!/\mathcal{A}_1 \otimes_{\mathbb{Z}/2} M_0, \mathbb{Z}/2) \cong \mathrm{Ext}^{*,*}_{\mathcal{A}_1}(M_0, \mathbb{Z}/2),$$

we only need to keep track of the $\mathcal{A}_1$-module structure.

REMARK 4.5.4. Let R be a graded exterior algebra on n generators over $\mathbb{Z}/2$

$$R = E(x_1, \dots, x_n) = \mathbb{Z}/2[x_1, \dots, x_n]/(x_1^2, \dots, x_n^2).$$

where x_i is in degree t_i. Then $\mathrm{Ext}_R^{*,*}(\mathbb{Z}/2, \mathbb{Z}/2)$ is a polynomial algebra on n generators

$$\mathrm{Ext}_R^{*,*}(\mathbb{Z}/2, \mathbb{Z}/2) \cong \mathbb{Z}/2[y_1, \ldots, y_n]$$

for $y_i \in \mathrm{Ext}^{1,t_i}(\mathbb{Z}/2, \mathbb{Z}/2)$. This is an example of a phenomenon called Koszul duality.

EXAMPLE 4.5.5. The module M_0 of Example 4.4.2 is isomorphic to $\mathcal{A}_1/\!/\mathcal{E}_0$, where $\mathcal{E}_0$ is the algebra generated by Sq^1. The algebra $\mathcal{E}_0$ is an exterior algebra on one generator in degree 1, so that

$$\mathrm{Ext}_{\mathcal{A}_1}^{*,*}(M_0, \mathbb{Z}/2) \cong \mathrm{Ext}_{\mathcal{E}_0}^{*,*}(\mathbb{Z}/2, \mathbb{Z}/2) \cong \mathbb{Z}/2[h_0]$$

where $h_0 \in \mathrm{Ext}_{\mathcal{E}_0}^{1,1}(\mathbb{Z}/2, \mathbb{Z}/2)$. The Adams chart for M_0 contains one h_0-tower starting in degree $(0, 0)$.

EXAMPLE 4.5.6. The $\mathcal{A}_1$-module $\mathcal{A}_1/\!/\mathcal{E}_1$ is the cohomology of $\Sigma^{-2}H^*(\mathbb{C}P^2)$, illustrated in Figure 21. By the change-of-rings isomorphism,

$$\mathrm{Ext}_{\mathcal{A}_1}^{*,*}(\mathcal{A}_1/\!/\mathcal{E}_1, \mathbb{Z}/2) \cong \mathrm{Ext}_{\mathcal{E}_1}^{*,*}(\mathbb{Z}/2, \mathbb{Z}/2).$$

Since $\mathcal{E}_1 = E(Q_0, Q_1)$, it follows that $\mathrm{Ext}_{\mathcal{A}_1}^{*,*}(\mathcal{A}_1/\!/\mathcal{E}_1, \mathbb{Z}/2)$ is a polynomial algebra on two generators. It is common to call the generator corresponding to $Q_0 = Sq^1$ by $h_0 \in \mathrm{Ext}_{\mathcal{A}_1}^{1,1}(\mathcal{A}_1/\!/\mathcal{E}_1, \mathbb{Z}/2)$. The generator corresponding to Q_1 is often called $v_1 \in \mathrm{Ext}_{\mathcal{A}_1}^{1,3}(\mathcal{A}_1/\!/\mathcal{E}_1, \mathbb{Z}/2)$, so that

$$\mathrm{Ext}_{\mathcal{A}_1}^{*,*}(\mathcal{A}_1/\!/\mathcal{E}_1, \mathbb{Z}/2) \cong \mathbb{Z}/2[h_0, v_1].$$

The Adams chart is depicted in Figure 22.

FIGURE 21. $\mathcal{A}_1/\!/\mathcal{E}_1$

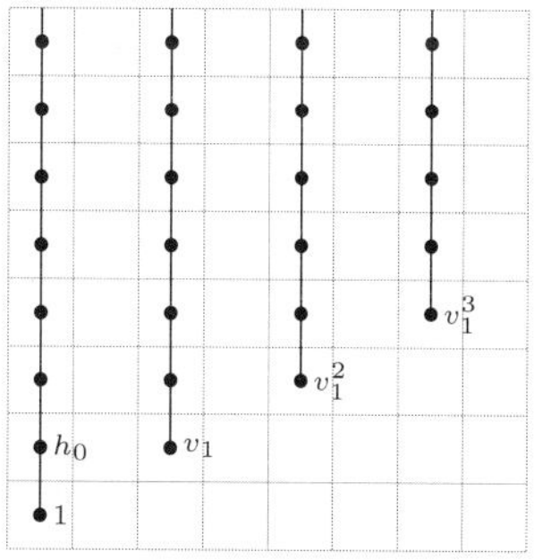

FIGURE 22. The Adams chart for $\mathrm{Ext}_{\mathcal{A}_1}^{s,t}(\mathcal{A}_1/\!/\mathcal{E}_1, \mathbb{Z}/2)$.

4.6. Long exact sequences. For some of the computations below we will need to use the long exact sequence induced on Ext from a short exact sequence of modules.

PROPOSITION 4.6.1. *Let $0 \to M \to N \to P \to 0$ be an exact sequence of $\mathcal{B}$-modules. Then there is a long exact sequence*

$$\cdots \longrightarrow \mathrm{Ext}_{\mathcal{B}}^{s,t}(P, \mathbb{Z}/2) \longrightarrow \mathrm{Ext}_{\mathcal{B}}^{s,t}(N, \mathbb{Z}/2) \longrightarrow \mathrm{Ext}_{\mathcal{B}}^{s,t}(M, \mathbb{Z}/2) \xrightarrow{\delta} \mathrm{Ext}_{\mathcal{B}}^{s+1,t}(P, \mathbb{Z}/2) \longrightarrow \mathrm{Ext}_{\mathcal{B}}^{s+1,t}(N, \mathbb{Z}/2) \longrightarrow \cdots$$

The map δ can be identified using the description of Ext in terms of extensions given in Section 4.1. Given an extension

$$0 \longrightarrow \Sigma^t\mathbb{Z}/2 \longrightarrow P_1 \longrightarrow \dots \longrightarrow P_s \longrightarrow M \longrightarrow 0 \tag{4.6.1}$$

we let $P_{s+1} = N$ and get an extension of length $s+1$

$$0 \longrightarrow \Sigma^t\mathbb{Z}/2 \longrightarrow P_1 \longrightarrow \dots \longrightarrow P_s \xrightarrow{} N = P_{s+1} \longrightarrow P \longrightarrow 0$$

$$P_s \searrow M \nearrow N$$

which corresponds to the boundary of the element of $\mathrm{Ext}_{\mathcal{B}}^{s,t}(M,\mathbb{Z}/2)$ represented by (4.6.1) in $\mathrm{Ext}_{\mathcal{B}}^{s+1,t}(P,\mathbb{Z}/2)$. See, e.g. [**McC01**, 9.6] for more details.

Computations using Proposition 4.6.1 can be done with the help of an Adams chart. The trick is to draw both $\mathrm{Ext}_{\mathcal{B}}^{s,t}(P,\mathbb{Z}/2)$ and $\mathrm{Ext}_{\mathcal{B}}^{s,t}(M,\mathbb{Z}/2)$ in the same chart and to treat the boundary map δ as a differential of slope $(-1,1)$. We illustrate this by an example.

EXAMPLE 4.6.2. We compute $\mathrm{Ext}_{\mathcal{A}_1}^{s,t}(R_0,\mathbb{Z}/2)$ for R_0 as depicted in Figure 23. The module R_0 sits in an exact sequence

$$0 \to \Sigma\mathbb{Z}/2 \to R_0 \to M_\infty \to 0, \tag{4.6.2}$$

so we use the long exact sequence of Proposition 4.6.1 to compute $\mathrm{Ext}_{\mathcal{A}_1}^{s,t}(R_0,\mathbb{Z}/2)$:

$$\dots \longrightarrow \mathrm{Ext}_{\mathcal{A}_1}^{s,t}(M_\infty,\mathbb{Z}/2) \longrightarrow \mathrm{Ext}_{\mathcal{A}_1}^{s,t}(R_0,\mathbb{Z}/2) \longrightarrow \mathrm{Ext}_{\mathcal{A}_1}^{s,t}(\Sigma\mathbb{Z}/2,\mathbb{Z}/2)$$

$$\xrightarrow[h_0]{\delta} \mathrm{Ext}_{\mathcal{A}_1}^{s+1,t}(M_\infty,\mathbb{Z}/2) \longrightarrow \mathrm{Ext}_{\mathcal{A}_1}^{s+1,t}(R_0,\mathbb{Z}/2) \longrightarrow \dots$$

The boundary is given by multiplication by h_0 since (4.6.2) is a representative extension for the element $h_0 \cdot 1 \in \mathrm{Ext}_{\mathcal{A}_1}^{1,1}(M_\infty,\mathbb{Z}/2)$.

In Figure 24, the classes of $\mathrm{Ext}_{\mathcal{A}_1}^{s,t}(\Sigma\mathbb{Z}/2,\mathbb{Z}/2)$ (blue online), which is illustrated in Figure 20, support boundaries (red online) to the classes of $\mathrm{Ext}_{\mathcal{A}_1}^{s+1,t}(M_\infty,\mathbb{Z}/2)$ (green online). The circled classes are the elements of $\mathrm{Ext}_{\mathcal{A}_1}^{s,t}(R_0,\mathbb{Z}/2)$ in this range. The dashed line indicates a multiplication by h_1 between a class coming from $\mathrm{Ext}_{\mathcal{A}_1}^{s,t}(\Sigma\mathbb{Z}/2,\mathbb{Z}/2)$ and a class coming from $\mathrm{Ext}_{\mathcal{A}_1}^{s+1,t}(M_\infty,\mathbb{Z}/2)$, which we have not justified. One way to do this is to compute a minimal resolution for R_0 and use the fact that multiplication by h_1 corresponds to the extension depicted in Figure 11.

EXAMPLE 4.6.3. Consider the $\mathcal{A}_1$-module depicted in Figure 25. Using Proposition 4.6.1, we get the Adams chart depicted in Figure 26.

REMARK 4.6.4. We present one last trick which is a variation on Proposition 4.6.1. It uses the fact that, although the module $\mathcal{A}_1/\!/\mathcal{E}_0$ is not projective, it has a nice periodic resolution as an $\mathcal{A}_1$-module. Given a module M, suppose that there is an exact complex

$$0 \longleftarrow M \xleftarrow{f_0} P_0 \xleftarrow{f_1} P_1 \xleftarrow{f_2} P_2 \longleftarrow \dots \tag{4.6.3}$$

where the P_s are direct sums of suspensions of copies of $\mathcal{A}_1$ and $\mathcal{A}_1/\!/\mathcal{E}_0$ and with the property that

$$f_s(P_s) \subseteq I(\mathcal{B})P_{s-1},$$

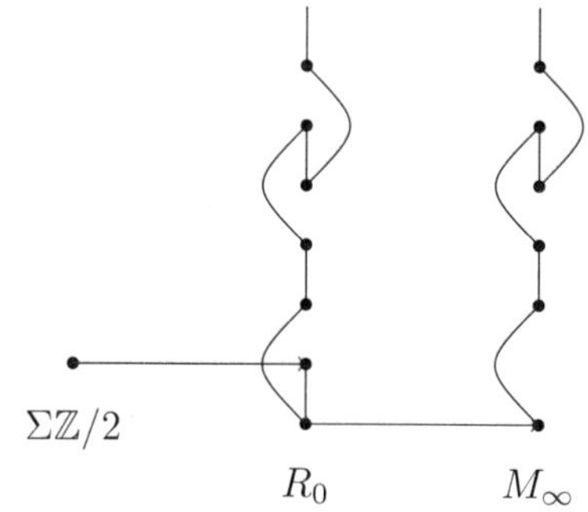

FIGURE 23. An extension exhibiting an $\mathcal{A}_1$-module we call R_0.

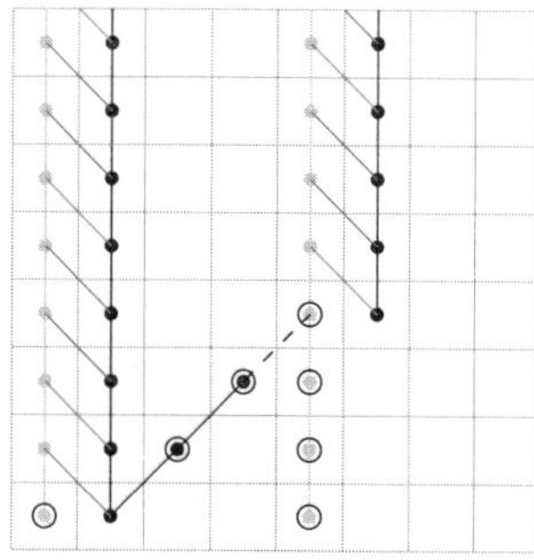

FIGURE 24. The computation of the Adams chart for $\operatorname{Ext}^{s,t}_{\mathcal{A}_1}(R_0, \mathbb{Z}/2)$ using the exact sequence of Figure 23.

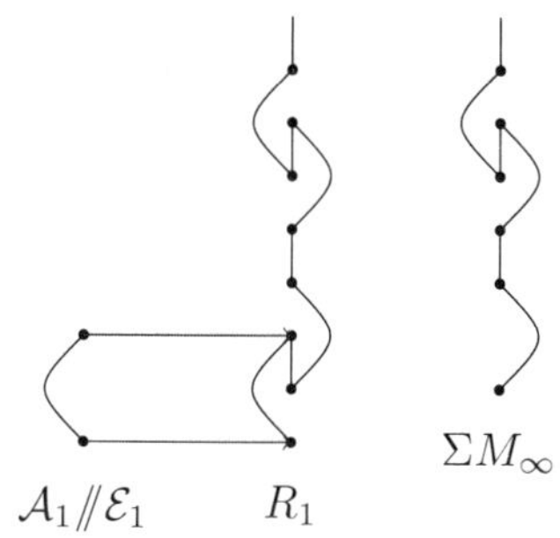

FIGURE 25. An exact sequence of $\mathcal{A}_1$-modules depicting R_1.

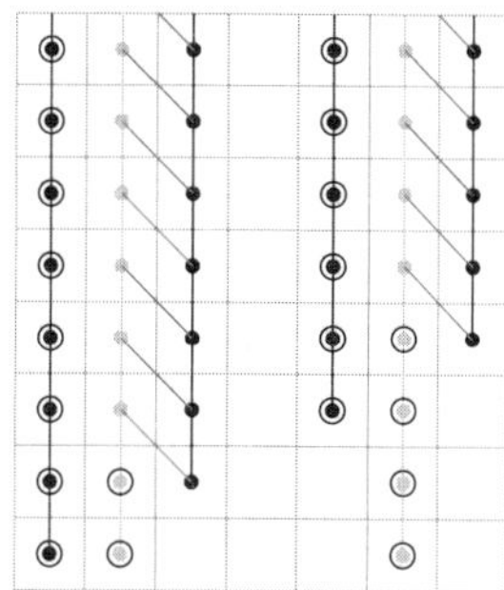

FIGURE 26. The computation of the Adams chart of $\operatorname{Ext}^{s,t}_{\mathcal{A}_1}(R_1, \mathbb{Z}/2)$ using the exact sequence of Figure 25.

so that $P_\bullet \to M$ is a "minimal resolution", but not by projective modules. We call this a "modified" minimal resolution. For each summand $\Sigma^t \mathcal{A}_1$ in P_s, there will be a generator of $\mathbb{Z}/2 \in \operatorname{Ext}^{s,t}_{\mathcal{A}_1}(M, \mathbb{Z}/2)$ and for each summand $\Sigma^t \mathcal{A}_1 /\!/ \mathcal{E}_0$ in P_s, there will be an h_0-tower whose generator is in $\operatorname{Ext}^{s,t}_{\mathcal{A}_1}(M, \mathbb{Z}/2)$.

The proof of this fact uses the collapsing of the spectral sequence of a double complex built from minimal resolutions.

EXAMPLE 4.6.5. We give a modified minimal resolution for $\mathcal{A}_1$ which is periodic in Figure 27. More precisely, the figure depicts the top row of (4.6.4). The periodic resolution is obtained by splicing copies of this complex together and is the bottom row of (4.6.4).

(4.6.4)

$$\begin{array}{ccccccccccccccc}
0 & \leftarrow & \mathbb{Z}/2 & \leftarrow & \mathcal{A}_1 & \leftarrow & \Sigma^2\mathcal{A}_1 \oplus \Sigma\mathcal{A}_1/\!/\mathcal{E}_0 & \leftarrow & \Sigma^4\mathcal{A}_1 & \leftarrow & \Sigma^7\mathcal{A}_1/\!/\mathcal{E}_0 & \leftarrow & \Sigma^{12}\mathbb{Z}/2 & \longleftarrow & 0 \\
 & & \| & & \| & & \| & & \| & & \| & & \uparrow & & \\
0 & \leftarrow & \mathbb{Z}/2 & \leftarrow & P_0 & \longleftarrow & P_1 & \longleftarrow & P_2 & \longleftarrow & P_3 & \longleftarrow & \Sigma^{12}P_0 & \leftarrow & \Sigma^{12}P_1 \leftarrow \dots
\end{array}$$

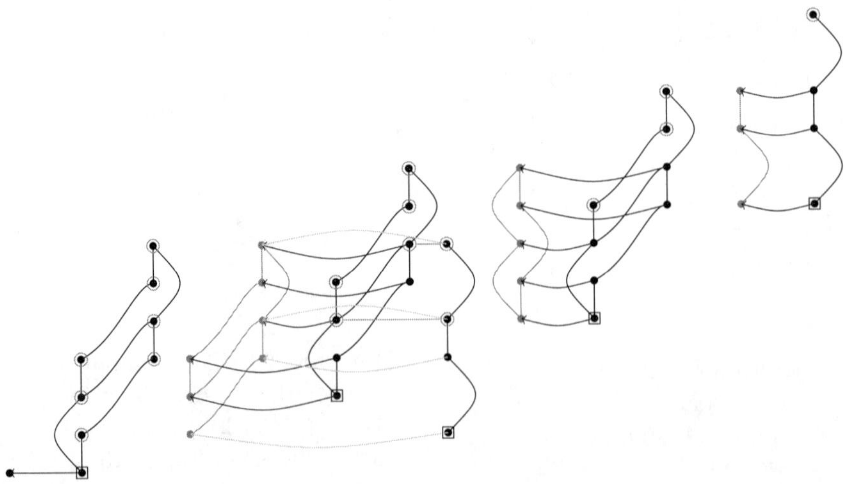

FIGURE 27. A modified minimal resolution for $\mathbb{Z}/2$. The dual of the boxed classes correspond to a copy of $\mathbb{Z}/2$ if they generate an $\mathcal{A}_1$, or an h_0-tower if they generate an $\mathcal{A}_1/\!\!/\mathcal{E}_0$.

EXAMPLE 4.6.6. Consider the module R_2 depicted in Figure 28. Using the resolution constructed in Example 4.6.5, we have a modified minimal resolution

$$0 \longleftarrow R_2 \longleftarrow \Sigma^{-1}P_1 \longleftarrow \Sigma^{-1}P_2 \longleftarrow \dots$$

So the Adams chart for $\mathrm{Ext}^{s,t}_{\mathcal{A}_1}(R_2, \mathbb{Z}/2)$ is a truncated version of that for $\mathrm{Ext}^{s,t}_{\mathcal{A}_1}(\mathbb{Z}/2, \mathbb{Z}/2)$ and is given in Figure 29. Similarly, the $\mathcal{A}_1$-modules J, called the *joker*, and Q called the *"upside down" question mark complex* also have Adams charts which are truncated versions of that for $\mathcal{A}_1$. These are depicted in Figure 29.

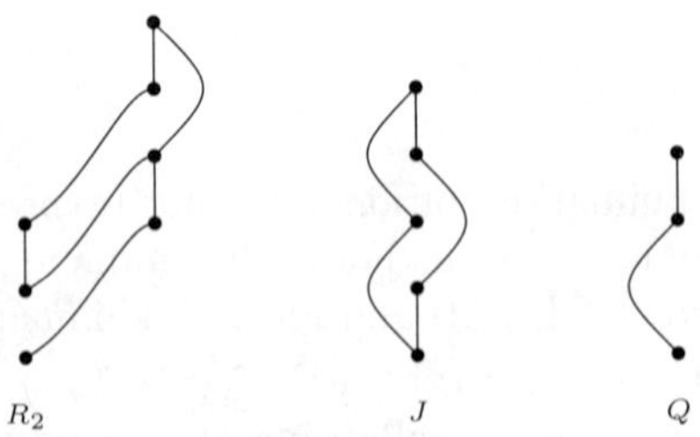

FIGURE 28. An $\mathcal{A}_1$-module we call R_2 (left), the joker J (center) and the "upside down" question mark complex Q (right).

4.7. The Adams spectral sequence. We turn to the construction of the spectral sequence. In this section, we make the following assumption:

ASSUMPTION 4.7.1. Let X be the suspension spectrum of a CW-complex that has finitely many cells in each dimension.

For example, the Thom spectra we are considering have this property since Grassmanians have cell structures with finitely many n-cells for each n. Some of

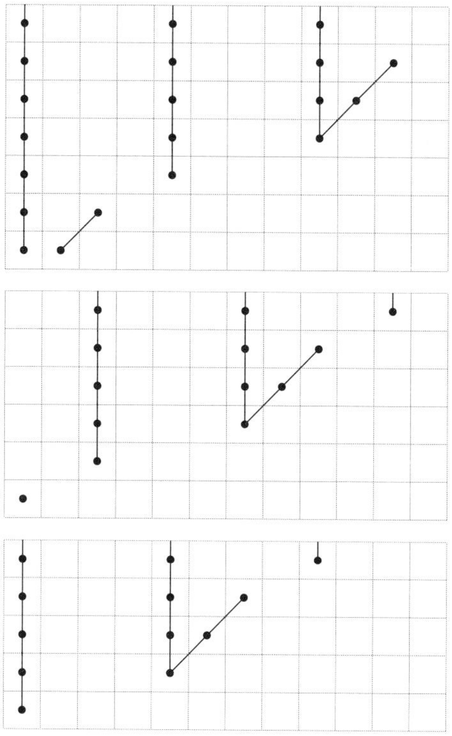

FIGURE 29. The Adams chart for $\operatorname{Ext}^{s,t}_{\mathcal{A}_1}(R_2, \mathbb{Z}/2)$ (top), $\operatorname{Ext}^{s,t}_{\mathcal{A}_1}(J, \mathbb{Z}/2)$ (center) and the Adams chart for $\operatorname{Ext}^{s,t}_{\mathcal{A}_1}(Q, \mathbb{Z}/2)$ (bottom).

this can be done in more generality, but all of our examples will have models of this form so we limit ourselves to this case. A friendly reference to spectral sequences is Hatcher's online notes [**Hat**]. Other great references are McCleary [**McC01**], Boardman [**Boa99**] and Miller [**Mil81**].

DEFINITION 4.7.2. The *Hurewicz homomorphism*

$$h\colon \pi_t X = [S^t, X] \to \operatorname{Hom}_{\mathcal{A}}(H^*(X), H^*(S^t)) \cong \operatorname{Hom}_{\mathcal{A}}(H^*(X), \Sigma^t\mathbb{Z}/2)$$

is defined by sending a map $f\colon S^t \to X$ to the induced map on cohomology, $f^*\colon H^*(X) \to H^*(S^t)$.

If h were an isomorphism, computing the homotopy groups of X would be as easy as understanding its cohomology. In certain cases, this does happen.

DEFINITION 4.7.3. A spectrum Z is a generalized Eilenberg–MacLane spectrum of finite type if

$$Z \simeq HV \simeq \bigvee_{i\in I} \Sigma^i H\mathbb{Z}/2$$

where V is a graded $\mathbb{Z}/2$ vector space which is finite in each degree.

The finiteness assumption in Definition 4.7.3 gives an isomorphism

$$\bigvee_{i\in I} \Sigma^i H\mathbb{Z}/2 \simeq \prod_{i\in I} \Sigma^i H\mathbb{Z}/2.$$

See (2.3.2).

EXAMPLE 4.7.4. Let X be a spectrum that satisfies Assumption 4.7.1. There is an isomorphism

$$H^*(H\mathbb{Z}/2 \wedge X) \cong \mathcal{A} \otimes_{\mathbb{Z}/2} H^*(X)$$

and a class $1 \otimes x \in H^{|x|}(H\mathbb{Z}/2 \wedge X)$ corresponds to a map $H\mathbb{Z}/2 \wedge X \to \Sigma^{|x|} H\mathbb{Z}/2$. By Assumption 4.7.1, the cohomology of X is finite in each degree, so

$$\prod_{x\in H^*(X)} \Sigma^{|x|} H\mathbb{Z}/2 \simeq \bigvee_{x\in H^*(X)} \Sigma^{|x|} H\mathbb{Z}/2$$

and the product of these maps is a weak equivalence:

$$H\mathbb{Z}/2 \wedge X \longrightarrow \bigvee_{x\in H^*(X)} \Sigma^{|x|} H\mathbb{Z}/2$$

So any spectrum of the form $H\mathbb{Z}/2 \wedge X$ for X satisfying Assumption 4.7.1 is a generalized Eilenberg–MacLane spectrum of finite type.

If Z is a generalized Eilenberg–MacLane spectrum of finite type, then the Hurewicz homomorphism is an isomorphism. So, the idea is to resolve X by generalized Eilenberg–MacLane spectra.

DEFINITION 4.7.5. Let X be a spectrum that satisfies Assumption 4.7.1. An *Adams resolution* is a sequence of spectra

$$(4.7.1) \qquad \begin{array}{ccccccccc} X = X_0 & \xleftarrow{i_0} & X_1 & \xleftarrow{i_1} & X_2 & \xleftarrow{i_2} & X_3 & \longleftarrow & \cdots \\ \downarrow{\scriptstyle j_0} & \nearrow{\scriptstyle \delta_0} & \downarrow{\scriptstyle j_1} & \nearrow{\scriptstyle \delta_1} & \downarrow{\scriptstyle j_2} & \nearrow{\scriptstyle \delta_2} & \downarrow{\scriptstyle j_3} & \nearrow{\scriptstyle \delta_3} & \\ K_0 & & K_1 & & K_2 & & K_3 & & \end{array}$$

where

$$X_{s+1} \xrightarrow{i_s} X_s \xrightarrow{j_s} K_s \xrightarrow{\delta_s} \Sigma X_{s+1}$$

are cofiber sequences (i.e. exact triangles) and such that

(a) $K_i \simeq \bigvee_{j\in I_i} \Sigma^j H\mathbb{Z}/2$ for some indexing set I_i, and
(b) $H^*(K_i) \to H^*(X_i)$ is surjective.

REMARK 4.7.6. From an Adams resolution, we obtain a sequence

$$X = X_0 \longrightarrow K_0 \longrightarrow \Sigma K_1 \longrightarrow \Sigma^2 K_2 \longrightarrow \cdots$$

where $\Sigma^s K_s \to \Sigma^{s+1} K_{s+1}$ is the composite $j_{s+1} \circ \delta_s$. Further, the resolution is constructed so that $H^*(\Sigma^\bullet K_\bullet) \to H^*(X)$ is a projective resolution of $H^*(X)$ as an $\mathcal{A}$-module.

REMARK 4.7.7. Let $\overline{H\mathbb{Z}/2}$ be defined by the fiber sequence $\overline{H\mathbb{Z}/2} \to S \to H\mathbb{Z}/2$. From Remark 4.7.4, it follows that

$$\begin{array}{ccccccc}
X & \xleftarrow{i_0} & \overline{H\mathbb{Z}/2} \wedge X & \xleftarrow{i_1} & \overline{H\mathbb{Z}/2}^{\wedge 2} \wedge X & \xleftarrow{i_2} & \cdots \\
\downarrow{\scriptstyle j_0} & \nearrow{\scriptstyle \delta_0} & \downarrow{\scriptstyle j_1} & \nearrow{\scriptstyle \delta_1} & \downarrow{\scriptstyle j_2} & \nearrow{\scriptstyle \delta_2} & \\
H\mathbb{Z}/2 \wedge X & & H\mathbb{Z}/2 \wedge \overline{H\mathbb{Z}/2} \wedge X & & H\mathbb{Z}/2 \wedge \overline{H\mathbb{Z}/2}^{\wedge 2} \wedge X & &
\end{array}$$

is an Adams resolution. So Adams resolutions always exist.

DEFINITION 4.7.8. Let

$$F^s = \operatorname{im}(\pi_* X_s \to \pi_* X).$$

Then $\alpha \in \pi_* X$ has *Adams filtration* s if $\alpha \in F^s \backslash F^{s+1}$.

The Adams filtration of an element is independent of the choice of Adams resolution.

LEMMA 4.7.9. *An element $f \in \pi_t X$ has Adams filtration $\geq s$ if and only if f factors as*

$$f : S^t = U_s \to U_{s-1} \to U_{s-2} \to \ldots \to U_1 \to U_0 = X$$

where the maps $U_i \to U_{i-1}$ induce the zero maps on mod-2 cohomology.

EXAMPLE 4.7.10. An element of $\pi_* X$ has Adams filtration 0 if and only if its image under the Hurewicz homomorphism is non-zero. The image of $\pi_t X_1 \to \pi_t X$ is the kernel of the map j_0. But j_0^* is surjective on cohomology, so

$$\operatorname{Hom}_{\mathcal{A}}(H^*(X), \Sigma^t \mathbb{Z}/2) \to \operatorname{Hom}_{\mathcal{A}}(H^*(K_0), \Sigma^t \mathbb{Z}/2)$$

is injective. In particular, i_0^* must be zero and the image of i_0 consists of elements of filtration $s \geq 1$.

Examples of elements of Adams filtration one are the Hopf maps

$$\eta \colon S^3 \to S^2, \qquad \nu \colon S^7 \to S^4, \qquad \sigma \colon S^{15} \to S^8.$$

We now turn to the construction of the Adams spectral sequence. Fix an Adams resolution of X as in (4.7.1). Applying $\pi_*(-)$, we get an unravelled exact couple

$$\begin{array}{ccccccccccc}
\pi_* X & = & \pi_* X_0 & \xleftarrow{i_0} & \pi_* X_1 & \xleftarrow{i_1} & \pi_* X_2 & \xleftarrow{i_2} & \pi_* X_3 & \longleftarrow & \cdots \\
 & & \downarrow{\scriptstyle j_0} & \nearrow{\scriptstyle \delta_0} & \downarrow{\scriptstyle j_1} & \nearrow{\scriptstyle \delta_1} & \downarrow{\scriptstyle j_2} & \nearrow{\scriptstyle \delta_2} & \downarrow{\scriptstyle j_3} & \nearrow & \\
 & & \pi_* K_0 & & \pi_* K_1 & & \pi_* K_2 & & \pi_* K_3 & &
\end{array}$$

from which we obtain a spectral sequence. More precisely, we let

(a) $E_1^{s,t} = \pi_{t-s} K_s \cong \pi_t \Sigma^s K_s$, and

(b) $d_1 \colon E_1^{s,t} \to E_1^{s+1,t}$ be given by $d_1 = \Sigma j_{s+1} \circ \delta_s$.

In general, $E_r^{*,*} = \ker(d_{r-1}) / \operatorname{im}(d_{r-1})$ and $d_r \colon E_r^{s,t} \to E_r^{s+r,t+r-1}$ is given by $d_r(x) = j(y)$ for any y such that $i^{r-1}(y) = \delta(x)$. Here, $i^n = i \circ \ldots \circ i$ iterated n-times and we have left out indices and suspensions.

PROPOSITION 4.7.11. *Let X satisfy Assumption 4.7.1. There is an isomorphism*

$$E_2^{s,t} \cong \operatorname{Ext}_{\mathcal{A}}^{s,t}(H^*(X), \mathbb{Z}/2).$$

PROOF. In degree t, the E_2 term is the cohomology of

$$0 \longrightarrow \pi_t K_0 \longrightarrow \pi_t \Sigma K_1 \longrightarrow \pi_t \Sigma^2 K_2 \longrightarrow \dots \tag{4.7.2}$$

However, the K_s are generalized Eilenberg MacLane spectra, so

$$\pi_t \Sigma^s K_s \cong \mathrm{Hom}_{\mathcal{A}}(H^*(\Sigma^s K_s), \mathbb{Z}/2).$$

The Adams resolutions are built so that $H^*(\Sigma^\bullet K_\bullet) \to H^*(X)$ is a projective resolution as $\mathcal{A}$-modules, so the homology of (4.7.2) is $\mathrm{Ext}_{\mathcal{A}}^{s,t}(H^*(X), \mathbb{Z}/2)$. □

In general, the Adams spectral sequence does not exactly compute the homotopy groups of the spectrum X. However, under Assumption 4.7.1, it does compute their 2-*completion*, a construction we review here.

DEFINITION 4.7.12. Let G be an abelian group. For each $s \in \mathbb{N}$, let

$$p_{s+1} \colon G/2^{s+1} \to G/2^s$$

be the map induced by reduction modulo 2^s. The 2-*completion* of G, denoted by $G_2^\wedge$, is the inverse limit of $G/2^s$ along the maps p_s. That is,

$$G_2^\wedge = \varprojlim_s G/2^s.$$

REMARK 4.7.13. Note that in the category of abelian groups, $\varprojlim_s G/2^s$ is isomorphic to the kernel of the map

$$p \colon \prod_s G/2^s \to \prod_s G/2^s$$

where p is the difference of the identity and the map to the product induced by the composites $\prod_s G/2^s \to G/2^{k+1} \xrightarrow{p_{k+1}} G/2^k$.

EXAMPLE 4.7.14. If $G = \mathbb{Z}$, then

$$\mathbb{Z}_2 := (\mathbb{Z})_2^\wedge = \varprojlim_s \mathbb{Z}/2^s$$

are the 2-adic integers. In general, if G is a finitely generated abelian group,

$$G_2^\wedge \cong G \otimes \mathbb{Z}_2.$$

In particular, if G is a finite abelian 2-group, then $G_2^\wedge \cong G$.

THEOREM 4.7.15. *Let X satisfy Assumption 4.7.1. Then the Adams spectral sequence for X computes the 2-completion of the homotopy groups of X. That is, the spectral sequence converges to $(\pi_* X)_2^\wedge$:*

$$\mathrm{Ext}_{\mathcal{A}}^{s,t}(H^*(X), \mathbb{Z}/2) \Longrightarrow (\pi_{t-s} X)_2^\wedge$$

REMARK 4.7.16. In fact, the Adams spectral sequence for X satisfying Assumption 4.7.1 computes the homotopy groups of a spectrum $X_2^\wedge$ that can be obtained using a construction analogous to completion for abelian group, and which has the property that $\pi_*(X_2^\wedge) \cong (\pi_* X)_2^\wedge$. In broad strokes, we define $X/2^s \in \mathrm{hSp}$ via the exact triangle:

$$X \xrightarrow{2^s} X \longrightarrow X/2^s \longrightarrow \Sigma X.$$

There are induced maps $p_{s+1}\colon X/2^{s+1} \to X/2^s$ and we define $\varprojlim_s X/2^s \in \mathrm{hSp}$ by the exact triangle

$$\varprojlim_s X/2^s \longrightarrow \prod_s X/2^s \xrightarrow{\ p\ } \prod_s X/2^s \longrightarrow \Sigma \varprojlim_s X/2^s$$

where p is the difference of the identity and the map to the product induced by the composites $\prod_s X/2^s \to X/2^{k+1} \xrightarrow{p_{k+1}} X/2^k$. This is called the *homotopy inverse limit.* For a spectrum X that satisfies Assumption 4.7.1, then

$$X_2^\wedge = \varprojlim_s X/2^s.$$

We refer the reader to Bousfield [**Bou79**, Section 2] and Ravenel [**Rav86**, II.2.1] for more details on this and related topics.

4.8. Using the Adams spectral sequence. In this section, we continue to assume that X satisfies Assumption 4.7.1, so that the Adams spectral sequence for X computes the 2-completion of the homotopy groups of X.

Computing Adams differentials is "an art not a science". There is no algorithm for determining them in general and it usually is a theorem when one computes a new differential in a spectral sequence of interest. However, there are rules to the game and the goal of this section is to share some of the tricks of the trade.

First, the Adams spectral sequence is depicted in an Adams chart as in Figure 30. In this grading, a d_r differential increases s by r and decreases $t-s$ by 1. If $d_r(x) = y$, we say that x hits, or *kills* y. The class x is the *source* and y the *target* of the differential. A class which is in the the kernel of d_r for every r is called a *permanent cycle.* A class which is hit by a differential is called a *boundary.* We say that x *survives* if it is a permanent cycle, but not a boundary.

The Adams spectral sequence for X is a module over the Adams spectral sequence for S^0. From this it follows that the differentials are h_0 and h_1-linear since h_0 detects $2 \in (\pi_0 S_0)_2^\wedge$ and h_1 detects the Hopf map $\eta \in (\pi_1 S_0)_2^\wedge$. That is, $d_r(h_i x) = h_i d_r(x)$ for $i = 0, 1$.

We draw each page $E_r^{*,*}$ of the spectral sequence in subsequent Adams charts, erasing pairs of classes x and y that are connected by a differential $d_r(x) = y$ as we "turn the pages". Letting the process go to infinity, or stopping when there are no possible differentials left, we get the last page, called $E_\infty^{*,*}$. The last page of the spectral sequence contains the information for $(\pi_* X)_2^\wedge$ in the form of an *associated graded.* That is, there is a filtration

$$(\pi_t X)_2^\wedge = F_\infty^{0,t} \supseteq F_\infty^{1,t+1} \supseteq F_\infty^{2,t+2} \supseteq \dots$$

related to $E_\infty^{*,*}$ by exact sequences

$$0 \to F_\infty^{s+1,t+s+1} \to F_\infty^{s,t+s} \to E_\infty^{s,t+s} \to 0. \tag{4.8.1}$$

So, each box in the $t-s$ column of the Adams chart at $E_\infty^{*,*}$ is a subquotient of the answer $(\pi_t X)_2^\wedge$ and the last problem is to reassemble them together. This is called *solving the extensions*, where the word "extension" refers to (4.8.1). We solve for $F_\infty^{0,t}/F^{s,t+s}$ inductively, starting with

$$0 \to E_\infty^{1,t+1} \to F_\infty^{0,t}/F_\infty^{2,t+2} \to E_\infty^{0,t} \to 0$$

and continuing on to

$$0 \to E_\infty^{s,t+s} \to F_\infty^{0,t}/F_\infty^{s+1,t+s+1} \to F_\infty^{0,t}/F_\infty^{s,t+s} \to 0.$$

We take the inverse limit once all of the terms $F_\infty^{0,t}/F^{s,t+s}$ have been determined.

An element $a \in E_\infty^{s,t+s}$ only represents a class $\alpha \in (\pi_t X)_2^\wedge$ modulo elements of *higher filtration*. The language used is that a *detects* the element α. Note that if a detects α, then it detects any class $\alpha + \beta$ where $\beta \in F_\infty^{s+1,t+s+1}$, so a class a may detect multiple elements.

For the Adams spectral sequence, if $a \in E_\infty^{s,t+s}$ and $b \in E_\infty^{s+1,t+s+1}$ are such that $h_0 a = b$, then a detects an element α and b detects an element β such that $2\alpha = \beta$. So multiplication by h_0 records multiplication by 2 and corresponds to a non-trivial, but easy to detect, extension as it comes from the module structure of the E_2-page. However, there can be non-trivial extensions coming from multiplications that do not come from the algebraic structure of the E_2-page. These are called *exotic extensions*. For example, if $2\zeta = \omega$, for $\zeta \in F_\infty^{s,t+s}$ and $\omega \in F_\infty^{s+\epsilon,t+s+\epsilon}$ where $\epsilon > 1$. Then ζ will be detected by some $z \in E_\infty^{s,t+s}$ and ω will be detected by some $w \in E_\infty^{s+\epsilon,t+s+\epsilon}$ so that these two classes are too far apart to be connected by an h_0. These situations are illustrated in Figure 30.

If there are no non-trivial differentials, we say that the spectral sequence *collapses*. We say that it *collapses at* E_r if $E_r^{*,*} = E_\infty^{*,*}$. Often, there will be no possibilities for non-trivial differentials as the target of any possible differential will be zero. In this case, we say that the spectral sequence *collapses for degree reasons*, or *is too sparse for differentials*. Finally, if there are no possibilities for exotic extensions because no two classes on the E_∞-page are aligned in a way that would allow for one to exist, we again say that there are *no exotic extensions for degree reasons* or that the spectral sequence is *too sparse for exotic extensions*. These are the best of all possible scenarios since differentials are hard to compute and exotic extensions are hard to solve. We will be in this situation in all of the examples in Section 5.

EXAMPLE 4.8.1. A typical example of solving extensions is when a column consists of a single h_0-tower, say starting in $E_\infty^{0,t}$. Then

$$\begin{array}{ccccccccc} 0 & \longrightarrow & E_\infty^{s,t+s} & \longrightarrow & F_\infty^{0,t}/F_\infty^{s+1,t+s+1} & \longrightarrow & F_\infty^{0,t}/F_\infty^{s,t+s} & \longrightarrow & 0 \\ & & \| & & \| & & \| & & \\ 0 & \longrightarrow & \mathbb{Z}/2 & \longrightarrow & \mathbb{Z}/2^{s+1} & \longrightarrow & \mathbb{Z}/2^s & \longrightarrow & 0 \end{array}$$

and $F_\infty^{0,t}/F_\infty^{s,t+s} \cong \mathbb{Z}/2^s\mathbb{Z}$ for all s. So

$$(\pi_t X)_2^\wedge \cong \varprojlim_s F_\infty^{0,t}/F_\infty^{s,t+s} \cong \varprojlim_s \mathbb{Z}/2^s \cong \mathbb{Z}_2$$

where $\mathbb{Z}_2$ are the 2-adic integers defined in Example 4.7.14.

5. Examples from the classification problems

In this section, we work out examples to illustrate the methodology. First, some notation. In [**FH16**], Freed and Hopkins give a uniform classification of fermionic symmetric groups ([**FH16**, 9.2]) in spacetime dimension n. There are two complex symmetry groups, denoted $H_n^c(s)$, and labelled by $s = 0, 1$ and eight real symmetry groups, denoted $H_n(s)$, and labelled by $s = 0, \pm1, \pm2, \pm3, 4$. They also show [**FH16**, 2.12] that in each case there are maps $H_n(s) \hookrightarrow H_{n+1}(s)$ stabilizing the groups, so that it makes sense to speak of $H(s)$ and $H^c(s)$ (this is precisely

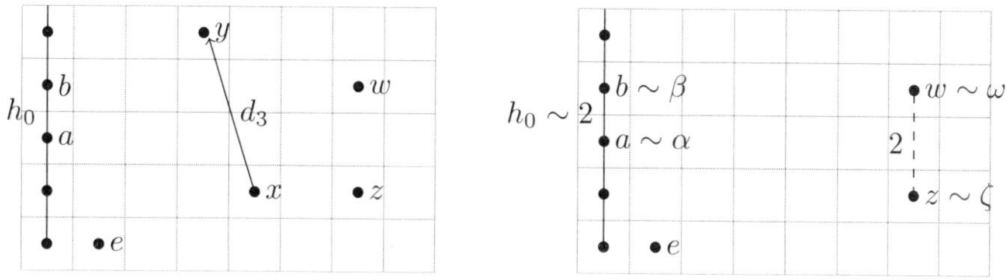

E_3 $\qquad$ E_4 or E_∞

FIGURE 30. Some phenomena in an Adams spectral sequence. The left chart is an example of an E_3-page and the right is the corresponding E_4-page, which in this case would be the E_∞-page as there is no possibilities for further differentials. (The class e cannot support a d_r differential to the h_0-tower since this would violate the h_0-linearity of the differentials.)

analogous to how $O(n)$ stabilizes to O). The Madsen-Tillman spectra (see Section 2.9) $MTH(s)$ are the cobordism theory of manifolds with stable tangential $H(s)$-structure. It is this cobordism theory that features in the Freed-Hopkins classification. This section will be devoted to computing the low dimensional homotopy groups of these cobordism spectra.

In [**FH16**], Freed and Hopkins produce the tables of Figure 31. The explanations in [**FH16**] are brief and some steps are left as exercises. In [**Cam17**], one of the authors gave a detailed explanation of the computation for $MT\,\mathrm{Pin}^-$, $MT\,\mathrm{Pin}^+$, $MT\,\mathrm{Pin}^{\tilde{c}-}$, $MT\,\mathrm{Pin}^{\tilde{c}+}$ and MTG^+. For this reason, we choose to apply the methods to explain the computations for MTG^0, MTG^-, $MT\,\mathrm{Spin}^c$ and $MT\,\mathrm{Pin}^c$, although we start by reproducing the computation for MTG^+ as a warm-up.

$*$	s	$X(H(s))$	π_0	π_1	π_2	π_3	π_4
$M\,\mathrm{Spin}$	0	$MO_0 \simeq S^0$	$\mathbb{Z}$	$\mathbb{Z}/2$	$\mathbb{Z}/2$	0	$\mathbb{Z}$
$MT\,\mathrm{Pin}^-$	$+1$	$\Sigma^{-1}MO_1$	$\mathbb{Z}/2$	$\mathbb{Z}/2$	$\mathbb{Z}/8$	0	0
$MT\,\mathrm{Pin}^+$	-1	ΣMTO_1	$\mathbb{Z}/2$	0	$\mathbb{Z}/2$	$\mathbb{Z}/2$	$\mathbb{Z}/16$
$MT\,\mathrm{Pin}^{\tilde{c}-}$	$+2$	$\Sigma^{-2}MO_2$	$\mathbb{Z}/2$	0	$\mathbb{Z}\times\mathbb{Z}/2$	0	$\mathbb{Z}/2$
$MT\,\mathrm{Pin}^{\tilde{c}+}$	-2	$\Sigma^{2}MTO_2$	$\mathbb{Z}/2$	0	$\mathbb{Z}$	$\mathbb{Z}/2$	$(\mathbb{Z}/2)^3$
MTG^+	$+3$	$\Sigma^{-3}MO_3$	$\mathbb{Z}/2$	0	$\mathbb{Z}/2$	0	$\mathbb{Z}/2\times\mathbb{Z}/4$
MTG^-	-3	$\Sigma^{3}MTO_3$	$\mathbb{Z}/2$	0	$\mathbb{Z}/2$	0	$(\mathbb{Z}/2)^3$
MTG^0	$+4$	$\Sigma^{-3}MSO_3$	$\mathbb{Z}$	0	0	0	$\mathbb{Z}^2$

$*$	s	$X(H^c(s))$	π_0	π_1	π_2	π_3	π_4
$MT\,\mathrm{Spin}^c$	0	$\Sigma^{-2}MU_1$	$\mathbb{Z}$	0	$\mathbb{Z}$	0	$\mathbb{Z}^2$
$MT\,\mathrm{Pin}^c$	1	$\Sigma^{-3}MU_1\wedge MO_1$	$\mathbb{Z}/2$	0	$\mathbb{Z}/4$	0	$\mathbb{Z}/8\times\mathbb{Z}/2$

FIGURE 31. The various real (top) and complex (bottom) symmetry groups studied in [**FH16**]. Note that $M\,\mathrm{Spin}\simeq MT\,\mathrm{Spin}$.

5.1. Reducing to computations over $\mathcal{A}_1$. Computations of $\mathcal{A}$ are in general difficult to perform without computer assistance. However, if one can reduce the computation to one over $\mathcal{A}_1$, constructing minimal resolutions becomes rather straightforward and computations can be done by hand, at least in some range.

The key to making the shift from computations over $\mathcal{A}$ to computations over $\mathcal{A}_1$ is the fact that the spectra MTH defined above satisfy

$$MTH \simeq \mathrm{MSpin} \wedge X(H)$$

where $X(H)$ are the Thom spectra of certain familiar vector bundles [**FH16**, 10.7]. The values of $X(H)$ for the groups H studied in [**FH16**] are given in Figure 31.

Since our cohomology is with field coefficients, namely $\mathbb{Z}/2$, the Künneth formula gives an isomorphism

$$H^*(MTH) \cong H^*(\mathrm{MSpin}) \otimes_{\mathbb{Z}/2} H^*(X(H)).$$

The key steps in the reductions of computations to an $\mathcal{A}_1$-module problem is the following theorem.

THEOREM 5.1.1 (Anderson, Brown, Peterson). *There is an isomorphism*

$$H^*(\mathrm{MSpin}) \cong \mathcal{A} \otimes_{\mathcal{A}_1} (\mathbb{Z}/2 \oplus M)$$

where M is a graded $\mathcal{A}_1$-module which is zero in degrees $t < 8$.

As a consequence of Theorem 5.1.1 and Remark 4.4.1, we have:

COROLLARY 5.1.2. *There is an isomorphism*

$$\mathrm{Ext}_{\mathcal{A}}^{s,t}(H^*(\mathrm{MSpin} \wedge X(H)), \mathbb{Z}/2) \cong \mathrm{Ext}_{\mathcal{A}_1}^{s,t}(H^*(X(H)), \mathbb{Z}/2)$$

if $t - s < 8$.

So low dimensional computations can be done over $\mathcal{A}_1$. We go through the following steps to compute $\pi_t MTH$ for $0 \le t \le 4$:

(1) Compute $H^*(X(H))$ as modules over the $\mathcal{A}_1$. See Section 3.4.
(2) Compute $\mathrm{Ext}_{\mathcal{A}_1}^{s,t}(H^*(X(H)), \mathbb{Z}/2)$ in the range $t - s \le 5$. See Section 4.4, Section 4.5 and Section 4.6.
(3) Compute the differentials and extensions. See Section 4.8. In all of our examples, the spectral sequences are too sparse for differentials and exotic extensions and this step is trivial.
(4) Read off $\pi_* MTH$.

We will do this one example at a time.

5.2. The case $s = 3$. This is the case of $H = G^+ = \mathrm{Pin}^+ \times_{\{\pm 1\}} SU_2$ and in this case,

$$MTG^+ \simeq \mathrm{MSpin} \wedge \Sigma^{-3} MO_3.$$

This example was stated in [**FH16**] and explicitly computed in [**Cam17**]. The cohomology of $H^*(\Sigma^{-3}MO_3)$ is illustrated in Figure 7. Let R_3 be the $\mathcal{A}_1$-module depicted in Figure 32, so that R_3 sits in an exact sequence

$$0 \to \Sigma Q \to R_3 \to M_\infty \to 0.$$

From Figure 7, we have that

$$H^*(\Sigma^{-3}MO_3) \approx R_3 \oplus \Sigma^2 \mathcal{A}_1 \oplus \Sigma^4 \mathcal{A}_1 \oplus \Sigma^5 \mathcal{A}_1$$

where we will use $\approx$ to denote that there is an isomorphism in the range necessary for computations of homotopy groups in degrees less than or equal to 4. We include the column $t-s=5$ to preclude the possibility of incoming differentials into the column $t-s=4$.

To compute the E_2-page of the spectral sequence

$$\mathrm{Ext}_{\mathcal{A}}^{s,t}(H^*(MTG^+),\mathbb{Z}/2) \approx \mathrm{Ext}_{\mathcal{A}_1}^{s,t}(H^*(\Sigma^{-3}MO_3),\mathbb{Z}/2) \Rightarrow \pi_{t-s}MTG^+,$$

we have to compute $\mathrm{Ext}_{\mathcal{A}_1}^{*,*}(R_3,\mathbb{Z}/2)$. Figure 32 and Figure 33 illustrate this computation.

The Adams spectral sequence computing π_*MTG^+ is depicted in Figure 34. The spectral sequence is too sparse for differentials and exotic extensions, so the homotopy groups are

$$\begin{aligned}
\pi_0 MTG^+ &= \mathbb{Z}/2\\
\pi_1 MTG^+ &= 0\\
\pi_2 MTG^+ &= \mathbb{Z}/2\\
\pi_3 MTG^+ &= 0\\
\pi_4 MTG^+ &= \mathbb{Z}/2\times\mathbb{Z}/4.
\end{aligned}$$

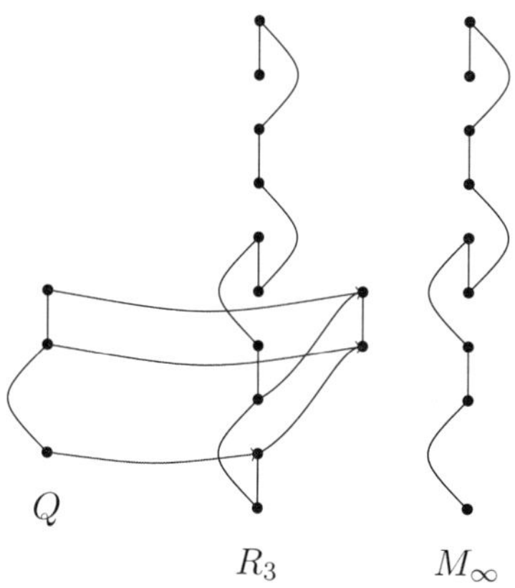

FIGURE 32. An $\mathcal{A}_1$-module we call R_3.

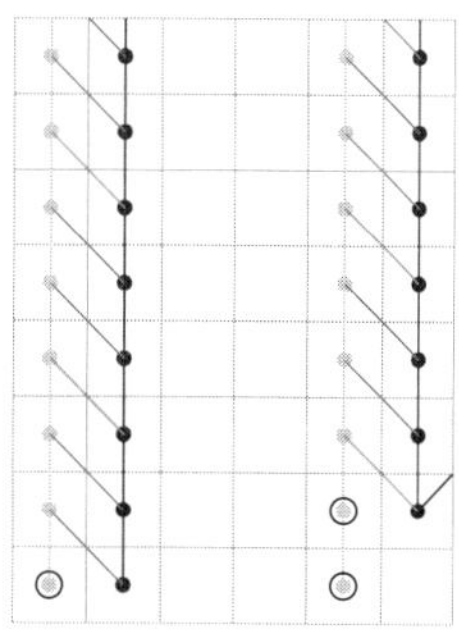

FIGURE 33. The computation of the Adams chart for $\mathrm{Ext}_{\mathcal{A}_1}^{s,t}(R_3,\mathbb{Z}/2)$ using the exact sequence of Figure 32. The Adams chart for Q is given in Figure 29 and M_∞ is discussed in Example 4.4.2.

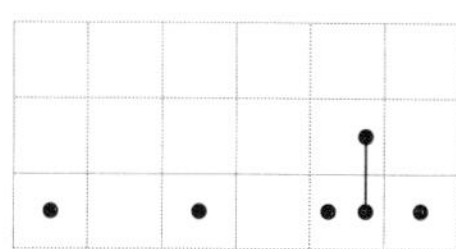

FIGURE 34. The Adams chart for $\mathrm{Ext}_{\mathcal{A}_1}^{s,t}(H^*(\Sigma^{-3}MO_3),\mathbb{Z}/2)$.

5.3. The case $s = -3$. This is the case of $H = G^- = \mathrm{Pin}^- \times_{\{\pm 1\}} SU_2$ and in this case,

$$MTG^- \simeq \mathrm{MSpin} \wedge \Sigma^3 MTO_3.$$

In the degrees relevant for us, the $\mathcal{A}_1$-module structure of $H^*(\Sigma^3 MTO_3)$ is given in Figure 8. We have that

$$H^*(\Sigma^3 MTO_3) \approx \mathcal{A}_1 \oplus \Sigma^2 R_0 \oplus \Sigma^4 \mathcal{A}_1 \oplus \Sigma^4 \mathcal{A}_1 \oplus \Sigma^5 R_5$$

where R_0 is the module depicted in Figure 23 and R_5 the module depicted in Figure 35. The module R_5 sits in a short exact sequence of $\mathcal{A}_1$-modules (pictured in Figure 35):

$$0 \to J \to R_5 \to \Sigma M_\infty \to 0.$$

Figure 24 gives $\mathrm{Ext}_{\mathcal{A}_1}^{*,*}(R_0, \mathbb{Z}/2)$ and Figure 36 gives $\mathrm{Ext}_{\mathcal{A}_1}^{*,*}(R_5, \mathbb{Z}/2)$. The Adams chart for $\mathrm{Ext}_{\mathcal{A}_1}^{s,t}(H^*(\Sigma^{-3} MTO_3), \mathbb{Z}/2)$ is depicted in Figure 37 in the range of interest. The spectral sequence is too sparse for differentials and exotic extensions and the homotopy groups of MTG^- are

$$\begin{aligned}
\pi_0 MTG^- &= \mathbb{Z}/2\\
\pi_1 MTG^- &= 0\\
\pi_2 MTG^- &= \mathbb{Z}/2\\
\pi_3 MTG^- &= 0\\
\pi_4 MTG^- &= (\mathbb{Z}/2)^3.
\end{aligned}$$

5.4. The case $s = 4$. This is the case of $H = G^0 = \mathrm{Spin} \times_{\{\pm 1\}} SU_2$ and in this case,

$$MTG^0 \simeq \mathrm{MSpin} \wedge \Sigma^{-3} MSO_3.$$

The $\mathcal{A}_1$-structure of $H^*(MSO_3)$ is depicted in Figure 9, and

$$H^*(\Sigma^{-3} MSO_3) \approx Q \oplus \Sigma^4 R_2.$$

The Adams chart for the modules Q and R_2 are depicted in Figure 29, and the Adams chart for $\mathrm{Ext}_{\mathcal{A}_1}^{s,t}(H^*(\Sigma^{-3} MSO_3), \mathbb{Z}/2)$ is in Figure 38. The spectral sequence is too sparse for differentials and exotic extensions and the homotopy groups of MTG^0 are

$$\begin{aligned}
\pi_0 MTG^0 &= \mathbb{Z}\\
\pi_1 MTG^0 &= 0\\
\pi_2 MTG^0 &= 0\\
\pi_3 MTG^0 &= 0\\
\pi_4 MTG^0 &= \mathbb{Z}^2.
\end{aligned}$$

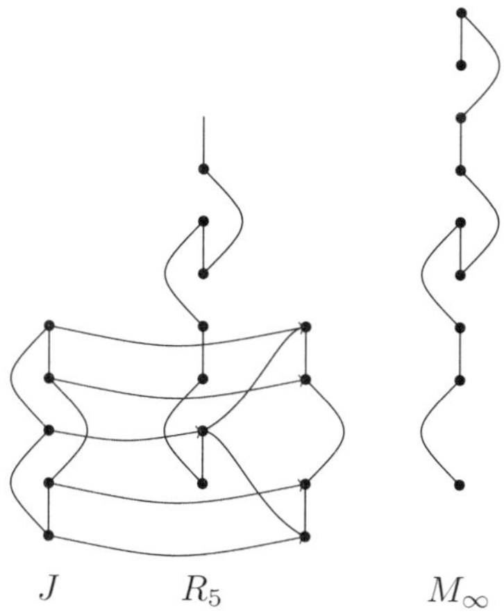

FIGURE 35. An $\mathcal{A}_1$-module we call R_5.

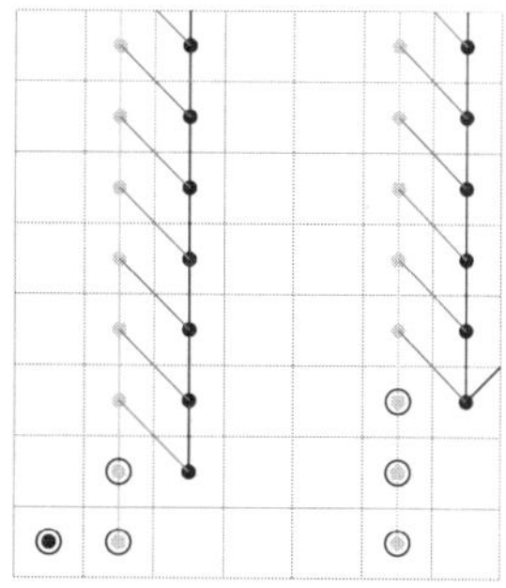

FIGURE 36. The computation of the Adams chart for $\mathrm{Ext}^{s,t}_{\mathcal{A}_1}(R_5, \mathbb{Z}/2)$ using the exact sequence of Figure 35. The Adams chart for J is given in Figure 29 and M_∞ is discussed in Example 4.4.2.

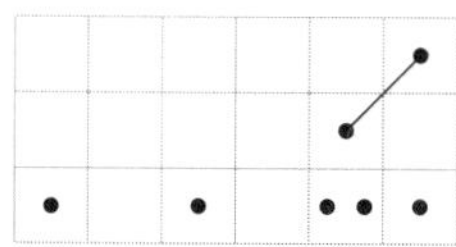

FIGURE 37. The Adams chart for $\mathrm{Ext}^{s,t}_{\mathcal{A}_1}(H^*(\Sigma^3 MTO_3), \mathbb{Z}/2)$.

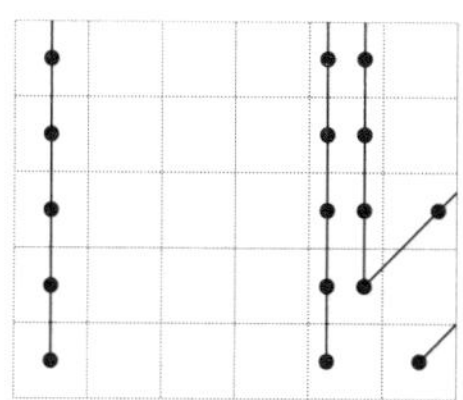

FIGURE 38. The Adams chart for $\mathrm{Ext}^{s,t}_{\mathcal{A}_1}(H^*(\Sigma^{-3} MSO_3), \mathbb{Z}/2)$.

5.5. The complex case $s = 0$. This is the case of $H^c = \mathrm{Spin}^c$ and in this case,

$$MTH^c(0) \simeq \mathrm{MSpin} \wedge \Sigma^{-2} MU_1.$$

The structure of $H^*(MU_1)$ as an $\mathcal{A}_1$-module is depicted in Figure 4. It is given by shifted sums of $\mathcal{A}_1 /\!\!/ \mathcal{E}_1$, so

$$H^*(\Sigma^{-2} MU_1) \approx \mathcal{A}_1 /\!\!/ \mathcal{E}_1 \oplus \Sigma^4 \mathcal{A}_1 /\!\!/ \mathcal{E}_1$$

In Example 4.5.6, we calculated that

$$\mathrm{Ext}^{*,*}_{\mathcal{A}_1}(\mathcal{A}_1 /\!\!/ \mathcal{E}_1, \mathbb{Z}/2) \cong \mathrm{Ext}^{*,*}_{\mathcal{E}_1}(\mathbb{Z}/2, \mathbb{Z}/2) \cong \mathbb{Z}/2[h_0, v_1]$$

for v_1 in degree $(s,t) = (1,3)$. The E_2-page of the Adams spectral sequence for $\pi_* MTH^c(0)$ is depicted in Figure 39. The spectral sequence is too sparse for differentials and exotic extensions and the homotopy groups of $MTH^c(0)$ are

$$\pi_0 MTH^c(0) = \mathbb{Z}$$
$$\pi_1 MTH^c(0) = 0$$
$$\pi_2 MTH^c(0) = \mathbb{Z}$$
$$\pi_3 MTH^c(0) = 0$$
$$\pi_4 MTH^c(0) = (\mathbb{Z})^2$$

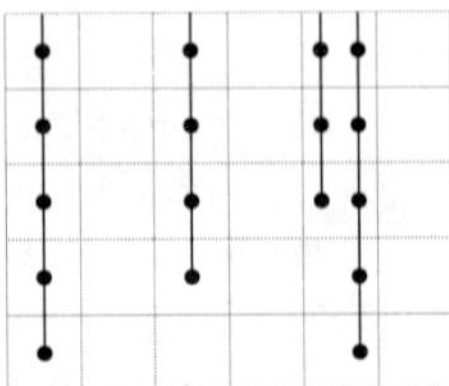

FIGURE 39. The Adams chart for $\mathrm{Ext}_{\mathcal{A}_1}^{s,t}(H^*(MU_1), \mathbb{Z}/2)$.

5.6. The complex case $s = 1$. This is the case of $H^c = \mathrm{Pin}^c$ and in this case,

$$MTH^c(1) \simeq \mathrm{MSpin} \wedge \Sigma^{-3} MU_1 \wedge MO_1.$$

The structure of $H^*(MU_1 \wedge MO_1)$ is depicted in Figure 5. We have

$$H^*(\Sigma^{-3} MU_1 \wedge MO_1) \approx R_6 \oplus \Sigma^4 R_6$$

for the module R_6 depicted in Figure 40. In order to compute the E_2-page of the Adams spectral sequence for $\pi_* MTH^c(1)$ we need to compute $\mathrm{Ext}_{\mathcal{A}_1}^{*,*}(R_6, \mathbb{Z}/2)$. The module R_6 sits in a short exact sequence of $\mathcal{A}_1$-modules (pictured in Figure 40):

$$0 \to \Sigma R_1 \to R_6 \to M_\infty \to 0$$

and $\mathrm{Ext}_{\mathcal{A}_1}^{*,*}(R_6, \mathbb{Z}/2)$ is computed in Figure 41. The E_2-page of the Adams spectral sequence for $\pi_* MTH^c(1)$ is depicted in Figure 42. The spectral sequence is too sparse for differentials and exotic extensions and the homotopy groups of $MTH^c(1)$ are

$$\pi_0 MTH^c(1) = \mathbb{Z}/2$$
$$\pi_1 MTH^c(1) = 0$$
$$\pi_2 MTH^c(1) = \mathbb{Z}/4$$
$$\pi_3 MTH^c(1) = 0$$
$$\pi_4 MTH^c(1) = \mathbb{Z}/2 \times \mathbb{Z}/8.$$

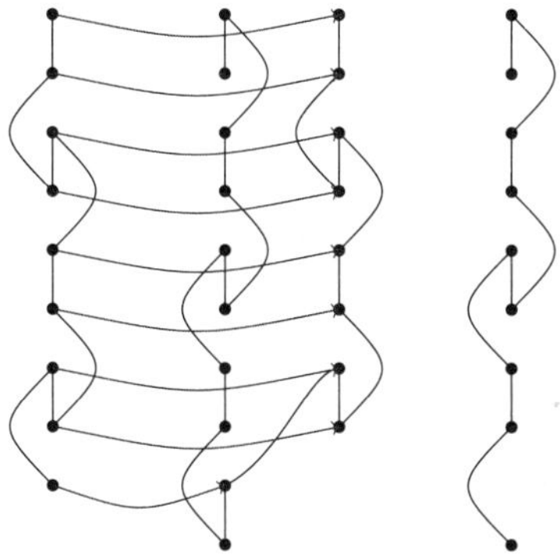

FIGURE 40. The exact sequence for R_6.

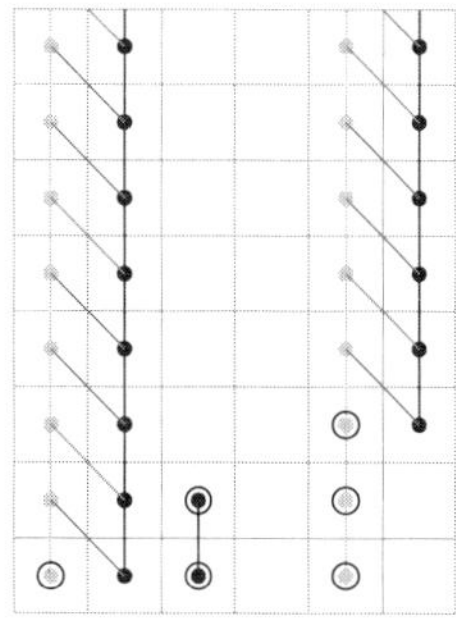

FIGURE 41. The computation of the Adams chart for $\mathrm{Ext}_{\mathcal{A}_1}^{s,t}(R_6, \mathbb{Z}/2)$ using the exact sequence of Figure 40. The Adams chart for R_1 is given in Figure 26 and M_∞ is discussed in Example 4.4.2.

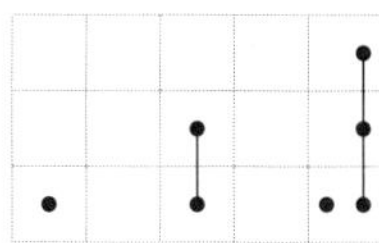

FIGURE 42. The Adams chart for $\mathrm{Ext}_{\mathcal{A}_1}^{s,t}(H^*(MU_1 \wedge MO_1), \mathbb{Z}/2)$.

Acknowledgments

The authors thank Andy Baker, Prasit Bhattacharya, Bert Guillou, and the referee for helpful comments.

References

[Ada58] J. F. Adams, *On the structure and applications of the Steenrod algebra*, Comment. Math. Helv. **32** (1958), 180–214, DOI 10.1007/BF02564578. MR0096219

[Ada60] J. F. Adams, *On the non-existence of elements of Hopf invariant one*, Ann. of Math. (2) **72** (1960), 20–104, DOI 10.2307/1970147. MR0141119

[Ada95] J. F. Adams, *Stable homotopy and generalised homology*, Chicago Lectures in Mathematics, University of Chicago Press, Chicago, IL, 1995. Reprint of the 1974 original. MR1324104

[AGP02] M. Aguilar, S. Gitler, and C. Prieto, *Algebraic topology from a homotopical viewpoint*, Universitext, Springer-Verlag, New York, 2002. Translated from the Spanish by Stephen Bruce Sontz. MR1908260

[Boa99] J. M. Boardman, *Conditionally convergent spectral sequences*, Homotopy invariant algebraic structures (Baltimore, MD, 1998), Contemp. Math., vol. 239, Amer. Math. Soc., Providence, RI, 1999, pp. 49–84, DOI 10.1090/conm/239/03597. MR1718076

[Bou79] A. K. Bousfield, *The localization of spectra with respect to homology*, Topology **18** (1979), no. 4, 257–281, DOI 10.1016/0040-9383(79)90018-1. MR551009

[Cam17] J. A. Campbell, *Homotopy Theoretic Classification of Symmetry Protected Phases*, ArXiv e-prints (2017).

[EKMM97] A. D. Elmendorf, I. Kriz, M. A. Mandell, and J. P. May, *Rings, modules, and algebras in stable homotopy theory*, Mathematical Surveys and Monographs, vol. 47, American Mathematical Society, Providence, RI, 1997. With an appendix by M. Cole. MR1417719

[FH16] D. S. Freed and M. J. Hopkins, *Reflection positivity and invertible topological phases*, ArXiv e-prints (2016).

[Hat] A. Hatcher, *Spectral sequences in algebraic topology*, Online book.

[Jam95] I. M. James (ed.), *Handbook of algebraic topology*, North-Holland, Amsterdam, 1995. MR1361885

[LMSM86] L. G. Lewis Jr., J. P. May, M. Steinberger, and J. E. McClure, *Equivariant stable homotopy theory*, Lecture Notes in Mathematics, vol. 1213, Springer-Verlag, Berlin, 1986. With contributions by J. E. McClure. MR866482

[Lur] J. Lurie, *Higher algebra*, Preprint.

[McC01] J. McCleary, *A user's guide to spectral sequences*, 2nd ed., Cambridge Studies in Advanced Mathematics, vol. 58, Cambridge University Press, Cambridge, 2001. MR1793722

[Mil81] H. R. Miller, *On relations between Adams spectral sequences, with an application to the stable homotopy of a Moore space*, J. Pure Appl. Algebra **20** (1981), no. 3, 287–312, DOI 10.1016/0022-4049(81)90064-5. MR604321

[MMSS01] M. A. Mandell, J. P. May, S. Schwede, and B. Shipley, *Model categories of diagram spectra*, Proc. London Math. Soc. (3) **82** (2001), no. 2, 441–512, DOI 10.1112/S0024611501012692. MR1806878

[MT68] R. E. Mosher and M. C. Tangora, *Cohomology operations and applications in homotopy theory*, Harper & Row, Publishers, New York-London, 1968. MR0226634

[Rav86] D. C. Ravenel, *Complex cobordism and stable homotopy groups of spheres*, Pure and Applied Mathematics, vol. 121, Academic Press, Inc., Orlando, FL, 1986. MR860042

[Sch] S. Schwede, *Symmetric spectra*, Online book.

[Wei94] C. A. Weibel, *An introduction to homological algebra*, Cambridge Studies in Advanced Mathematics, vol. 38, Cambridge University Press, Cambridge, 1994. MR1269324

[Whi62] G. W. Whitehead, *Generalized homology theories*, Trans. Amer. Math. Soc. **102** (1962), 227–283, DOI 10.2307/1993676. MR0137117

Department of Mathematics, University of Colorado at Boulder, Campus Box 395, Boulder, Colorado 80309

Email address: agnes.beaudry@colorado.edu

Department of Mathematics, Vanderbilt University, 1326 Stevenson Center, Nashville, Tennessee 37240

Email address: j.campbell@vanderbilt.edu

Contemporary Mathematics
Volume **718**, 2018
https://doi.org/10.1090/conm/718/14489

Flagged higher categories

David Ayala and John Francis

Abstract. We introduce *flagged* (∞, n)*-categories* and prove that they are equivalent to Segal sheaves on Joyal's category $\mathbf{\Theta}_n$. As such, flagged (∞, n)-categories provide a model-independent formulation of Segal sheaves. This result generalizes the statement that n-groupoid objects in spaces are effective, as we explain and contextualize as an instance of Koszul duality. Along the way, we establish a useful expression for the univalent-completion of such a Segal sheaf. Finally, we conjecture a characterization of flagged (∞, n)-categories as stacks on (∞, n)-categories that satisfy descent with respect to colimit diagrams that do not generate invertible i-morphisms for any i.

Contents

Introduction

In [**Re2**], Rezk introduced a definition an (∞, n)-category as a presheaf

$$\mathcal{C}\colon \mathbf{\Theta}_n^{\mathsf{op}} \longrightarrow \mathcal{S}\mathsf{paces}$$

on a combinatorial category of Joyal's (developed in [**Be**]) that satisfies two families of conditions:

- **"Completeness"**, or, in this paper, **"univalent-completeness"**, as per the use of the term in Homotopy Type theory, as to distinguish from other notions of completion;
- **"Segal"**, or, in this paper, **"Segal-sheaf"**, so as to evoke a locality condition.

2010 *Mathematics Subject Classification.* Primary 18A05. Secondary 55U35, 55P65.

Key words and phrases. (∞, n)-categories, higher categories, flagged higher categories, Segal spaces, univalence, groupoid objects, Cech nerve.

The first author was supported by the National Science Foundation under award 1507704. The second author was supported by the National Science Foundation under award 1508040.

(We will review this notions shortly, as Definition 0.5.) For formal reasons, each presheaf on $\mathbf{\Theta}_n$ determines an (∞, n)-category via *Segal-sheafification* followed by *univalent-completion*. Many examples of (∞, n)-categories of especial interest, even for $n = 1$, are univalent-completions of naturally presented Segal sheaves on the category $\mathbf{\Theta}_n$. For instance, the following disparate constructions occur as univalent-completions:

- Replace a group by its moduli space of torsors, which loses conjugation information within the group. (See Example 0.15.)
- Replace a ring by its category of modules, which remembers only its Morita-type. (See Example 0.17.)
- Replace a category by its idempotent completion. (See Example 0.18.)
- Replace a suitably connective sequence $(X_0 \to \cdots \to X_n)$ of spaces by the space X_n alone. (See Example 0.21.)
- Replace a smooth closed manifold by its smooth h-cobordism-type, which loses simple-homotopy-type. (See Example 0.22.)

Each of these constructions losses information, as indicated. In light of this lossy univalent-completion construction, it is in order to find a conceptual, model-independent, formulation of Segal sheaves on $\mathbf{\Theta}_n$. Our purpose for such a formulation is to accommodate native examples of such entities, and also to house such entities in a framework that can borrow results from established (∞, n)-category theory (even for $n = 1$).

In this paper we give a model-independent formulation of Segal sheaves on $\mathbf{\Theta}_n$, a corollary concerning higher groupoid objects, and conjecture another model-independent formulation; we state these three assertions informally here.

(1) (Theorem 0.27) A Segal sheaf on $\mathbf{\Theta}_n$ is equivalent to a flag $\mathcal{C}_0 \to \mathcal{C}_1 \to \cdots \to \mathcal{C}_n$ in which each $\mathcal{C}_i$ is an (∞, i)-category and, for each $0 \leq k \leq i \leq j \leq n$, the functor $\mathcal{C}_i \to \mathcal{C}_j$ is surjective on spaces of k-morphisms with specified source-target.
(2) (Corollary 0.33) An n-groupoid object in $\mathcal{S}\mathsf{paces}$ is precisely a flag $X_0 \to X_1 \to \cdots \to X_n$ of spaces for which, for each $0 \leq i \leq j \leq n$, the map $X_i \to X_j$ is i-connective.
(3) (Conjecture 0.45) A Segal sheaf on $\mathbf{\Theta}_n$ is a stack on the ∞-category Cat_n of (∞, n)-category that satisfies descent with respect to those colimit diagrams that do not generate invertible i-morphisms for any $0 \leq i \leq n$.

TQFT context. This technical article is a first step in an anticipated duality between certain topological quantum field theories. Specifically, in an earlier work [**AF2**] the authors express a duality between two classes of topological quantum field theories. One side of this duality is a state-sum theory, instantiated as factorization homology of a partially commutative $\Bbbk$-algebra, specifically an $\mathcal{E}_n$-algebra in $\mathsf{Mod}_{\Bbbk}$, where $\Bbbk$ is a fixed field of characteristic 0. The other side of the duality is a sigma-model with target X, perturbative about a point $\mathsf{Spec}(\Bbbk) \xrightarrow{p} X$. (Here, X is an $\mathcal{E}_n$-stack over $\Bbbk$, such as the putative output of a deformation quantization of a shifted symplectic $\Bbbk$-scheme in the sense of [**PTVV**].) The essential relationship between these two theories is Koszul duality, which articulates how deformations of $p \in X$ is governed by its shifted tangent space $\mathbb{T}_p^{\mathcal{E}_n} X[-n]$, as it is endowed with the structure of an $\mathcal{E}_n$-algebra. Corollary 0.33 in this article is a first step toward establishing a duality between general state-sum theories, instantiated as factorization homology of (∞, n)-categories [**AFR**], and sigma-models that are not necessarily perturbative about a point, but about a certain flag $X_0 \to X_1 \to \cdots \to X_n = X$. Specifically, Corollary 0.33 articulates this Koszul duality in the case that each X_i is a homotopy type, which can be interpreted as an infinitesimal algebraic stack.

This is accomplished through a technical result about higher category theory, which is of independent interest. Specifically, the main Theorem 0.27 of this article characterizes Segal sheaves on $\mathbf{\Theta}_n$, which may be regarded as not-necessarily univalent-complete (∞, n)-categories, as *flagged* (∞, n)*-categories*, which are flags $\mathcal{C}_0 \to \mathcal{C}_1 \to \cdots \to \mathcal{C}_n$ in which each $\mathcal{C}_i$ is an (∞, i)-category and each arrow is suitable connected. A notable example of such being $\mathsf{Bord}_0 \to \cdots \to \mathsf{Brod}_n$, the flag of bordism (∞, i)-categories: as a flagged (∞, n)-category, isomorphism are diffeomorphisms; as a (∞, n)-category, isomorphism are h-cobordisms.

Conventions. We make use of Lurie's work [**Lu1**], as well as Joyal's work [**Jo1**], for the foundations of ∞-category theory – there, quasi-category theory. This includes a comprehensive theory of colimits, limits, (space-valued) presheaves, the Yoneda embedding as a colimit completion, the unstraightening construction, and Bousfield localizations among presentable ∞-categories. We assume the reader has operational, though not necessarily technical, acquaintance with these features of ∞-category theory. We also call on some more specific features of the ∞-category $\mathcal{S}\mathsf{paces}$, which are consequences of the fact that it is an ∞-topos in the sense of §6 of [**Lu1**]. We assume the reader has a working acquaintance with Joyal's category $\mathbf{\Theta}_n$, as it is presented in Berger's work [**Be**] as well as Rezk's work [**Re2**]. We assume the reader has a working acquaintance with (∞, n)-categories as developed by Rezk in [**Re2**].

We make use of the following notation.

NOTATION 0.1.

- We may denote the colimit of a presheaf $\mathcal{F}\colon \mathcal{C}^{\mathsf{op}} \to \mathcal{S}\mathsf{paces}$ as $|\mathcal{F}| := \mathsf{colim}(\mathcal{F})$.
- Let $\mathcal{C}$ be an ∞-category. By right Kan extension, each presheaf $(\mathcal{C}^{\mathsf{op}} \xrightarrow{\mathcal{F}} \mathcal{S}\mathsf{paces}) \in \mathsf{PShv}(\mathcal{C})$ extends along the Yoneda embedding as a functor

$$\mathcal{F}\colon \mathsf{PShv}(\mathcal{C})^{\mathsf{op}} \xrightarrow{\mathsf{Map}(-,\mathcal{F})} \mathcal{S}\mathsf{paces}~, \qquad \mathcal{E} \mapsto \mathcal{F}(\mathcal{E}) := \mathsf{Map}(\mathcal{E}, \mathcal{F})~.$$

0.1. Setup and main results. We give a definition of Joyal's category $\mathbf{\Theta}_n$ ([**Jo2**]), which follows Definition 3.9 in [**Be**].

DEFINITION 0.2. The category $\mathbf{\Theta}_n$, and its subcategory $\mathbf{\Theta}_n^{\mathsf{cls}} \hookrightarrow \mathbf{\Theta}_n$ of *closed* morphisms, are defined by induction on n as follows.

- For $n < 0$, $\mathbf{\Theta}_n := \emptyset$. Assume $n \geq 0$. An object in the category $\mathbf{\Theta}_n$ is a pair of objects $[p] \in \mathbf{\Delta}$ and $(S_1, \dots, S_p) \in (\mathbf{\Theta}_{n-1})^{\times p}$; such an object is typically denoted $[p](S_1, \dots, S_p)$. A morphism in $\mathbf{\Theta}_n$ from $[p](S_1, \dots, S_p)$ to $[q](T_1, \dots, T_q)$ is a morphism $[p] \xrightarrow{\sigma} [q]$ in $\mathbf{\Delta}$ together with, for each $0 < i \leq p$ and $\sigma(i-1) < j \leq \sigma(i)$, a morphism $S_i \xrightarrow{\tau_{ij}} T_j$ in $\mathbf{\Theta}_n$. Composition of morphisms in $\mathbf{\Theta}_n$ is given by composing morphisms in $\mathbf{\Delta}$ and composing collections of morphisms in $\mathbf{\Theta}_{n-1}$.
- For each $n \geq 0$, the subcategory $\mathbf{\Theta}_n^{\mathsf{cls}} \hookrightarrow \mathbf{\Theta}_n$ contains all objects, and for $n > 0$, only those morphisms $[p](S_1, \dots, S_p) \xrightarrow{(\sigma, (\tau_{ij}))} [q](T_1, \dots, T_q)$ for which σ is a consecutive inclusion and each τ_{ij} is a morphism in $\mathbf{\Theta}_{n-1}^{\mathsf{cls}}$.

For $0 \leq k \leq n$, the *k-cell* is $c_k := [1](c_{k-1})$ for $k > 0$ and $c_0 = [0]$. Its *boundary* is the sub-presheaf of that represented by c_k,

$$\partial c_k \colon \mathbf{\Theta}_n^{\mathsf{op}} \longrightarrow \mathcal{S}\mathsf{paces} ,$$

whose value on $T \in \mathbf{\Theta}_n$ consists of those morphisms $T \to c_k$ that factor through a morphism $c_j \to c_k$ for some $0 \leq j < k$.

REMARK 0.3. Consider the $(2,1)$-category $\mathsf{Cat}_n^{\mathsf{strict}}$ of *strict n-categories*, by which it is meant the $(2,1)$-category of ordinary categories enriched over the Cartesian monoidal $(2,1)$-category of strict $(n-1)$-categories. There is a fully faithful functor $\mathbf{\Theta}_n \hookrightarrow \mathsf{Cat}_n^{\mathsf{strict}}$, as established in [**Be**]. The *nerve* functor is the restricted Yoneda functor along this fully faithful functor:

$$\mathsf{Cat}_n^{\mathsf{strict}} \longrightarrow \mathsf{PShv}(\mathbf{\Theta}_n) , \qquad \mathsf{C} \mapsto \big(T \mapsto \mathsf{Cat}_n^{\mathsf{strict}}(T, \mathsf{C})\big) .$$

NOTATION 0.4. Let $0 \leq i \leq n$. The strict i-category corepresenting an invertible i-morphism in a strict n-groupoid is denoted $c_{i-1} \wr \mathsf{E}(c_1)$. The presheaf on $\mathbf{\Theta}_n$ which is its nerve is given the same notation. (This is a special case of Definition 2.1, to come.)

Notice the morphism between presheaves on $\mathbf{\Theta}_n$,

$$c_{i-1} \wr \mathsf{E}(c_1) \longrightarrow c_{i-1} , \tag{1}$$

corepresenting identity i-morphisms as invertible i-morphisms within strict n-categories. This morphism between presheaves on $\mathbf{\Theta}_n$ determines a morphism between ∞-overcategories:

$$\mathbf{\Theta}_{n/c_{i-1} \wr \mathsf{E}(c_1)} \longrightarrow \mathbf{\Theta}_{n/c_{i-1}} , \tag{2}$$

where the domain ∞-category is the base change of the ∞-overcategory $\mathsf{PShv}(\mathbf{\Theta}_n)_{/c_{i-1} \wr \mathsf{E}(c_1)}$ along the Yoneda functor. The functor (2), in turn, defines a functor

$$\big(\mathbf{\Theta}_{n/c_{i-1} \wr \mathsf{E}(c_1)}\big)^{\triangleright} \longrightarrow \mathbf{\Theta}_n \tag{3}$$

whose value on the cone point is $c_{i-1} \in \mathbf{\Theta}_n$.

DEFINITION 0.5.

- A *Segal cover in* $\mathbf{\Theta}_n$ is a colimit diagram $\mathcal{J}^{\triangleright} \to \mathbf{\Theta}_n^{\mathsf{cls}}$. The ∞-category of *Segal sheaves (on* $\mathbf{\Theta}_n$*)* is the full ∞-subcategory

  $$\mathsf{Shv}(\mathbf{\Theta}_n) \subset \mathsf{PShv}(\mathbf{\Theta}_n)$$

  consisting of those presheaves that carry (the opposites of) Segal covers to limit diagrams.
- For $0 < i \leq n$, the *i-univalence diagram in* $\mathbf{\Theta}_n$ is the functor $\big(\mathbf{\Theta}_{n/c_{i-1} \wr \mathsf{E}(c_1)}\big)^{\triangleright} \xrightarrow{(3)} \mathbf{\Theta}_n$. The ∞-category of *univalent Segal sheaves (on* $\mathbf{\Theta}_n$*)* is the full ∞-subcategory

  $$\mathsf{Shv}^{\mathsf{unv}}(\mathbf{\Theta}_n) \subset \mathsf{Shv}(\mathbf{\Theta}_n)$$

  consisting of those Segal sheaves that carry (the opposites of) univalence diagrams to limit diagrams.

REMARK 0.6. We use the notation $\mathsf{Shv}(\boldsymbol{\Theta}_n)$ for Segal sheaves on $\boldsymbol{\Theta}_n$ to suggestively regard the Segal condition on a presheaf as a descent condition with respect to a notion of a cover. We warn the reader, however, that Segal covers do *not* form a Grothendieck topology on $\boldsymbol{\Theta}_n$. Likewise, the ∞-category $\mathsf{Shv}(\boldsymbol{\Theta}_n)$ is not an ∞-topos; therefore, there is no Grothendieck site for which $\mathsf{Shv}(\boldsymbol{\Theta}_n)$ is its ∞-category of sheaves. Nevertheless, the Segal covers of Definition 0.5 *do* define a Grothendieck topology on the subcategory $\boldsymbol{\Theta}_n^{\mathsf{cls}} \subset \boldsymbol{\Theta}_n$. Said another way, the pullback ∞-category in the diagram

$$\begin{array}{ccc} \mathsf{Shv}(\boldsymbol{\Theta}_n^{\mathsf{cls}}) & \longrightarrow & \mathsf{Shv}(\boldsymbol{\Theta}_n) \\ \downarrow & & \downarrow \\ \mathsf{PShv}(\boldsymbol{\Theta}_n^{\mathsf{cls}}) & \longrightarrow & \mathsf{PShv}(\boldsymbol{\Theta}_n) \end{array}$$

is an ∞-topos. In fact, this ∞-topos is free on its infinitesimal basics:

$$\mathsf{PShv}(\mathbb{G}_n^{\mathsf{cls}}) \simeq \mathsf{Shv}(\boldsymbol{\Theta}_n^{\mathsf{cls}}) ,$$

where $\mathbb{G}_n^{\mathsf{cls}} \subset \boldsymbol{\Theta}_n^{\mathsf{cls}}$ is the full subcategory consisting of the cells.

We recall the following culminating definition of [**Re2**].

DEFINITION 0.7 ([**Re2**]). The ∞-category Cat_n of (∞, n)-categories is initial among presentable ∞-categories under $\boldsymbol{\Theta}_n$,

$$\boldsymbol{\Theta}_n \longrightarrow \mathsf{Cat}_n ,$$

that carry Segal covers to colimit diagrams and carry univalence diagrams to colimit diagrams.

OBSERVATION 0.8. From their defining universal properties, there is a canonical identification between ∞-categories under $\boldsymbol{\Theta}_n$:

$$\mathsf{Cat}_n \simeq \mathsf{Shv}^{\mathsf{unv}}(\boldsymbol{\Theta}_n) .$$

DEFINITION 0.9. Let $0 \leq i \leq n$. A functor $\mathcal{C} \to \mathcal{D}$ between (∞, n)-categories is *i-connective* if, for each $0 \leq k \leq i$, each solid diagram among (∞, n)-categories

$$\begin{array}{ccc} \partial c_k & \longrightarrow & \mathcal{C} \\ \downarrow & \nearrow & \downarrow \\ c_k & \longrightarrow & \mathcal{D} \end{array}$$

can be filled.

EXAMPLE 0.10. Let $\mathcal{X} \to \mathcal{Y}$ be a map between ∞-groupoids, and let $n \geq 0$ be an integer. Regarded as a functor between (∞, n)-categories, it is i-connective if and only if it is i-connective as a map between spaces. This is to say that, for each choice $x \mapsto y$ of compatible base points and for each $0 \leq k \leq i$, the induced map $\pi_k(X; x) \to \pi_k(Y; y)$ is a bijection for $k < i$ and is surjective for $k = i$.

REMARK 0.11. Let $\mathcal{C} \to \mathcal{D}$ be a functor between (∞, n)-categories. One might say this functor is *k-surjective* if it is surjective on spaces of k-morphisms with specified source-target. More precisely, if, for each functor $\partial c_k \to \mathcal{C}$, the resulting map between spaces $\mathsf{Map}^{\partial c_k/}(c_k, \mathcal{C}) \to \mathsf{Map}^{\partial c_k/}(c_k, \mathcal{D})$ is surjective on path components. Through this terminology, the functor $\mathcal{C} \to \mathcal{D}$ being i-connective is equivalent to it being k-surjective for each $0 \leq k \leq i$.

DEFINITION 0.12. A *flagged* (∞, n)*-category* is a sequence of morphisms among (∞, n)-categories

$$\mathcal{C}_0 \longrightarrow \mathcal{C}_1 \longrightarrow \cdots \longrightarrow \mathcal{C}_n$$

satisfying the following conditions:

- for each $0 \leq i \leq n$, the (∞, n)-category $\mathcal{C}_i$ is actually an (∞, i)-category;
- for each $0 \leq i \leq j \leq n$, the functor $\mathcal{C}_i \to \mathcal{C}_j$ is i-connective.

The ∞-category of *flagged* (∞, n)*-categories* is the full ∞-subcategory

$$\mathsf{fCat}_n \subset \mathsf{Fun}\big([n], \mathsf{Cat}_n\big)$$

consisting of the flagged (∞, n)-categories.

NOTATION 0.13. For $\mathcal{C}$ an (∞, n)-category, we denote by $\mathcal{C}^{\sim} := \mathcal{C}_{\leq 0}$ its maximal ∞-subgroupoid.

EXAMPLE 0.14. In general, a flagged $(\infty, 1)$-category is an $(\infty, 1)$-category $\mathcal{C}$, together with a *surjective* functor $\mathcal{G} \to \mathcal{C}$ from an ∞-groupoid. Here *surjective* means *essentially surjective*, or equivalently it means π_0-surjective on spaces of objects.

EXAMPLE 0.15. Let A be an associative algebra in the Cartesian symmetric monoidal ∞-category $\mathcal{S}\mathsf{paces}$. Its deloop $* \to \mathfrak{B}A$ is an ∞-category equipped with a functor from the terminal ∞-groupoid which is surjective on maximal ∞-subgroupoids.

EXAMPLE 0.16. More generally, let A be an $\mathcal{E}_n$-algebra A in the Cartesian symmetric monoidal ∞-category $\mathcal{S}\mathsf{paces}$. Its n-fold deloop $* \to \cdots \to * \to \mathfrak{B}^n A$ is an (∞, n)-category equipped with a functor from the terminal $(\infty, n-1)$-category which is $(n-1)$-connected. For $A \simeq \Omega^n X$ an n-fold loop space on a based space X, then the above flagged (∞, n)-category is $* \to \cdots \to * \to X$, so long as X is n-connective.

EXAMPLE 0.17. Consider the ordinary category Morita whose objects are associative rings, whose morphisms from A to B are (B, A)-bimodules, and whose composition rule is given as follows: for P a (B, A)-bimodule, and for Q a (C, B)-bimodule, the composition $Q \circ P$ is the (C, A)-bimodule $P \underset{B}{\otimes} Q$. Equivalences in Morita are Morita equivalences between rings. In particular, for each commutative ring R, the objects $\mathsf{Mat}_{2\times 2}(R)$ and R are equivalent in Morita. Consider the flagged $(\infty, 1)$-category

$$\mathsf{Rings}^{\sim} \longrightarrow \mathsf{Morita}\ .$$

The underlying ∞-groupoid of this flagged $(\infty, 1)$-category is, by design, that of isomorphisms between associative rings.

EXAMPLE 0.18. More generally, consider the $(2, 1)$-category Corr whose objects are ordinary categories and whose morphisms from $\mathcal{C}$ to $\mathcal{D}$ are $(\mathcal{D}, \mathcal{C})$-bimodules, and whose composition rule is given by coend. Two categories are equivalent in Corr if their idempotent completions are equivalent as categories. In particular, the ordinary category corepresenting an idempotent and that corepresenting a retraction are equivalent in Corr yet they are not equivalent as categories – there is a unique fully faithful epimorphism between them, and it is not surjective. Consider the flagged $(\infty, 1)$-category

$$\mathsf{Cat}^{\sim} \longrightarrow \mathsf{Corr}\ .$$

The underlying ∞-groupoid of this flagged $(\infty,1)$-category is, by design, that of equivalences between ordinary categories.

EXAMPLE 0.19. We make use of the terminology introduced in Remark 0.11. In general, a flagged $(\infty,2)$-category is an $(\infty,2)$-category $\mathcal{C}_2$, together with a functor $\mathcal{C}_1 \to \mathcal{C}_2$ from an $(\infty,1)$-category that is 0-surjective and 1-surjective, together with a 0-surjective functor $\mathcal{C}_0 \to \mathcal{C}_1$ from an ∞-groupoid.

EXAMPLE 0.20. Let $\mathcal{G}$ be an $(\infty,0)$-category, which is simply a space. Denote by $\mathcal{G}_{\leq -1}$ its (-1)-truncation: this is a space which is initial if $\mathcal{G}$ is empty, and is final otherwise. Then $\mathcal{G} \to \mathcal{G}_{\leq -1}$ is a flagged $(\infty,1)$-category. This construction is present in the definition of an enriched ∞-category, as developed in [**GH**]. Namely, for $\mathcal{V}$ a monoidal ∞-category, the canonical functor to its deloop $* \to \mathfrak{B}\mathcal{V}$ is a flagged $(\infty,1)$-category (internal to Cat). A $\mathcal{V}$-enriched ∞-category, with underlying ∞-groupoid $\mathcal{C}_0$, is a lax functor between flagged $(\infty,1)$-categories $\mathsf{hom}_{\mathcal{C}}\colon (\mathcal{C}_0 \to (\mathcal{C}_0)_{\leq -1}) \xrightarrow{\mathsf{lax}} (* \to \mathfrak{B}\mathcal{V})$ satisfying a certain univalent-condition.

EXAMPLE 0.21. A flagged (∞,n)-category in which each constituent (∞,i)-category is, in fact, an ∞-groupoid is precisely a flag of spaces $(X_0 \to \cdots \to X_n)$ in which each map $X_i \to X_j$ is i-connective. (For an interesting example, consider a knot $K \subset S^3$, and take $X_0 = X_1 = K$ and $X_2 = X_3 = S^3$. For another interesting example, consider a point $* \in X$ in an n-connective space and take $X_0 = \cdots = X_{n-1} = *$ and $X_n = X$.) The underlying (∞,n)-category of this flagged (∞,n)-category is the ∞-groupoid X_n, which is blind to the maps $X_i \to X_n$. In the example coming from a point $* \in X$ in an n-connective space, this flagged (∞,n)-category is the $\mathcal{E}_n$-algebra $\Omega^n X$, whereas the (∞,n)-category associated to this flagged (∞,n)-category is its n-fold deloop X, as an unpointed space. In the case $n=1$, the space of automorphisms of X is the space of outer automorphisms of the $\mathcal{E}_1$-algebra ΩX.

EXAMPLE 0.22. Two closed $(n-1)$-manifolds are equivalent as $(n-1)$-morphisms in the (∞,n)-category $\mathcal{B}\mathsf{ord}_n$ if and only if they are h-cobordant (see §2.2 of [**Lu3**] for a discussion of this). Consider the fantastic example of a flagged (∞,n)-category

$$\mathcal{B}\mathsf{ord}_0 \longrightarrow \mathcal{B}\mathsf{ord}_1 \longrightarrow \cdots \longrightarrow \mathcal{B}\mathsf{ord}_n$$

given by the standard functors. The underlying (∞,i)-category in this flagged (∞,n)-category, by design, codifies diffeomorphisms between compact i-manifolds (with corner structure).

OBSERVATION 0.23. Consider the standard sequence of fully faithful right adjoint functors

$$\mathbf{\Theta}_0 \hookrightarrow \mathbf{\Theta}_1 \hookrightarrow \ldots \hookrightarrow \mathbf{\Theta}_{n-1} \hookrightarrow \mathbf{\Theta}_n\ . \tag{4}$$

Each of the functors in this sequence, as well as their left adjoints, preserves Segal covers and univalence diagrams. Therefore, left Kan extension along each functor in the above sequence determines a sequence of fully faithful left adjoint functors

$$\mathsf{Cat}_0 \hookrightarrow \mathsf{Cat}_1 \hookrightarrow \ldots \hookrightarrow \mathsf{Cat}_{n-1} \hookrightarrow \mathsf{Cat}_n\ ;$$

the right adjoint to each of these functors is given by restriction along the corresponding functor in (4).

TERMINOLOGY 0.24. For each $0 \leq i \leq j \leq n$, the value of the right adjoint to $\mathsf{Cat}_i \hookrightarrow \mathsf{Cat}_j$ on an (∞,j)-category $\mathcal{C}$ is its *maximal (∞,i)-subcategory* $\mathcal{C}_{\leq i} \subset \mathcal{C}$.

OBSERVATION 0.25. Evaluation at the target defines a left adjoint

$$\mathsf{fCat}_n \longrightarrow \mathsf{Cat}_n$$

in a localization between ∞-categories. The right adjoint carries an (∞, n)-category $\mathcal{C}$ to the flagged (∞, n)-category

$$\mathcal{C}_{\leq 0} \longrightarrow \mathcal{C}_{\leq 1} \longrightarrow \cdots \longrightarrow \mathcal{C}_{\leq n-1} \longrightarrow \mathcal{C}_{\leq n} = \mathcal{C}\ .$$

The fully faithful functors

$$\mathbf{\Theta}_n \hookrightarrow \mathsf{Cat}_n \hookrightarrow \mathsf{fCat}_n \tag{5}$$

determine the restricted Yoneda functors

$$\mathsf{fCat}_n \longrightarrow \mathsf{PShv}(\mathsf{Cat}_n) \qquad \text{and} \qquad \mathsf{fN}\colon \mathsf{fCat}_n \longrightarrow \mathsf{PShv}(\mathbf{\Theta}_n)\ . \tag{6}$$

In light of the factorization (5), the functor fN extends the standard nerve functor:

$$\mathsf{N}\colon \mathsf{Cat}_n \hookrightarrow \mathsf{fCat}_n \xrightarrow{\mathsf{fN}} \mathsf{PShv}(\mathbf{\Theta}_n)\ .$$

REMARK 0.26. Let $\underline{\mathcal{C}} = (\mathcal{C}_0 \to \mathcal{C}_1 \to \cdots \to \mathcal{C}_n)$ be a flagged (∞, n)-category. The value of the presheaf $\mathsf{fN}(\underline{\mathcal{C}})$ on $T \in \mathbf{\Theta}_n$ is the space of fillers in the commutative diagram among (∞, n)-categories:

$$\begin{array}{ccccccccc} T_{\leq 0} & \longrightarrow & T_{\leq 1} & \longrightarrow & \cdots & \longrightarrow & T_{\leq n-1} & \longrightarrow & T_{\leq n} \\ \downarrow & & \downarrow & & & & \downarrow & & \downarrow \\ \mathcal{C}_0 & \longrightarrow & \mathcal{C}_1 & \longrightarrow & \cdots & \longrightarrow & \mathcal{C}_{n-1} & \longrightarrow & \mathcal{C}_n. \end{array}$$

In the case that each canonical functor to the maximal (∞, i)-subcategory $\mathcal{C}_i \to (\mathcal{C}_n)_{\leq i}$ is an equivalence, such a diagram is just the data of its rightmost vertical arrow.

Here is our main result, which we prove in §3.4.

THEOREM 0.27. *The restricted Yoneda functor*

$$\mathsf{fN}\colon \mathsf{fCat}_n \xrightarrow{(6)} \mathsf{PShv}(\mathbf{\Theta}_n)$$

is fully faithful, with image consisting of those presheaves that carry (the opposites of) Segal covers to limit diagrams:

$$\mathsf{fN}\colon \mathsf{fCat}_n \xrightarrow{\simeq} \mathsf{Shv}(\mathbf{\Theta}_n)\ .$$

REMARK 0.28. Theorem 0.27 offers a model-independent description of Segal sheaves on $\mathbf{\Theta}_n$, as we explain. By Definition 0.7, the ∞-category Cat_n of (∞, n)-categories is defined via a universal property that references $\mathbf{\Theta}_n$. The work of Barwick–Schommer-Pries ([**BS**]) articulates a sense in which this dependence on $\mathbf{\Theta}_n$ can be relieved, or rather replaced by an assortment of other basic categories $\mathcal{T}_n$. In this way, we regard the ∞-category Cat_n of (∞, n)-categories as *model-independent* – it can be described as a full ∞-subcategory of presheaves on an assortment of basic categories $\mathcal{T}_n$, not just $\mathcal{T}_n = \mathbf{\Theta}_n$. Supported by this, the Definition 0.12 of the ∞-category of flagged (∞, n)-categories, then, is a model-independent notion. Theorem 0.27 therefore gives a model-independent description of Segal sheaves on $\mathbf{\Theta}_n$.

0.2. A corollary. We draw a corollary of Theorem 0.27, as it specializes to the case of groupoids. To state this corollary, we give two auxiliary definitions.

Recall from Notation 0.4 the strict n-category $c_{i-1} \wr \mathsf{E}(c_1)$. Consider the functor between strict n-categories,

$$c_i \longrightarrow c_{i-1} \wr \mathsf{E}(c_1) \ , \tag{7}$$

corepresenting invertible i-morphisms as examples of i-morphsims within strict n-categories.

DEFINITION 0.29. The ∞-category of *n-groupoid objects (in $\mathcal{S}$paces)* is the full ∞-subcategory

$$n\,\mathsf{Gpd}[\mathcal{S}] \ \subset \ \mathsf{PShv}(\mathbf{\Theta}_n)$$

consisting of those presheaves $\mathcal{G}\colon \mathbf{\Theta}_n^{\mathsf{op}} \to \mathcal{S}\mathsf{paces}$ that satisfy the following conditions.

(1) $\mathcal{G}$ carries (the opposites of) Segal diagrams to limit diagrams.
(2) For each $0 < i \leq n$, each solid diagram among presheaves on $\mathbf{\Theta}_n$ admits a filler:

$$\begin{array}{ccc} c_i & \longrightarrow & \mathcal{G} \\ {\scriptstyle (7)}\big\downarrow & \overset{\exists}{\nearrow} & \\ c_{i-1} \wr \mathsf{E}(c_1) & & \end{array} .$$

REMARK 0.30. Let $0 < i \leq n$. The morphism $c_i \xrightarrow{(7)} c_{i-1} \wr \mathsf{E}(c_1)$ in $\mathsf{Shv}(\mathbf{\Theta}_1)$ is an epimorphism. So, for each Segal sheaf $\mathcal{F}$ on $\mathbf{\Theta}_n$, the map between spaces induced by $\mathcal{F}\big(c_{i-1} \wr \mathsf{E}(c_1)\big) \xrightarrow{\mathcal{F}(7)} \mathcal{F}(c_i)$ is a monomorphism. Therefore, condition (2) in Definition 0.29 is equivalent to the condition that, for each $0 < i \leq n$, this monomorphism $\mathcal{G}\big(c_{i-1} \wr \mathsf{E}(c_1)\big) \xrightarrow{\simeq} \mathcal{G}(c_i)$ is surjective on components, and is therefore an equivalence. In the case that $\mathcal{G}$ is the nerve of a strict n-category, this condition (2) is exactly the condition that $\mathcal{G}$ is, in fact, a strict n-groupoid. This justifies the terminology of Definition 0.29.

The next definition isolates the examples discussed in Example 0.21.

DEFINITION 0.31. The ∞-category of *n-flagged ∞-groupoids* is the full ∞-subcategory

$$\mathsf{fGpd}_n \ \subset \ \mathsf{Fun}([n], \mathcal{S}\mathsf{paces})$$

of sequences $(X_0 \to \cdots \to X_n)$ for which the map $X_i \to X_j$ is i-connective for each $0 \leq i \leq j \leq n$.

OBSERVATION 0.32. There are evident fully faithful functors

$$n\,\mathsf{Gpd}[\mathcal{S}] \ \hookrightarrow \ \mathsf{Shv}(\mathbf{\Theta}_n) \qquad \text{and} \qquad \mathsf{fGpd}_n \ \hookrightarrow \ \mathsf{fCat}_n \ .$$

We isolate the following consequence of Theorem 0.27, which is of independent interest. Its proof occupies §3.5.

COROLLARY 0.33. *The equivalence of Theorem* 0.27 *restricts as an equivalence between* ∞*-categories:*

$$\begin{array}{ccc} \mathsf{fGpd}_n & \overset{\mathsf{fN}}{\underset{\simeq}{\dashrightarrow}} & n\,\mathsf{Gpd}[\mathcal{S}] \\ {\scriptstyle \text{Obs } 0.32}\downarrow & & \downarrow{\scriptstyle \text{Obs } 0.32} \\ \mathsf{fCat}_n & \underset{\text{Thm } 0.27}{\overset{\mathsf{fN}}{\underset{\simeq}{\longrightarrow}}} & \mathsf{Shv}(\mathbf{\Theta}_n). \end{array}$$

REMARK 0.34. Let $\underline{X} = (X_0 \to X_1 \to \cdots \to X_n)$ be an n-flagged ∞-groupoid. Through Corollary 0.33, this is equivalent data to the n-groupoid object $\mathsf{fN}(\underline{X})$. This n-groupoid object $\mathsf{fN}(\underline{X})$ is the Segal sheaf on $\mathbf{\Theta}_n$ with the following values on cells and their boundaries.

- $\mathsf{fN}(\underline{X})(c_0) \simeq X_0$. Also, $\mathsf{fN}(\underline{X})(\partial c_1) \simeq X_0 \times X_0$.
- $\mathsf{fN}(\underline{X})(c_1) \simeq X_0 \underset{X_1}{\times} X_0$. Also, $\mathsf{fN}(\underline{X})(\partial c_2) \simeq \big(X_0 \underset{X_1}{\times} X_0\big) \underset{X_0 \times X_0}{\times} \big(X_0 \underset{X_1}{\times} X_0\big)$.
- $$\begin{aligned} \mathsf{fN}(\underline{X})(c_2) &\simeq X_2 \underset{\left(X_2 \underset{X_2}{\times} X_2\right) \underset{X_2\times X_2}{\times} \left(X_2 \underset{X_2}{\times} X_2\right)}{\times} \big(X_0 \underset{X_1}{\times} X_0\big) \underset{X_0 \times X_0}{\times} \big(X_0 \underset{X_1}{\times} X_0\big) \\ &\simeq X_2 \underset{X^{S^1}}{\times} \mathsf{fN}(\underline{X})(\partial c_2)\ . \end{aligned}$$
- In general, there are pullback squares among spaces:

$$\begin{array}{ccc} \mathsf{fN}(\underline{X})(c_i) & \longrightarrow & X_i \\ \downarrow & & \downarrow{\scriptstyle \mathsf{diag}} \\ \mathsf{fN}(\underline{X})(\partial c_i) & \longrightarrow & X_i^{S^{i-1}} \end{array} \qquad \text{and} \qquad \begin{array}{ccc} \mathsf{fN}(\underline{X})(\partial c_j) & \longrightarrow & \mathsf{fN}(\underline{X})(c_{j-1}) \\ \downarrow & & \downarrow \\ \mathsf{fN}(\underline{X})(c_{j-1}) & \longrightarrow & \mathsf{fN}(\underline{X})(\partial c_{j-1}). \end{array}$$

Informally, an object in $\mathsf{fN}(\underline{X})$ is a point in X_0. A 1-morphism in $\mathsf{fN}(\underline{X})$ is a path in X_1 equipped with lifts of its endpoints to X_0. A 2-morphism in $\mathsf{fN}(\underline{X})$ is a 2-disk in X_2, equipped with compatible lifts of its hemispheres to X_1 and lifts of its poles to X_0. Continuing, an i-morphism is an i-disk in X_i equipped with compatible lifts of its hemispherical j-strata to X_j for $j < i$. Informally, composition is given by concatenating disks.

REMARK 0.35. Let $\underline{X}$ be an n-flagged ∞-groupoid. The connectivity assumptions on $\underline{X}$ ensure that the space X_n can be recovered as the colimit of the n-groupoid object $\mathsf{fN}(\underline{X})$. Without these connectivity assumptions, this colimit $|\,\mathsf{fN}(\underline{X})|$ would report a suitable connective cover of X_n.

REMARK 0.36. Let $\underline{X} = (X_0 \to X_1)$ be a 1-flagged ∞-groupoid. The 1-groupoid object $\mathsf{fN}(\underline{X})$, which is in particular a simplicial space, is the *Cech nerve* of the map $X_0 \to X_1$, in the sense of §6.1.2 of [**Lu1**]. In this sense, Corollary 0.33 generalizes the fact that every 1-groupoid object $\mathcal{G}_\bullet$ in $\mathcal{S}\mathsf{paces}$ is the Cech nerve of the canonical map $\mathcal{G}_0 \to |\mathcal{G}_\bullet|$ to its colimit.

REMARK 0.37. Let $\underline{X} = (X_0 \to \cdots \to X_n)$ be an n-flagged ∞-groupoid. After Remark 0.36, the n-groupoid object $\mathsf{fN}(\underline{X})$, which is in particular a presheaf on $\mathbf{\Theta}_n$, can be interpreted as the n-*Cech nerve* of the given flag $\underline{X}$. In this way, Corollary 0.33 states that every n-groupoid object $\mathbf{\Theta}_n^{\mathsf{op}} \xrightarrow{\mathcal{G}_\bullet} \mathcal{S}\mathsf{paces}$ in $\mathcal{S}\mathsf{paces}$ is the n-Cech nerve of the canonical flag of maps $\mathcal{G}_0 \to |\mathcal{G}_{\bullet\leq 1}| \to |\mathcal{G}_{\bullet\leq 2}| \to \cdots \to |\mathcal{G}_{\bullet\leq n}|$.

REMARK 0.38. Remark 0.36 can be interpreted as an instance of unstable Koszul duality over the $\mathcal{E}_1$-operad, which we expand on now. Let $\underline{X} = (X_0 \to X_1)$ be a 1-flagged ∞-groupoid. Fix a field $\Bbbk$ of characteristic 0; consider the presentable ∞-category $\mathsf{Stack}(\Bbbk)$ of (commutative) $\Bbbk$-stacks. The functor $* \xrightarrow{\langle \mathsf{Spec}(\Bbbk) \rangle} \mathsf{Stack}(\Bbbk)$, which selects the terminal (commutative) $\Bbbk$-stack, uniquely extends as a colimit preserving functor $\mathcal{S}\mathsf{paces} \to \mathsf{Stack}(\Bbbk)$. In this way, each space, and each diagram among spaces, determines a (commutative) $\Bbbk$-stack, and a diagram of (commutative) $\Bbbk$-stacks, respectively. In particular, the given map $X_0 \to X_1$ between spaces determines a map between (commutative) $\Bbbk$-stacks. Let us suppose that $X_0 \simeq *$ is terminal. $\mathcal{E}_1$-deformations of this map between (commutative) $\Bbbk$-stacks is organized as a functor from local Artin $\mathcal{E}_1$-$\Bbbk$-algebras. Koszul duality asserts that this functor is represented by an augmented $\mathcal{E}_1$-algebra $\mathbb{T}_*^{\mathcal{E}_1} X_1$. For formal reasons, this representing augmented $\mathcal{E}_1$-$\Bbbk$-algebra $\mathbb{T}_*^{\mathcal{E}_1} X_1$ is the universal enveloping $\mathcal{E}_1$-algebra of the Lie algebra $\mathbb{T}_* X_1$, which is the tangent space of the (commutative) $\Bbbk$-stack X_1 at its point. For other formal reasons, this representing augmented $\mathcal{E}_1$-$\Bbbk$-algebra $\mathbb{T}_*^{\mathcal{E}_1} X_1$ is the group ring $\Bbbk[\Omega X] := \mathsf{C}_*(\Omega X_1; \Bbbk)$ on the group ΩX_1 which $\mathsf{fN}(\underline{X})$ codifies, as it is equipped with its standard augmentation. Conversely, the assumption that the map between spaces $* \to X_1$ is 0-connective gives that the canonical map from the Maurer–Cartan $\mathcal{E}_1$-$\Bbbk$-stack,

$$\mathsf{MC}_{\Bbbk[\Omega X_1]} =: \mathsf{Spec}(\Bbbk)/\Bbbk[\Omega X_1] \xrightarrow{\simeq} X_1 ,$$

is an equivalence (as $\mathcal{E}_1$-$\Bbbk$-stacks). Though less developed, we anticipate a similar interpretation of Corollary 0.33 for the general $n = 1$ case (in which X_0 is general). Specifically,

- the 1-groupoid $\mathsf{fN}(\underline{X})$, which we regard as an unstable version of an $\mathcal{E}_1$-algebroid over X_0, represents $\mathcal{E}_1$-deformations of the map $X_0 \to X_1$ between (commutative) $\Bbbk$-stacks;
- the connectivity of the map $X_0 \to X_1$ between spaces ensures that X_1, as an $\mathcal{E}_1$-$\Bbbk$-stack, is the Maurer–Cartan $\mathcal{E}_1$-$\Bbbk$-stack of this $\mathcal{E}_1$-algebroid over X_0.

REMARK 0.39. We follow-up on Remark 0.38. Though even less developed, we speculate a further interpretation of Corollary 0.33 for the case of general n. Specifically, for $\underline{X} = (X_0 \to \cdots \to X_n)$ an n-flagged ∞-groupoid,

- the n-groupoid $\mathsf{fN}(\underline{X})$, which we regard as an unstable version of an $\mathcal{E}_1$-algebroid over an $\mathcal{E}_1$-algebroid over ... over an $\mathcal{E}_1$-algebroid over X_0, represents compatible $\mathcal{E}_1$-deformations of each map $X_i \to X_{i+1}$ in the given flag $\underline{X}$;
- the connectivity of each map $X_i \to X_j$ ensures that X_n, as an $\mathcal{E}_n$-$\Bbbk$-stack, is the Maurer–Cartan $\mathcal{E}_n$-$\Bbbk$-stack of this iterated $\mathcal{E}_1$-algebroid over X_0.

0.3. A conjecture. We state a conjecture, and some related problems, that are prompted by this work. To state our conjecture we single out a class of diagrams in Cat_n.

DEFINITION 0.40. A *gaunt colimit diagram in* Cat_n is a functor $\mathcal{J}^{\triangleright} \to \mathsf{Cat}_n$ for which, for each $0 \leq i \leq n$, the composite functor $\mathcal{J}^{\triangleright} \to \mathsf{Cat}_n \xrightarrow{(-)_{\leq i}} \mathsf{Cat}_i$ is a colimit diagram.

EXAMPLE 0.41. For each $0 < i < p$, the diagram

$$\begin{array}{ccc} \{i\} & \longrightarrow & \{i < \cdots < p\} \\ \downarrow & & \downarrow \\ \{0 < \cdots < i\} & \longrightarrow & \{0 < \cdots < p\} \end{array}$$

is a gaunt colimit diagram in Cat_1. More generally, for each Segal cover $\mathcal{J}^{\triangleright} \to \boldsymbol{\Theta}_n$, the composite functor

$$\mathcal{J}^{\triangleright} \longrightarrow \boldsymbol{\Theta}_n \longrightarrow \mathsf{Cat}_n$$

is a gaunt colimit diagram.

EXAMPLE 0.42. While the diagram in Cat_1

$$\begin{array}{ccc} \{0 < 2\} \amalg \{1 < 3\} & \longrightarrow & \{0 < 1 < 2 < 3\} \\ \downarrow & & \downarrow \\ * \amalg * & \longrightarrow & * \end{array}$$

is a colimit diagram, it is *not* a gaunt colimit diagram. Note, however, that the map from the colimit of maximal ∞-subgroupoids to the maximal ∞-subgroupoid of $*$, ∞-groupoids

$$\Big(\{0,1,2,3\} \coprod_{\{0,2\}\amalg\{1,3\}} * \amalg *\Big) \simeq \{-\} \amalg \{+\} \longrightarrow * ~,$$

is 0-connective, which is to say that it is surjective on path components.

EXAMPLE 0.43. Consider the ordinary 1-category $\mathsf{E}(c_1)$ corepresenting an isomorphism. Consider the functor $\boldsymbol{\Delta}_{/\mathsf{E}(c_1)} := \boldsymbol{\Delta} \underset{\mathsf{Cat}_1}{\times} (\mathsf{Cat}_1)/\mathsf{E}(c_1) \to \boldsymbol{\Delta} \hookrightarrow \mathsf{Cat}_1$ from the slice category. Consider its terminal extension

$$(\boldsymbol{\Delta}_{/\mathsf{E}(c_1)})^{\triangleright} \longrightarrow \mathsf{Cat}_1 ~.$$

While this functor is a colimit diagram, it is *not* a gaunt colimit diagram. More generally, for each $0 < i \leq n$, consider the strict n-groupoidification $\mathsf{E}(c_i)$ of the i-cell. While the composite functor

$$(\boldsymbol{\Theta}_{n/\mathsf{E}(c_i)})^{\triangleright} \longrightarrow \boldsymbol{\Theta}_n \hookrightarrow \mathsf{Cat}_n$$

is a colimit diagram, it is *not* a gaunt colimit diagram.

REMARK 0.44. Heuristically, a colimit diagram $\mathcal{J}^{\triangleright} \to \mathsf{Cat}_n$ is *gaunt* if it does not generate invertible i-morphisms for any i.

We make the following.

CONJECTURE 0.45. *The restricted Yoneda functor*

$$\mathsf{fCat}_n \xrightarrow{(6)} \mathsf{PShv}(\mathsf{Cat}_n)$$

is fully faithful, with image consisting of those presheaves that carry (the opposites of) gaunt colimit diagrams to limit diagrams.

REMARK 0.46. We reflect on the statement of Conjecture 0.45. By definition, the ∞-category Cat_n is presentable. Therefore, the image of the Yoneda functor $\mathsf{Cat}_n \to \mathsf{PShv}(\mathsf{Cat}_n)$, which is fully faithful, consists of those presheaves that carry (the opposites of) *all* colimit diagrams to limit diagrams.

PROBLEM 0.47. Identify checkable criteria for when a colimit diagram $\mathcal{J}^{\triangleright} \to \mathsf{Cat}_n$ is in fact a gaunt colimit diagram.

PROBLEM 0.48. Let $0 \leq i \leq n$. For $\partial c_i \to \mathcal{C}$ a functor to a finite gaunt n-category, give an explicit description (for instance as a presheaf on $\mathbf{\Theta}_n$, or even as a presheaf on finite gaunt n-categories) of the pushout (∞, n)-category

$$\mathcal{C} \underset{\partial c_i}{\amalg} c_i \ .$$

1. Comparing Segal sheaves and flagged higher categories

We establish adjunctions connecting the ∞-categories $\mathsf{Shv}(\mathbf{\Theta}_n)$ and fCat_n. In doing so, we introduce some interpolating ∞-categories.

1.1. The adjunction. Recall the sequence of functors (4); for each $0 \leq i \leq j \leq n$, denote that functor as

$$\iota := \iota_{ij} \colon \mathbf{\Theta}_i \hookrightarrow \mathbf{\Theta}_j \tag{8}$$

where the subscripts are omitted if the indices are understood from context. The sequence (4) of fully faithful right adjoint functors is selected by a functor:

$$\mathbf{\Theta}_\bullet \colon [n] \xrightarrow{\langle(4)\rangle} \mathsf{Cat} \ , \qquad i \mapsto \mathbf{\Theta}_i \ .$$

Postcomposing this functor with the functor $\mathsf{PShv} \colon \mathsf{Cat} \to \mathsf{Pr}^{\mathsf{L}}$, to presentable ∞-categories and left adjoint functors among them, results in a functor $[n] \xrightarrow{\mathbf{\Theta}_\bullet} \mathsf{Cat} \xrightarrow{\mathsf{PShv}} \mathsf{Pr}^{\mathsf{L}}$. The unstraightening of this functor is an ∞-category over $[n]$,

$$\mathsf{PShv}(\mathbf{\Theta}_\bullet) \longrightarrow [n] \ , \tag{9}$$

which is both a coCartesian fibration and a Cartesian fibration.

REMARK 1.1. Let $0 \leq i \leq j \leq n$. Consider the morphism $c_1 \xrightarrow{\langle i \leq j\rangle} [n]$. The coCartesian monodromy functor of (9) over the morphism is the unique colimit preserving functor $\iota_! \colon \mathsf{PShv}(\mathbf{\Theta}_i) \to \mathsf{PShv}(\mathbf{\Theta}_j)$ extending the composite functor $\mathbf{\Theta}_i \hookrightarrow \mathbf{\Theta}_j \hookrightarrow \mathsf{PShv}(\mathbf{\Theta}_j)$. The Cartesian monodromy functor of (9) over this same morphism is the functor $\mathsf{PShv}(\mathbf{\Theta}_i) \leftarrow \mathsf{PShv}(\mathbf{\Theta}_j) \colon \iota^*$ given by pullback along $\mathbf{\Theta}_i \hookrightarrow \mathbf{\Theta}_j$. Notice that both of these monodromy functors preserve colimits, and that the coCartesian monodromy functor is left adjoint to the Cartesian monodromy functor.

Consider the ∞-category of sections of (9):

$$\Gamma\big(\mathsf{PShv}(\mathbf{\Theta}_\bullet)\big) \ := \ \mathsf{Fun}_{/[n]}\big([n], \mathsf{PShv}(\mathbf{\Theta}_\bullet)\big) \ .$$

Explicitly, an object of $\Gamma\big(\mathsf{PShv}(\mathbf{\Theta}_\bullet)\big)$ is, for each $0 \leq i \leq n$, a presheaf $\mathcal{F}_i \in \mathsf{PShv}(\mathbf{\Theta}_i)$, together with, for each $0 < i \leq n$, a morphism $\iota_! \mathcal{F}_{i-1} \to \mathcal{F}_i$ between presheaves on $\mathbf{\Theta}_i$. Now, because $n \in [n]$ is a final object, Cartesian monodromy of the unique morphisms in $[n]$ to this final object define a functor from the fiber over n to this ∞-category of sections:

$$\mathsf{PShv}(\mathbf{\Theta}_n) \longrightarrow \Gamma\big(\mathsf{PShv}(\mathbf{\Theta}_\bullet)\big) \ . \tag{10}$$

This functor (10) is fully faithful, and its image consists of the *Cartesian* sections, which are those sections that carry morphisms to (9)-Cartesian morphisms. Precomposing with the Yonda functor $\boldsymbol{\Theta}_n \hookrightarrow \mathsf{PShv}(\boldsymbol{\Theta}_n)$ determines the solid diagram among ∞-categories:

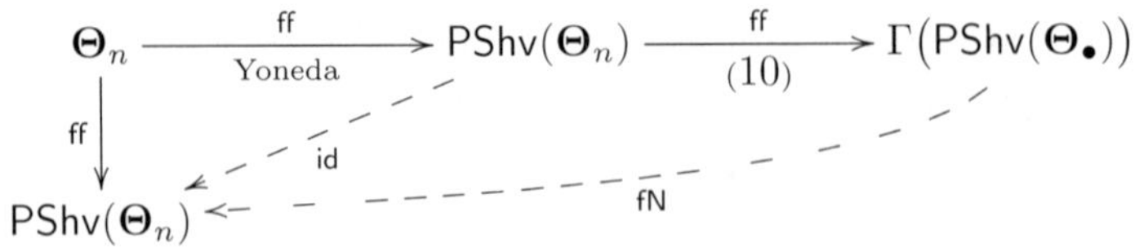

Left Kan extensions define the fillers in this diagram, which is indeed a commutative diagram because each of the solid arrows is a fully faithful functor. From the universal property of the Yoneda functor, the inner filler is the identity functor on $\mathsf{PShv}(\boldsymbol{\Theta}_n)$, as indicated. As is always the case for left Kan extensions through a Yoneda functor, the outer filler is the restricted Yoneda functor. (We give this left Kan extension the same notation as (6) because it extends that functor, as we will see.) From the universal property of left Kan extensions, the resulting triangle among presentable ∞-categories

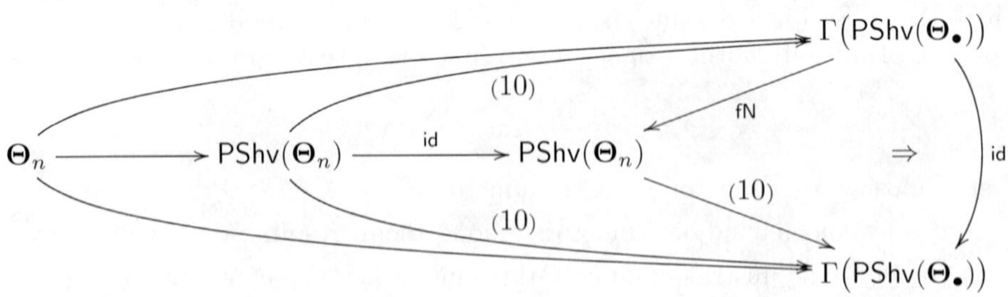

lax-commutes, as indicated. By construction, the resulting outer lax-commutative triangle among ∞-categories is, in fact, a commutative triangle. From the universal property of the Yoneda functor $\boldsymbol{\Theta}_n \to \mathsf{PShv}(\boldsymbol{\Theta}_n)$ as a colimit completion, it follows that the second-to-outer lax-commutative triangle is also, in fact, a commutative triangle. This concludes the construction of an adjunction

$$(10)\colon \mathsf{PShv}(\boldsymbol{\Theta}_n) \rightleftarrows \Gamma\big(\mathsf{PShv}(\boldsymbol{\Theta}_\bullet)\big)\colon \mathsf{fN}~. \tag{11}$$

Explicitly, the left adjoint evaluates on a presheaf $\mathcal{F} \in \mathsf{PShv}(\boldsymbol{\Theta}_n)$ as the section $i \mapsto \mathcal{F}_{|\boldsymbol{\Theta}_i^{\mathsf{op}}}$; in the case that $\mathcal{F}$ is represented by an object $T \in \boldsymbol{\Theta}_n$, we implement the other notation $T_{\leq\bullet}$ in place of $\mathcal{F}_{|\boldsymbol{\Theta}_\bullet^{\mathsf{op}}}$. Explicitly, the right adjoint evaluates on a section $\mathcal{F}_\bullet$ as the presheaf $T \mapsto \mathsf{Map}_\Gamma(T_{\leq\bullet}, \mathcal{F}_\bullet)$, whose values are spaces of morphisms in the ∞-category $\Gamma\big(\mathsf{PShv}(\boldsymbol{\Theta}_\bullet)\big)$.

Inspecting the definition of the restricted Yoneda functor $\mathsf{fN}\colon \Gamma\big(\mathsf{PShv}(\boldsymbol{\Theta}_\bullet)\big) \to \mathsf{PShv}(\boldsymbol{\Theta}_n)$, as n varies, reveals the following.

Observation 1.2. For each $0 \leq i \leq j \leq n$, the diagram among ∞-categories

$$\begin{array}{ccc} \mathsf{PShv}(\boldsymbol{\Theta}_j) & \xleftarrow{\ \mathsf{fN}\ } & \Gamma\big(\mathsf{PShv}(\boldsymbol{\Theta}_{\bullet\leq j})\big) \\ \downarrow & & \downarrow \\ \mathsf{PShv}(\boldsymbol{\Theta}_i) & \xleftarrow{\ \mathsf{fN}\ } & \Gamma\big(\mathsf{PShv}(\boldsymbol{\Theta}_{\bullet\leq i})\big) \end{array}$$

canonically commutes.

1.2. Restricting the adjunction. We now show that the adjunction (11) restricts to Segal objects.

Recall the ∞-category of (9). Consider the full ∞-subcategories

$$\mathsf{PShv}(\mathbf{\Theta}_\bullet) \supset \mathsf{Shv}(\mathbf{\Theta}_\bullet) \supset \mathsf{Shv}^{\mathsf{unv}}(\mathbf{\Theta}_\bullet) \tag{12}$$

consisting of those pairs $\big(i \in [n], \mathcal{F} \in \mathsf{PShv}(\mathbf{\Theta}_i)\big)$ for which $\mathcal{F}_i \in \mathsf{Shv}(\mathbf{\Theta}_i)$, and for which $\mathcal{F}_i \in \mathsf{Shv}^{\mathsf{unv}}(\mathbf{\Theta}_i)$, respectively.

LEMMA 1.3. *In the commutative diagram among ∞-categories,*

$$\begin{array}{ccccc} \mathsf{PShv}(\mathbf{\Theta}_\bullet) & \longleftarrow & \mathsf{Shv}(\mathbf{\Theta}_\bullet) & \longleftarrow & \mathsf{Shv}^{\mathsf{unv}}(\mathbf{\Theta}_\bullet) \\ & \searrow & \downarrow & \swarrow & \\ & & [n] & & \end{array},$$

each of the vertical functors is both a coCartesian fibration and a Cartesian fibration, and each of the horizontal functors is fully faithful and preserves coCartesian morphisms and Cartesian morphisms over $[n]$.

PROOF. By Definition 0.5, both Segal covers and univalence diagrams are, in particular, limit diagrams in $\mathbf{\Theta}_n$. By direct inspection, for each $0 \leq i \leq j \leq n$, the fully faithful functor $\mathbf{\Theta}_i \hookrightarrow \mathbf{\Theta}_j$ carries Segal covers to Segal covers and carries univalence diagrams to univalence diagrams. From these two points, it follows that the adjunction $\iota_!\colon \mathsf{PShv}(\mathbf{\Theta}_i) \rightleftarrows \mathsf{PShv}(\mathbf{\Theta}_j)\colon \iota^*$ restricts as an adjunction

$$\iota_!\colon \mathsf{Shv}(\mathbf{\Theta}_i) \rightleftarrows \mathsf{Shv}(\mathbf{\Theta}_j)\colon \iota^* ,$$

which further restricts as an adjunction

$$\iota_!\colon \mathsf{Shv}^{\mathsf{unv}}(\mathbf{\Theta}_i) \rightleftarrows \mathsf{Shv}^{\mathsf{unv}}(\mathbf{\Theta}_j)\colon \iota^* ,$$

The lemma follows. □

For each $0 \leq i \leq n$, via Bousfield localization, the fully faithful inclusion between presentable ∞-categories, $\mathsf{Shv}^{\mathsf{unv}}(\mathbf{\Theta}_i) \hookrightarrow \mathsf{Shv}(\mathbf{\Theta}_i)$, is a right adjoint:

$$(-)^{\widehat{}}_{\mathsf{unv}}\colon \mathsf{Shv}(\mathbf{\Theta}_i) \rightleftarrows \mathsf{Shv}^{\mathsf{unv}}(\mathbf{\Theta}_i) . \tag{13}$$

The left adjoint is *univalent-completion.*

LEMMA 1.4. *The fully faithful inclusion*

$$\mathsf{Shv}(\mathbf{\Theta}_\bullet) \hookleftarrow \mathsf{Shv}^{\mathsf{unv}}(\mathbf{\Theta}_\bullet)$$

is a right adjoint functor. Its left adjoint functor lies over $[n]$, and the adjunction is given on fibers over $i \in [n]$ as the Bousfield localization $(-)^{\widehat{}}_{\mathsf{unv}}\colon \mathsf{Shv}(\mathbf{\Theta}_i) \rightleftarrows \mathsf{Shv}^{\mathsf{unv}}(\mathbf{\Theta}_i)$ implementing univalent-completion.

PROOF. The existence of a left adjoint is a condition on the given fully faithful functor. Because the given fully faithful functor is a coCartesian functor between coCartesian fibrations over $[n]$, the result is proved upon showing that the following two points.

(1) For each $0 \leq i \leq n$, the functor between fibers over i is a right adjoint in an adjunction,

$$\mathsf{L}_i\colon \mathsf{Shv}(\mathbf{\Theta}_i) \rightleftarrows \mathsf{Shv}^{\mathsf{unv}}(\mathbf{\Theta}_i) .$$

(2) For each $0 < i \leq n$, the diagram

$$\begin{array}{ccc} & \overset{\mathsf{L}_{i-1}}{\curvearrowright} & \\ \mathsf{Shv}(\boldsymbol{\Theta}_{i-1}) & \Downarrow & \mathsf{Shv}^{\mathsf{unv}}(\boldsymbol{\Theta}_{i-1}) \\ \downarrow & & \downarrow \\ \mathsf{Shv}(\boldsymbol{\Theta}_i) & \xrightarrow{\mathsf{L}_i} & \mathsf{Shv}^{\mathsf{unv}}(\boldsymbol{\Theta}_i) \end{array}$$

lax-commutes.

The first point is exactly the adjunction (13), in which $\mathsf{L}_i = \widehat{(-)}_{\mathsf{unv}}$ is univalent-completion. The fully faithful functor $\mathsf{Shv}^{\mathsf{unv}}(\boldsymbol{\Theta}_{i-1}) \hookrightarrow \mathsf{Shv}^{\mathsf{unv}}(\boldsymbol{\Theta}_i)$ is a left adjoint, with right adjoint given by restriction: $\mathcal{C} \mapsto \mathcal{C}_{<i}$. Therefore, the sought lax-commutative diagram is implemented from the commutativity of the diagram involving right adjoints to the sought lax-commutative diagram:

$$\begin{array}{ccc} \mathsf{Shv}(\boldsymbol{\Theta}_{i-1}) & \longleftarrow & \mathsf{Shv}^{\mathsf{unv}}(\boldsymbol{\Theta}_{i-1}) \\ \uparrow & & \uparrow \\ \mathsf{Shv}(\boldsymbol{\Theta}_i) & \longleftarrow & \mathsf{Shv}^{\mathsf{unv}}(\boldsymbol{\Theta}_i) \end{array}$$

□

Taking sections, Lemma 1.4 has the following useful consequence.

COROLLARY 1.5. *The fully faithful inclusion*

$$\Gamma\big(\mathsf{Shv}(\boldsymbol{\Theta}_\bullet)\big) \hookleftarrow \Gamma\big(\mathsf{Shv}^{\mathsf{unv}}(\boldsymbol{\Theta}_\bullet)\big)$$

is a right adjoint functor. Its left adjoint carries a section $\mathcal{F}_\bullet$ to the section $\widehat{(\mathcal{F}_\bullet)}_{\mathsf{unv}}$, whose value on $i \in [n]$ is the univalent-completion of the Segal sheaf $\mathcal{F}_i \in \mathsf{Shv}(\boldsymbol{\Theta}_i)$.

REMARK 1.6. Verifying that a presheaf on $\boldsymbol{\Theta}_n$ satisfies the Segal condition, as defined in Definition 0.5, can be reduced to a simpler problem, as we now explain. Each closed morphism in $\boldsymbol{\Theta}_n$ is a monomorphism. Therefore, for each $T \in \boldsymbol{\Theta}_n^{\mathsf{cls}}$, the overcategory $\boldsymbol{\Theta}^{\mathsf{cls}}_{n/T}$ is a a poset. Inspecting the definition of the category $\boldsymbol{\Theta}_n$, this poset $\boldsymbol{\Theta}^{\mathsf{cls}}_{n/T}$ is, in fact, finite. Therefore, each colimit in $\boldsymbol{\Theta}_n^{\mathsf{cls}}$ can be expressed as a finite iteration of pushouts. It follows that a presheaf $\mathcal{F} \in \mathsf{PShv}(\boldsymbol{\Theta}_n)$ is *Segal* if and only if it carries (the opposites of) pushout diagrams in $\boldsymbol{\Theta}_n^{\mathsf{cls}}$ to pullback diagrams among spaces. We make implicit use of this reduction as we proceed.

LEMMA 1.7. *The adjunction* (11) *restricts as an adjunction*

$$(10)\colon \mathsf{Shv}(\boldsymbol{\Theta}_n) \rightleftarrows \Gamma\big(\mathsf{Shv}(\boldsymbol{\Theta}_\bullet)\big)\colon \mathsf{fN}\ . \tag{14}$$

PROOF. Because the asserted restriction is to fully faithful ∞-subcategories in the adjunction (11), we need only show that the left and the right adjoint functors restrict as desired.

From its definition, Lemma 1.3 gives that the functor (10) restricts as a functor

$$(10)\colon \mathsf{Shv}(\boldsymbol{\Theta}_n) \rightleftarrows \Gamma\big(\mathsf{Shv}(\boldsymbol{\Theta}_\bullet)\big)\ ,$$

which is necessarily fully faithful.

It remains to prove that fN restricts likewise. Let $\mathcal{F}_\bullet \in \Gamma\big(\mathsf{Shv}(\boldsymbol{\Theta}_\bullet)\big)$. We must show, then, that the presheaf $\mathsf{fN}(\mathcal{F}_\bullet)\colon \boldsymbol{\Theta}_n^{\mathsf{op}} \to \mathcal{S}\mathsf{paces}$ carries (opposites) of Segal covers to limit diagrams. By definition of fN as a restricted Yoneda functor, this is implied by the functor $\mathbf{\Theta}_n \to \mathsf{fCat}_n$ carrying Segal covers to colimit diagrams. By definition of a Segal cover, this is implied by the functor $\mathbf{\Theta}_n^{\mathsf{cls}} \to \mathsf{fCat}_n$ preserving colimit diagrams. By definition of fCat_n as a full ∞-subcategory of $\mathsf{Fun}([n], \mathsf{Cat}_n)$, this is implied by each of the forgetful functors $(-)_{\leq i}\colon \mathbf{\Theta}_n^{\mathsf{cls}} \to \mathsf{Cat}_n \xrightarrow{\text{forget}} \mathsf{Cat}_i$ preserving colimit diagrams.

Now let $T_\bullet\colon \mathcal{J}^{\triangleright} \to \mathbf{\Theta}_n^{\mathsf{cls}}$ be a colimit diagram. Denote the value on the cone point as $T := T_* \in \mathbf{\Theta}_n$. We must show that the composite functor $(T_\bullet)_{\leq i}\colon \mathcal{J}^{\triangleright} \to \mathsf{Cat}_i$ is a colimit diagram. In general, for each k, the full subcategory of $\mathbf{\Theta}_k^{\mathsf{cls}}$ consisting of the cells strongly generates; also, by definition, the functor $\mathbf{\Theta}_k \hookrightarrow \mathsf{Cat}_k$ preserves such colimit diagrams. We can therefore reduced to the case that the functor $\mathcal{J} \simeq \mathcal{E}(T) \hookrightarrow \mathbf{\Theta}_{n/T}^{\mathsf{cls}}$ is the inclusion of the full subcategory $\mathcal{E}(T) \subset \mathbf{\Theta}_{n/T}^{\mathsf{cls}}$ consisting of those closed morphisms $T_j \to T$ for which $T_j \simeq c_k$ is a k-cell for some $0 \leq k \leq n$. This, a priori, ∞-category $\mathcal{E}(T)$ is in fact a finite poset (see Remark 1.6). The lemma is proved once we establish the following sequence of equivalences among (∞, i)-categories:

$$\begin{aligned}
\underset{C\in\mathcal{E}(T)}{\mathsf{colim}}\, C_{\leq i} &\xleftarrow{\simeq} \underset{C\in\mathcal{E}(T)}{\mathsf{colim}}\ \underset{C'\in\mathcal{E}(C)_{\leq i}}{\mathsf{colim}}\, C' \\
&\xleftarrow{\simeq} \underset{C'\in\mathcal{E}(T)_{\leq i}}{\mathsf{colim}}\, C' \\
&\xrightarrow{\simeq} T_{\leq i}\ .
\end{aligned}$$

From the universal property of the right adjoint functor $(-)_{\leq i}\colon \mathsf{Cat}_n \to \mathsf{Cat}_i$, for each $S \in \mathbf{\Theta}_n$, there is a canonical identification as a colimit:

$$S_{\leq i} \simeq \mathsf{colim}\big(\mathbf{\Theta}_{i/S} \to \mathbf{\Theta}_i \hookrightarrow \mathsf{Cat}_i\big)\ .$$

Consider the subcategory $\mathbf{\Theta}_{i/S}^{\mathsf{cls}} \subset \mathbf{\Theta}_{i/S}$ consisting of the closed morphisms to S and closed morphisms among them. Via the active-closed factorization system on the category $\mathbf{\Theta}_i$, the inclusion of this subcategory is a final functor. Consider the full subcategory $\mathcal{E}(S)_{\leq i} \subset \mathcal{E}(S) \subset \mathbf{\Theta}_{i/S}^{\mathsf{cls}}$ consisting of those $(C \to S)$ for which $C \simeq c_k$ is a k-cell for some $0 \leq k \leq i$. Now, for each k, the full subcategory of $\mathbf{\Theta}_k^{\mathsf{cls}}$ consisting of the cells strongly generates; also, by definition, the functor $\mathbf{\Theta}_k \hookrightarrow \mathsf{Cat}_k$ preserves such colimit diagrams. We conclude from these observations an identification

$$S_{\leq i} \simeq \mathsf{colim}\big(\mathcal{E}(S)_{\leq i} \hookrightarrow \mathbf{\Theta}_{i/S}^{\mathsf{cls}} \hookrightarrow \mathbf{\Theta}_{i/S} \to \mathbf{\Theta}_i \hookrightarrow \mathsf{Cat}_i\big)\ .$$

Applying this to $S = T$ gives the final equivalence in the above string. Notice that the assignments $S \mapsto \mathcal{E}(S)$ and $S \mapsto \mathcal{E}(S)_{\leq i}$ each evidently extend as a functors $\mathbf{\Theta}_n^{\mathsf{cls}} \to \mathsf{Poset}_{/\mathbf{\Theta}_n^{\mathsf{cls}}}$. This gives the first of the equivalences. The second equivalence follows from Quillen's Theorem A, using the following observation.

> Let $C' \to T$ be a closed morphism in $\mathbf{\Theta}_n$ from an i-cell. Then the poset of factorizations of this closed morphism through a closed morphism $C \to T$ from a cell, has an initial object.

Finally, the composition of this string of equivalence is evidently the canonical morphism we intended to show is an equivalence. □

Concatenating Corollary 1.5 and Lemma 1.7 results in the composite adjunction

$$(15)\qquad \widehat{(-_{|\mathbf{\Theta}_\bullet^{\mathsf{op}}})}_{\mathsf{unv}}\colon \mathsf{Shv}(\mathbf{\Theta}_n) \rightleftarrows \Gamma\big(\mathsf{Shv}(\mathbf{\Theta}_\bullet)\big) \rightleftarrows \Gamma\big(\mathsf{Shv}^{\mathsf{unv}}(\mathbf{\Theta}_\bullet)\big)\colon \mathsf{fN}\ .$$

Unwinding definitions reveals the next observation.

OBSERVATION 1.8. There is a canonical fully faithful functor between ∞-categories:

$$\mathsf{fCat}_n \hookrightarrow \Gamma\big(\mathsf{Shv}^{\mathsf{unv}}(\boldsymbol{\Theta}_\bullet)\big) . \tag{16}$$

The image consists of those sections $\mathcal{C}_\bullet = (\mathcal{C}_0 \to \mathcal{C}_1 \to \cdots \to \mathcal{C}_n)$ for which, for each $0 \leq i \leq j \leq n$, the functor $\mathcal{C}_i \to \mathcal{C}_j$ between (∞, j)-categories is i-connective.

The proof of the next result occupies §3.1.

LEMMA 1.9. *The value of the left adjoint of the adjunction (15) on a Segal sheaf $\mathcal{F} \in \mathsf{Shv}(\boldsymbol{\Theta}_n)$ lies in the image of the fully faithful functor of Observation 1.8:*

$$(\mathcal{F}_{|\boldsymbol{\Theta}_\bullet^{\mathsf{op}}})\widehat{\ }_{\mathsf{unv}} \in \mathsf{fCat}_n .$$

Through Observation 1.8, Lemma 1.9 has the following consequence.

COROLLARY 1.10. *The adjunction (15) restricts as an adjunction*

$$(-_{|\boldsymbol{\Theta}_\bullet^{\mathsf{op}}})\widehat{\ }_{\mathsf{unv}} \colon \mathsf{Shv}(\boldsymbol{\Theta}_n) \rightleftarrows \mathsf{fCat}_n \colon \mathsf{fN} . \tag{17}$$

1.3. Explicating the adjunction. After Corollary 1.10, our main result (Theorem 0.27) is implied by showing that both the unit and the counit transformations of the adjunction (17) are equivalences. So we explicate the values of left and right adjoints, as well as the unit and the counit, of the adjunction (17).

1.3.1. *The left adjoint.* The value of the left adjoint $(-_{|\boldsymbol{\Theta}_\bullet^{\mathsf{op}}})\widehat{\ }_{\mathsf{unv}}$ of the adjunction (15) on a Segal sheaf $\mathcal{F} \in \mathsf{Shv}(\boldsymbol{\Theta}_n)$ is the section $(\mathcal{F}_{|\boldsymbol{\Theta}_\bullet^{\mathsf{op}}})\widehat{\ }_{\mathsf{unv}}$ of the functor $\mathsf{Shv}^{\mathsf{unv}}(\boldsymbol{\Theta}_\bullet) \to [n]$ that is the assignment

$$[n] \ni i \mapsto (\mathcal{F}_{|\boldsymbol{\Theta}_i^{\mathsf{op}}})\widehat{\ }_{\mathsf{unv}} \in \mathsf{Shv}^{\mathsf{unv}}(\boldsymbol{\Theta}_i) ,$$

which is the univalent-completion of the restriction $\mathcal{F}_{|\boldsymbol{\Theta}_i^{\mathsf{op}}} \in \mathsf{Shv}(\boldsymbol{\Theta}_i)$.

1.3.2. *The right adjoint.* The value of the right adjoint fN of the adjunction (15) on a section $\mathcal{C}_\bullet \in \Gamma\big(\mathsf{Shv}^{\mathsf{unv}}(\boldsymbol{\Theta}_\bullet)\big)$ is the Segal sheaf $\mathsf{fN}(\mathcal{C}_\bullet) \in \mathsf{Shv}(\boldsymbol{\Theta}_n)$ that is the assignment

$$\boldsymbol{\Theta}_n^{\mathsf{op}} \ni T \mapsto \mathsf{Map}_{\Gamma_{\leq n}}(T_{\leq\bullet}, \mathcal{C}_\bullet) \in \mathcal{S}\mathsf{paces} ,$$

which is the space of morphisms in $\Gamma\big(\mathsf{Shv}^{\mathsf{unv}}(\boldsymbol{\Theta}_\bullet)\big)$ from $T_{\leq\bullet} = (T_{\leq 0} \to T_{\leq 1} \to \cdots \to T_{\leq n-1} \to T)$ to $\mathcal{C}_\bullet$. The next result (Corollary 1.12) makes the values of the Segal sheaf $\mathsf{fN}(\mathcal{C}_\bullet)$ more explicit.

OBSERVATION 1.11. For $\mathcal{C}_\bullet$ and $\mathcal{D}_\bullet$ sections of the functor $\mathsf{PShv}(\boldsymbol{\Theta}_\bullet) \to [n]$, the canonical square among spaces of morphisms

$$\begin{array}{ccc} \mathsf{Map}_{\Gamma_{\leq n}}(\mathcal{C}_\bullet, \mathcal{D}_\bullet) & \longrightarrow & \mathsf{Map}_{\mathsf{PShv}(\boldsymbol{\Theta}_n)}(\mathcal{C}_n, \mathcal{D}_n) \\ \downarrow & & \downarrow \\ \mathsf{Map}_{\Gamma_{< n}}(\mathcal{C}_{\bullet<n}, \mathcal{D}_{\bullet<n}) & \longrightarrow & \mathsf{Map}_{\mathsf{PShv}(\boldsymbol{\Theta}_n)}(\mathcal{C}_{n-1}, \mathcal{D}_n) \end{array}$$

is a pullback.

PROOF. The coCartesian monodromy functors of the coCartesian fibration $\mathsf{PShv}(\boldsymbol{\Theta}_\bullet) \to [n]$ are given by left Kan extensions along fully faithful functors. Therefore, these coCartesian monodromy functors are fully faithful. These coCartesian monodromy functors thusly define a fully faithful functor to the fiber over the final object $n \in [n]$:

$$\Gamma\big(\mathsf{PShv}(\boldsymbol{\Theta}_\bullet)\big) \hookrightarrow \mathsf{Fun}\big([n], \mathsf{PShv}(\boldsymbol{\Theta}_n)\big) =: \mathsf{PShv}(\boldsymbol{\Theta}_n)^{[n]} .$$

The canonical functor $[n-1] \coprod_{\{n-1\}} \{n-1<n\} \to [n]$ between ∞-categories is an equivalence from the pushout. Consequently, for $\mathcal{X}$ an ∞-category, and for $x_\bullet, y_\bullet \in \mathsf{Fun}([n],\mathcal{X}) =: \mathcal{X}^{[n]}$ two functors $[n] \to \mathcal{X}$, the canonical square among spaces of morphisms

$$\begin{array}{ccc} \mathcal{X}^{[n]}(x_\bullet, y_\bullet) & \longrightarrow & \mathcal{X}(x_n, y_n) \\ \downarrow & & \downarrow \\ \mathcal{X}^{[n-1]}(x_{\bullet<n}, y_{\bullet<n}) & \longrightarrow & \mathcal{X}(x_{n-1}, y_n) \end{array}$$

is a pullback. Apply this to the case $\mathcal{X} = \mathsf{PShv}(\boldsymbol{\Theta}_n)$. □

Corollary 1.12. *Let $\mathcal{C}_\bullet$ be a section of the functor $\mathsf{PShv}(\boldsymbol{\Theta}_\bullet) \to [n]$. Let $T \in \boldsymbol{\Theta}_n$. There is a canonical pullback diagram among spaces:*

$$\begin{array}{ccc} \mathsf{fN}(\mathcal{C}_\bullet)(T) & \longrightarrow & \mathcal{C}_n(T) \\ \downarrow & & \downarrow \\ \mathsf{fN}(\mathcal{C}_{\bullet<n})(T_{<n}) & \longrightarrow & \mathcal{C}_n(T_{<n}) . \end{array}$$

Alternatively, there is a canonical limit diagram among spaces:

$$\begin{array}{ccccccccc} & & & & \mathsf{fN}(\mathcal{C}_\bullet)(T) & & & & \\ & \swarrow & & \swarrow & \downarrow & \searrow & & \searrow & \\ \mathcal{C}_0(T_{\leq 0}) & & \mathcal{C}_1(T_{\leq 1}) & & \cdots & & \mathcal{C}_{n-1}(T_{\leq n-1}) & & \mathcal{C}_n(T) \\ & \searrow & \downarrow & \searrow & \downarrow & \searrow & \downarrow & \searrow & \downarrow \\ & & \mathcal{C}_1(T_{\leq 0}) & & \mathcal{C}_2(T_{\leq 1}) & & \mathcal{C}_{n-1}(T_{\leq n-2}) & & \mathcal{C}_n(T_{\leq n-1}). \end{array}$$

Corollary 1.12 makes apparent the following.

Observation 1.13. Let $0 \leq i \leq n$. Let $\mathcal{F}_\bullet$ be a section of the functor $\mathsf{PShv}(\boldsymbol{\Theta}_\bullet) \to [n]$. Restricting this section over $[i] = \{0 < \cdots < i\} \subset [n]$ determines the section $\mathcal{F}_{\bullet\leq i}$ of $\mathsf{PShv}(\boldsymbol{\Theta}_\bullet) \to [i]$. There is a canonical equivalence between presheaves on $\boldsymbol{\Theta}_i$:

$$\mathsf{fN}(\mathcal{F}_{\bullet\leq i}) \simeq \mathsf{fN}(\mathcal{F}_\bullet)_{|\boldsymbol{\Theta}_i^{\mathsf{op}}} .$$

1.3.3. *The unit.* The unit of the adjunction (15) evaluates on each Segal sheaf on $\boldsymbol{\Theta}_n$ as the morphism between presheaves

$$\text{unit}\colon \mathcal{F} \longrightarrow \mathsf{fN}\big((\mathcal{F}_{|\boldsymbol{\Theta}_\bullet^{\mathsf{op}}})^\wedge_{\mathsf{unv}}\big)$$

whose value on $T \in \boldsymbol{\Theta}_n$ is described inductively (via Observation 1.13) through Corollary 1.12 as the square among spaces

$$\begin{array}{ccccc} \mathcal{F}(T) & & \longrightarrow & & \mathcal{F}^\wedge_{\mathsf{unv}}(T) \\ \downarrow & & & & \downarrow \\ \mathcal{F}(T_{<n}) & \xrightarrow{\text{induction}} & \mathsf{fN}\big((\mathcal{F}_{|\boldsymbol{\Theta}_{\bullet<n}})^\wedge_{\mathsf{unv}}\big) & \longrightarrow & \mathcal{F}^\wedge_{\mathsf{unv}}(T_{<n}) . \end{array} \tag{18}$$

Without compressing this description via induction, the unit transformation can be described through Corollary 1.12 as the canonical diagram among spaces

(19)

$$\begin{array}{ccccccc} & & & \mathcal{F}(T) & & & \\ & \swarrow & \swarrow & \downarrow & \searrow & & \\ (\mathcal{F}_{|\boldsymbol{\Theta}_0^{\mathsf{op}}})^{\widehat{}}_{\mathsf{unv}}(T_{\leq 0}) & & (\mathcal{F}_{|\boldsymbol{\Theta}_1^{\mathsf{op}}})^{\widehat{}}_{\mathsf{unv}}(T_{\leq 1}) & & (\mathcal{F}_{|\boldsymbol{\Theta}_{n-1}^{\mathsf{op}}})^{\widehat{}}_{\mathsf{unv}}(T_{\leq n-1}) & & (\mathcal{F}_{|\boldsymbol{\Theta}_n^{\mathsf{op}}})^{\widehat{}}_{\mathsf{unv}}(T) \\ & \searrow & \downarrow & \searrow & \downarrow & \searrow & \downarrow \\ & & (\mathcal{F}_{|\boldsymbol{\Theta}_1^{\mathsf{op}}})^{\widehat{}}_{\mathsf{unv}}(T_{\leq 0}) & & \cdots & & (\mathcal{F}_{|\boldsymbol{\Theta}_n^{\mathsf{op}}})^{\widehat{}}_{\mathsf{unv}}(T_{\leq n-1}). \end{array}$$

The proof of the next result occupies §1.3.3.

LEMMA 1.14. *For each Segal sheaf $\mathcal{F}$ on $\boldsymbol{\Theta}_n$, and for each $T \in \boldsymbol{\Theta}_n$, both of the diagrams among spaces (18) and (19) are limit diagrams.*

1.3.4. *The counit.* The counit of the adjunction (15) evaluates on each section $\mathcal{C}_\bullet$ of the functor $\mathsf{Shv}^{\mathsf{unv}}(\boldsymbol{\Theta}_\bullet^{\mathsf{op}}) \to [n]$ as the morphism between sections

$$\text{counit} \colon (\mathsf{fN}(\mathcal{C}_\bullet)_{|\boldsymbol{\Theta}_\bullet^{\mathsf{op}}})^{\widehat{}}_{\mathsf{unv}} \longrightarrow \mathcal{C}_\bullet$$

whose value on $i \in [n]$ is described, through Observation 1.13, as the canonical functor between (∞, i)-categories

$$\big(\mathsf{fN}(\mathcal{C}_{\bullet \leq i})\big)^{\widehat{}}_{\mathsf{unv}} \longrightarrow \mathcal{C}_i$$

from the univalent-completion of the Segal sheaf $\mathsf{fN}(\mathcal{C}_{\bullet \leq i})$ on $\boldsymbol{\Theta}_i$ that evaluates on each $T \in \boldsymbol{\Theta}_i$ as the canonical map

$$\mathsf{fN}(\mathcal{C}_{\bullet \leq i})(T) \longrightarrow \mathcal{C}_i(T) \tag{20}$$

as in Corollary 1.12.

OBSERVATION 1.15. Let $\underline{\mathcal{C}}$ be a flagged (∞, n)-category. Through the fully faithful functor (16), regard $\underline{\mathcal{C}}$ as a section of the functor $\mathsf{Shv}^{\mathsf{unv}}(\boldsymbol{\Theta}_\bullet) \to [n]$. The connectivity assumptions on each $\mathcal{C}_i \to \mathcal{C}_j$ ensure that, for each $0 \leq i \leq n$ and each $T \in \boldsymbol{\Theta}_i$, the map (20) between spaces is surjective (on path components). From the 2-out-of-3 property for surjections, it follows that the counit evaluates as a surjection

$$\text{counit} \colon \big(\mathsf{fN}(\mathcal{C}_{\bullet \leq i})\big)^{\widehat{}}_{\mathsf{unv}}(T) \longrightarrow \mathcal{C}_i(T)$$

is surjective (on path components).

The proof of the next result occupies §1.3.4.

LEMMA 1.16. *Let $\underline{\mathcal{C}} \in \mathsf{fCat}_n$ be a flagged (∞, n)-category. For each $0 \leq i \leq n$, the canonical morphism between Segal sheaves on $\boldsymbol{\Theta}_i$*

$$\text{counit} \colon \mathsf{fN}(\mathcal{C}_{\bullet \leq i}) \longrightarrow \mathcal{C}_i$$

witnesses a univalent-completion: $(\mathsf{fN}(\mathcal{C}_{\bullet \leq i}))^{\widehat{}}_{\mathsf{unv}} \xrightarrow{\simeq} \mathcal{C}_i$.

2. Univalent-completion

We establish a formula for univalent-completion. Section §2.1 establishes the formula for univalent-completion. Section §2.2 justifies that this formula indeed implements univalent-completion.

2.1. The formula. We establish the somewhat explicit formula for univalent-completion.

Notation 0.4 is an instance of the following.

DEFINITION 2.1. The functor

$$\mathsf{E}\colon \mathbf{\Theta}_n \longrightarrow \mathsf{PShv}(\mathbf{\Theta}_n) \tag{21}$$

carries an object $T = [p](T_1, \ldots, T_p) \in \mathbf{\Theta}_n$ to the presheaf of sets $\mathsf{E}(T)$ which sends $S = [q](S_1, \ldots, S_q)$ to

$$\Big(\{0, \ldots, q\} \xrightarrow{f} \{0, \ldots, p\}\ ,\ \big(g_{ij} \in \mathsf{E}(T_i)(S_j)\big)_{0<i\leq p;\ j\in\mathsf{Hull}\{f(i-1), f(i)\}}\Big)\ .$$

Each value of the functor (21) is an explicit description of strict n-groupoid-completion, as we now observe.

OBSERVATION 2.2. The functor (21) is identical to the composite functor

$$\mathsf{E}\colon \mathbf{\Theta}_n \hookrightarrow \mathsf{Cat}_n^{\mathsf{strict}} \xrightarrow{n\text{-groupoid completion}} \mathsf{Gpd}_n^{\mathsf{strict}} \hookrightarrow \mathsf{Cat}_n^{\mathsf{strict}} \xrightarrow{\text{nerve}} \mathsf{PShv}(\mathbf{\Theta}_n)\ ,$$

which we briefly explain. Here, $\mathsf{Gpd}_n^{\mathsf{strict}} \subset \mathsf{Cat}_n^{\mathsf{strict}}$ is its full $(2,1)$-subcategory consisting of those categories enriched in $\mathsf{Gpd}_{n-1}^{\mathsf{strict}}$ for which each (1-)morphism is invertible. The first functor is the fully faithful functor of Remark 0.3. The second functor is the left adjoint to the inclusion of the strict n-groupoids. The third functor is the inclusion of the strict n-groupoids. The fourth functor is the *nerve* functor of Remark 0.3.

Observation 2.2 yields the following.

OBSERVATION 2.3. The functor (21) factors through n-groupoids, and in particular Segal sheaves: $\mathsf{E}\colon \mathbf{\Theta}_n \xrightarrow{(21)} n\,\mathsf{Gpd}[\mathcal{S}] \underset{\text{Obs } 0.32}{\subset} \mathsf{Shv}(\mathbf{\Theta}_n)$.

Using that n-groupoid completion is a left adjoint, Observation 2.2 also yields the following.

OBSERVATION 2.4. There is a canonical natural transformation

$$\mathbf{\Theta}_n \ \underset{\mathsf{E}}{\overset{\text{Yoneda}}{\Downarrow}}\ \mathsf{Shv}(\mathbf{\Theta}_n)\ ,$$

which is by epimorphisms.

We next observe how Observation 2.4 is useful for evaluating the right adjoint of the fully faithful functor $n\,\mathsf{Gpd}[\mathcal{S}] \hookrightarrow \mathsf{Shv}(\mathbf{\Theta}_n)$.

OBSERVATION 2.5. The functor resulting from Observation 2.4,

$$\mathsf{Shv}(\mathbf{\Theta}_n) \longrightarrow \mathsf{PShv}(\mathbf{\Theta}_n)\ , \qquad \mathcal{F} \mapsto \Big(T \mapsto \mathcal{F}\big(\mathsf{E}(T)\big)\Big) \tag{22}$$

factors through the full ∞-subcategory $n\,\mathsf{Gpd}[\mathcal{S}] \subset \mathsf{Shv}(\mathbf{\Theta}_n) \subset \mathsf{PShv}(\mathbf{\Theta}_n)$ the right adjoint to the inclusion $n\,\mathsf{Gpd}[\mathcal{S}] \hookrightarrow \mathsf{Shv}(\mathbf{\Theta}_n)$.

Using the adjoint functor theorem, there is a Bousfield localization

$$\mathsf{Shv}(\mathbf{\Theta}_n) \rightleftarrows \mathsf{Shv}^{\mathsf{unv}}(\mathbf{\Theta}_n)~, \qquad \mathcal{F} \mapsto \mathcal{F}^{\widehat{}}_{\mathsf{unv}}~. \tag{23}$$

We now give a somewhat explicit formula for this left adjoint, which is inspired the *classifying diagram* construction of [**Re1**], which is elaborated in [**JT**]. Consider the composite functor

$$L\colon \mathsf{PShv}(\mathbf{\Theta}_n) \xrightarrow{\mathcal{F}\mapsto\mathcal{F}(-\times\mathsf{E}(\bullet))} \mathsf{PShv}(\mathbf{\Theta}_n \times \mathbf{\Theta}_n) \xrightarrow{\mathcal{H}\mapsto|\mathcal{H}|_\bullet} \mathsf{PShv}(\mathbf{\Theta}_n) \tag{24}$$

taking values

$$L(\mathcal{F}) := \big(T \mapsto \big|\mathcal{F}(T\times \mathsf{E}(\bullet))|_\bullet\big)~.$$

The initial functor $\ast \xrightarrow{\langle[0]\rangle} \mathbf{\Theta}_n^{\mathsf{op}}$ determines the initial functor $\mathbf{\Theta}_n^{\mathsf{op}} = \mathbf{\Theta}_n^{\mathsf{op}}\times \ast \to \mathbf{\Theta}_n^{\mathsf{op}} \times \mathbf{\Theta}_n^{\mathsf{op}}$. This, in turn, determines the natural transformation

$$\mathsf{id} \longrightarrow L~, \qquad \mathcal{F}\mapsto \Big(\mathcal{F} = \mathcal{F}(-\times \mathsf{E}([0])) \to \big|\mathcal{F}(-\times\mathsf{E}(\bullet))|_\bullet =: L(\mathcal{F})\Big)~. \tag{25}$$

The proof of the next result occupies §2.2.

PROPOSITION 2.6. *There is a canonical commutative diagram among ∞-categories:*

$$\begin{array}{ccc}
\mathsf{Shv}(\mathbf{\Theta}_n) & \xrightarrow{(-)^{\widehat{}}_{\mathsf{unv}}} & \mathsf{Shv}^{\mathsf{unv}}(\mathbf{\Theta}_n) \\
\downarrow & & \downarrow \\
\mathsf{PShv}(\mathbf{\Theta}_n) & \xrightarrow{L} & \mathsf{PShv}(\mathbf{\Theta}_n),
\end{array}$$

which extends as a commutative diagram among ∞-categories:

$$\begin{array}{ccc}
c_1\times\mathsf{Shv}(\mathbf{\Theta}_n) & \xrightarrow{\text{unit}} & \mathsf{Shv}(\mathbf{\Theta}_n) \\
\downarrow & & \downarrow \\
c_1\times\mathsf{PShv}(\mathbf{\Theta}_n) & \xrightarrow{(25)} & \mathsf{PShv}(\mathbf{\Theta}_n).
\end{array}$$

In other words, for each Segal sheaf $\mathcal{F}\in\mathsf{PShv}(\mathbf{\Theta}_n)$, and for each $T\in\mathbf{\Theta}_n$, there is a canonical identification between spaces:

$$\big|\mathsf{Map}(T\times\mathsf{E}(\bullet),\mathcal{F})\big| \xrightarrow{\simeq} \mathcal{F}^{\widehat{}}_{\mathsf{unv}}(T)~;$$

with respect to this identification, the value of the unit of the adjunction (23) on $\mathcal{F}$ evaluates on $T\in\mathbf{\Theta}_n$ as the canonical map between spaces

$$\mathcal{F}(T)\simeq \mathsf{Map}\big(T\times\mathsf{E}([0]),\mathcal{F}\big) \longrightarrow \big|\mathsf{Map}(T\times\mathsf{E}(\bullet),\mathcal{F})\big| \simeq \mathcal{F}^{\widehat{}}_{\mathsf{unv}}(T)~.$$

2.2. The formula is correct. We prove Proposition 2.6. This result is proved upon establishing the following features of the functor L of (24) and the natural transformation $\mathsf{id}\to L$ of (25).

(1) Should $\mathcal{F}$ carry (the opposites of) Segal covers to limit diagrams, then the presheaf $L(\mathcal{F})$ also carries (the opposites of) Segal covers to limit diagrams. In other words, there is a factorization of the restriction:

$$\begin{array}{ccc}
\mathsf{Shv}(\mathbf{\Theta}_n) & \overset{L}{\dashrightarrow} & \mathsf{Shv}(\mathbf{\Theta}_n) \\
\downarrow & & \downarrow \\
\mathsf{PShv}(\mathbf{\Theta}_n) & \xrightarrow{L} & \mathsf{PShv}(\mathbf{\Theta}_n).
\end{array}$$

(2) Should $\mathcal{F}$ carry (the opposites of) Segal covers to limit diagrams, then the presheaf $L(\mathcal{F})$ carries (the opposites of) univalence diagrams to limit diagrams. In other words, there is a further factorization of the restriction:

$$\begin{array}{ccc} \mathsf{Shv}(\Theta_n) & \overset{L}{\dashrightarrow} & \mathsf{Shv}^{\mathsf{unv}}(\Theta_n) \\ & {\scriptstyle L}\searrow \quad \swarrow & \\ & \mathsf{PShv}(\Theta_n) & \end{array} .$$

(3) Should $\mathcal{F}$ carry (the opposites of) Segal covers and univalence diagrams to limit diagrams, then the natural transformation $\mathsf{id} \to L$ evaluates as an equivalence. In other words, the lax-commutative diagram

$$\begin{array}{ccc} & \overset{\mathsf{id}}{\frown\!\!\to} & \\ \mathsf{Shv}^{\mathsf{unv}}(\Theta_n) & \Downarrow & \mathsf{Shv}^{\mathsf{unv}}(\Theta_n) \\ & \searrow \quad \nearrow {\scriptstyle L} & \\ & \mathsf{PShv}(\Theta_n) & \end{array} .$$

is, in fact, a commutative diagram.

- **Proof of (1):** Let $\mathcal{F} \in \mathsf{Shv}(\Theta_n)$ be a Segal sheaf. Let

$$\begin{array}{ccc} T^0 & \longrightarrow & T^+ \\ \downarrow & & \downarrow \\ T^- & \longrightarrow & T \end{array}$$

be a pushout diagram on Θ_n^{cls}. This diagram determines the diagram in $\mathsf{Fun}\big(\Theta_n, \mathsf{PShv}(\Theta_n)\big)$:

$$\begin{array}{ccc} T^0 \times \mathsf{E}(\bullet) & \longrightarrow & T^+ \times \mathsf{E}(\bullet) \\ \downarrow & & \downarrow \\ T^- \times \mathsf{E}(\bullet) & \longrightarrow & T \times \mathsf{E}(\bullet). \end{array}$$

Through Observation 2.2, using that the Segal condition is closed under products of presheaves, this is in fact a diagram in $\mathsf{Fun}\big(\Theta_n, \mathsf{Shv}(\Theta_n)\big) \subset \mathsf{Fun}\big(\Theta_n, \mathsf{PShv}(\Theta_n)\big)$. Established in [**Re2**], products in $\mathsf{Shv}(\Theta_n)$ distribute over colimits. Therefore the above diagram is a pushout diagram in $\mathsf{Fun}\big(\Theta_n, \mathsf{Shv}(\Theta_n)\big)$. Applying the Segal sheaf $\mathcal{F}$ results in the pullback diagram in $\mathsf{Fun}\big(\Theta_n^{\mathsf{op}}, \mathcal{S}\mathsf{paces}\big)$:

(26)
$$\begin{array}{ccc} \mathcal{F}\big(T \times \mathsf{E}(\bullet)\big) & \longrightarrow & \mathcal{F}\big(T^+ \times \mathsf{E}(\bullet)\big) \\ \downarrow & & \downarrow \\ \mathcal{F}\big(T^- \times \mathsf{E}(\bullet)\big) & \longrightarrow & \mathcal{F}\big(T^0 \times \mathsf{E}(\bullet)\big). \end{array}$$

Now, as in [**Re2**], the ∞-category $\mathsf{Shv}(\mathbf{\Theta}_n)$ is Cartesian-closed. It follows that, for each $S \in \mathbf{\Theta}_n$, the presheaf

$$\mathcal{F}(S \times \bullet)\colon \mathbf{\Theta}_n^{\mathsf{op}} \longrightarrow \mathcal{S}\mathsf{paces}$$

is a Segal-sheaf. Observation 2.5 gives that the maximal n-subgroupoid of this Segal-sheaf $\mathcal{F}(S \times \bullet)$ is the presheaf $\mathcal{F}\big(S \times \mathsf{E}(\bullet)\big)$. We conclude, in particular, that the pullback diagram (26) is in fact a diagram in the full ∞-subcategory $n\,\mathsf{Gpd}[\mathcal{S}] \subset \mathsf{Fun}(\mathbf{\Theta}_n^{\mathsf{op}}, \mathcal{S}\mathsf{paces})$ of Observation 0.32. Because this full ∞-subcategory is closed under limits, the above diagram is a pullback diagram in $n\,\mathsf{Gpd}[\mathcal{S}]$. The functor $|-|\colon n\,\mathsf{Gpd}[\mathcal{S}] \to \mathcal{S}\mathsf{paces}$ carries the above diagram to a diagram among spaces:

$$\begin{array}{ccc} L(\mathcal{F})(T) := \big|\mathcal{F}\big(T \times \mathsf{E}(\bullet)\big)\big| & \longrightarrow & \big|\mathcal{F}\big(T^+ \times \mathsf{E}(\bullet)\big)\big| =: L(\mathcal{F})(T^+) \\ \downarrow & & \downarrow \\ L(\mathcal{F})(T^-) := \big|\mathcal{F}\big(T^- \times \mathsf{E}(\bullet)\big)\big| & \longrightarrow & \big|\mathcal{F}\big(T^0 \times \mathsf{E}(\bullet)\big)\big| =: L(\mathcal{F})(T^0). \end{array} \tag{27}$$

This proof that $L(\mathcal{F})$ is a Segal sheaf whenever $\mathcal{F}$ is a Segal sheaf is complete once we show (27) is a pullback among spaces.

Restricting the diagram (26) in $\mathsf{PShv}(\mathbf{\Theta}_n)$ along the standard diagonal functor $\mathbf{\Delta}^{\times n} \to \mathbf{\Theta}_n$ determines a pullback diagram

$$\begin{array}{ccc} \mathcal{G}([\bullet],\dots,[\bullet]) & \longrightarrow & \mathcal{G}_+([\bullet],\dots,[\bullet]) \\ \downarrow & & \downarrow \\ \mathcal{G}_-([\bullet],\dots,[\bullet]) & \longrightarrow & \mathcal{G}_0([\bullet],\dots,[\bullet]) \end{array} \tag{28}$$

in $\mathsf{PShv}(\mathbf{\Delta}^{\times n})$. Because the diagram (26) belongs to $n\,\mathsf{Gpd}[\mathcal{S}]$, this restricted diagram (28) belongs to the ∞-category $\mathsf{Gpd}^n[\mathcal{S}]$ of groupoid objects in the ∞-category $\mathsf{Gpd}^{n-1}[\mathcal{S}]$, where $\mathsf{Gpd}^1[\mathcal{S}] = 1\,\mathsf{Gpd}[\mathcal{S}]$ is the ∞-category of groupoid objects in $\mathcal{S}\mathsf{paces}$. Now, let $P := \big([p_1],\dots,[p_{n-1}]\big) \in \mathbf{\Delta}^{n-1}$. Consider the functor $\mathbf{\Delta} \xrightarrow{(\langle P\rangle,\mathsf{id})} \mathbf{\Delta}^{\times n-1} \times \mathbf{\Delta} = \mathbf{\Delta}^{\times n}$. Consider the restriction

$$\begin{array}{ccc} \mathcal{G}^P(\bullet) & \longrightarrow & \mathcal{G}^P_+(\bullet) \\ \downarrow & & \downarrow \\ \mathcal{G}^P_-(\bullet) & \longrightarrow & \mathcal{G}^P_0(\bullet) \end{array} \tag{29}$$

of the diagram (28) along this functor. This diagram (29) is a pullback diagram among simplicial spaces. Because $\mathcal{F} \in \mathsf{Shv}(\mathbf{\Theta}_n)$ is a Segal sheaf, inspecting the standard diagonal functor $\mathbf{\Delta}^{\times n} \to \mathbf{\Theta}_n$ gives that the vertical arrows in the pullback diagram (29) are Cartesian fibrations as well as coCartesian fibrations among Segal sheaves on $\mathbf{\Delta}$. Quillen's Theorem B (of [**AF1**]) applied to (29) grants that the resulting diagram among spaces

$$\begin{array}{ccc} |\mathcal{G}^P(\bullet)| & \longrightarrow & |\mathcal{G}^P_+(\bullet)| \\ \downarrow & & \downarrow \\ |\mathcal{G}^P_-(\bullet)| & \longrightarrow & |\mathcal{G}^P_0(\bullet)| \end{array} \tag{30}$$

is a pullback. Proceeding by induction on n gives that the diagram among spaces

$$\begin{array}{ccc} \big|\mathcal{G}\big([\bullet],\dots,[\bullet]\big)\big| & \longrightarrow & \big|\mathcal{G}_+\big([\bullet],\dots,[\bullet]\big)\big| \\ \downarrow & & \downarrow \\ \big|\mathcal{G}_-\big([\bullet],\dots,[\bullet]\big)\big| & \longrightarrow & \big|\mathcal{G}_0\big([\bullet],\dots,[\bullet]\big)\big| \end{array} \tag{31}$$

is a pullback diagram. Now, there is a canonical morphism among square diagrams in $\mathcal{S}\mathsf{paces}$ from (31) to diagram (27). By direct examination using Quillen's Theorem A, the diagonal functor $\mathbf{\Delta}^{\times n} \to \mathbf{\Theta}_n$ is final. It follows that this comparison morphism from (31) to (27) is an equivalence between square diagrams. Finally, because (31) is a pullback diagram, then so too is (27).

- **Proof of (2):** Let $\mathcal{F} \in \mathsf{Shv}(\mathbf{\Theta}_n)$ be a Segal sheaf. By Definition 0.5 of the univalenct-condition, we must show, for each $0 \leq i < n$, that the canonical map between spaces

$$L(\mathcal{F})(c_i) \longrightarrow L(\mathcal{F})\big(c_i \wr \mathsf{E}(c_1)\big) \tag{32}$$

is an equivalence. Consider the diagram in $\mathsf{Shv}(\mathbf{\Theta}_n)$:

$$\begin{array}{ccccccc} \partial c_i \times \mathsf{E}(c_{i+1}) & & & \longrightarrow & & & c_i \times \mathsf{E}(c_{i+1}) \\ \downarrow & \searrow & & & & \swarrow & \downarrow \\ & & \partial c_i & \longrightarrow & c_i & & \\ & & \downarrow & & \downarrow & & \\ & & \partial c_i & \longrightarrow & c_i & & \\ \downarrow & \nearrow & & & & \nwarrow & \downarrow \\ \partial c_i & & & \longrightarrow & & & c_i \wr \mathsf{E}(c_1). \end{array}$$

In this diagram, the outer square is a pushout, and the inner square is trivially a pushout. Furthermore, the diagonal arrows in this square are carried to equivalences by the localization $\mathsf{Shv}(\mathbf{\Theta}_n) \to \mathsf{Shv}^{\mathsf{unv}}(\mathbf{\Theta}_n)$. Applying the Segal sheaf $L(\mathcal{F})$ to this diagram results in the diagram among spaces,

$$\begin{array}{ccccccc} L(\mathcal{F})(c_i) & & & \longrightarrow & & & L(\mathcal{F})(c_i) \\ \downarrow & \searrow & & & & \swarrow & \downarrow \\ & & L(\mathcal{F})\big(c_i \wr \mathsf{E}(c_1)\big) & \longrightarrow & L(\mathcal{F})\big(c_i \times \mathsf{E}(c_{i+1})\big) & & \\ & & \downarrow & & \downarrow & & \\ & & L(\mathcal{F})\big(\partial c_i \times \mathsf{E}(c_{i+1})\big) & \longrightarrow & L(\mathcal{F})(\partial c_i) & & \\ \downarrow & \nearrow & & & & \nwarrow & \downarrow \\ L(\mathcal{F})(\partial c_i) & & & \longrightarrow & & & L(\mathcal{F})(\partial c_i), \end{array}$$

in which both the inner and the outer squares are pullback squares. Therefore, the down-rightward arrow, which is (32), is an equivalence provided each of the down-leftward arrow and the up-rightward arrow is an equivalence. This is implied by showing, for each $S, T \in \boldsymbol{\Theta}_n$, that the canonical map between spaces,

$$L(\mathcal{F})(T) := \big|\mathcal{F}\big(T \times \mathsf{E}(\bullet)\big)\big| \longrightarrow \big|\mathcal{F}\big(T \times \mathsf{E}(S) \times \mathsf{E}(\bullet)\big)\big| =: L(\mathcal{F})\big(T \times \mathsf{E}(S)\big) ,$$

is an equivalence. So fix $S, T \in \boldsymbol{\Theta}_n$.

As established in the **Proof of (1)** above, the presheaf $\mathcal{G}_T(\bullet) := \mathcal{F}\big(T \times \mathsf{E}(\bullet)\big)$ on $\boldsymbol{\Theta}_n$ belongs to the full ∞-subcategory $n\,\mathsf{Gpd}[\mathcal{S}] \underset{\text{Obs } 0.32}{\subset} \mathsf{PShv}(\boldsymbol{\Theta}_n)$. Inspecting the Definition 2.1 of the functor $\mathsf{E}\colon \boldsymbol{\Theta}_n \to n\,\mathsf{Gpd}[\mathcal{S}]$, its left Kan extension to $\mathsf{Cat}_n^{\mathsf{strict}}$ preserves products. Therefore $\mathsf{E}(S \times \bullet) = \mathsf{E}(S) \times \mathsf{E}(\bullet)$. We conclude an identification of the exponential presheaf,

$$\mathcal{G}_T(\bullet)^S := \mathcal{G}_T(S \times \bullet) = \mathcal{F}\big(T \times \mathsf{E}(S \times \bullet)\big) = \mathcal{F}\big(T \times \mathsf{E}(S) \times \mathsf{E}(\bullet)\big) ,$$

and that this is also an n-groupoid object in $\mathcal{S}\mathsf{paces}$. So it remains to establish the following general claim:

(†): Let $\boldsymbol{\Theta}_n^{\mathsf{op}} \xrightarrow{\mathcal{G}} \mathcal{S}\mathsf{paces}$ be an n-groupoid object. Let $S \in \boldsymbol{\Theta}_n$ be an object. Consider the morphism $\mathcal{G} \to \mathcal{G}^S$ between n-groupoid objects induced by the unique morphism $S \to *$ in $\boldsymbol{\Theta}_n$. The resulting map between colimits

(33) $$|\mathcal{G}| \longrightarrow |\mathcal{G}^S|$$

is an equivalence between spaces.

Consider the morphism $s\colon * \to S$ that selects the unique object for which there are no non-identity morphisms to it in the n-category S. This morphism is a section of the unique morphism $S \to *$ in $\boldsymbol{\Theta}_n$. This determines a retraction $|\mathcal{G}^S| \xrightarrow{s^*} |\mathcal{G}|$, for which (33) is a section. The claim (†) then follows upon showing the composite map among spaces

$$|\mathcal{G}^S| \xrightarrow{s^*} |\mathcal{G}| \xrightarrow{(33)} |\mathcal{G}^S|$$

is equivalent to the identity map. We do this representably. More precisely, we construct a filler among strict n-categories

(34) $$\begin{array}{ccccccccc} S & \xrightarrow{\cong} & * \times S & \xrightarrow{0 \times \mathsf{id}} & c_n \times S & \xleftarrow{1 \times \mathsf{id}} & * \times S & \xleftarrow{\cong} & S \\ \downarrow{\scriptstyle !} & & & & \downarrow{\scriptstyle H_S} & & & \swarrow{\scriptstyle \mathsf{id}} & \\ * & & \xrightarrow{\quad s \quad} & & S & & & & \end{array} .$$

Such a filler determines a filler in the diagram among presheaves on $\boldsymbol{\Theta}_n$:

$$\begin{array}{ccccc} \mathcal{G}^S & \xrightarrow{\quad 0 \quad} & c_n \times \mathcal{G}^S & \xleftarrow{\quad 1 \quad} & \mathcal{G}^S \\ \uparrow{\scriptstyle !^*} & & \downarrow & \nearrow{\scriptstyle \mathsf{id}} & \\ \mathcal{G} & \xleftarrow[\quad s^* \quad]{} & \mathcal{G}^S & & \end{array} .$$

The colimit functor $|-|\colon \mathsf{PShv}(\boldsymbol{\Theta}_n) \to \mathcal{S}\mathsf{paces}$ preserves products, and carries c_n to $*$. Applying this colimit functor to the above diagram results in the sought equivalence between the composite map $|\mathcal{G}^S| \to |\mathcal{G}| \to |\mathcal{G}^S|$ and the identity map.

So we are left to construct a filler, $H_S \colon c_n \times S \to S$, in the diagram (34). We do this for all $S \in \boldsymbol{\Theta}_n$, by induction on n. The base case that $n = 0$ is tautological. So assume $n > 0$. Write $S = [p](S_1, \dots, S_p)$. If $p = 0$, then $S = *$ is terminal and there is a unique H_S. So assume $p > 0$. Because S is a colimit of its maximal Segal cover, it is enough to define H_S on each term of this colimit, compatibly. So the functor H_S is determined by declaring, for each consecutive morphism $c_n \to S$ in $\boldsymbol{\Theta}_n$, what is the composite functor among strict n-categories:

$$H_S \circ (\mathsf{id}, f) \colon c_n \xrightarrow{(\mathsf{id},f)} c_n \times S \xrightarrow{H_S} S\ .$$

So let $c_n \xrightarrow{f} S$ be a consecutive morphism in $\boldsymbol{\Theta}_n$ that is injective on sets of objects. So f evaluates on objects as $f(0) = i-1$ and $f(1) = i$ for some $0 < i \leq p$, and f evaluates on strict $(n-1)$-categories of morphisms as $\mathsf{Hom}_f(0,1) \colon \mathsf{Hom}_{c_n}(0,1) = c_{n-1} \xrightarrow{f_i} S_i = \mathsf{Hom}_S(i-1,i)$ for some consecutive morphism $f_i \colon c_{n-1} \to S_i$ in $\boldsymbol{\Theta}_{n-1}$. We declare the composite functor among strict n-categories

$$H_S \circ (\mathsf{id}, f) \colon c_n \xrightarrow{(\mathsf{id},f)} c_n \times S \xrightarrow{H_S} S$$

as follows. On objects, this composite functor is

$$H_S \circ (\mathsf{id}, f)(0) = 0 \qquad \text{and} \qquad H_S \circ (\mathsf{id}, f)(1) = f(1).$$

On $(n-1)$-categories of morphisms, this composite functor,

$$\begin{aligned}
\mathsf{Hom}_{c_n}(0,1) \quad & = \quad c_{n-1} \\
& \xrightarrow{(\mathsf{id},f_i)} \quad c_{n-1} \times S_i \\
& = \quad \mathsf{Hom}_{c_n}(0,1) \times \mathsf{Hom}_S\big(f(0), f(1)\big) \\
& \xrightarrow{\mathsf{Hom}_{H_S}(0,1)} \quad \mathsf{Hom}_S\big(0, f(1)\big) = \prod_{0<j\leq i} S_j\ ,
\end{aligned}$$

is defined by declaring the composition

$$c_{n-1} \xrightarrow{(\mathsf{id},f_i)} c_{n-1} \times S_i \xrightarrow{\mathsf{Hom}_{H_S}(0,1)} \prod_{0<j\leq i} S_j \xrightarrow{\mathsf{pr}} S_j$$

to be the following. If $j < i$, this latter composition is constant at the object $0 \in S_j$. If $j = i$, this latter composition is $H_{S_i} \circ (\mathsf{id}, f_i)$; note that, because $S_i \in \boldsymbol{\Theta}_{n-1}$, this has been defined by induction on n. This completes the construction of H_S, for each $S \in \boldsymbol{\Theta}_n$, and therefore completes the proof of (2).

- **Proof of (3):** Let $\mathcal{F} \in \mathsf{Shv}^{\mathsf{unv}}(\boldsymbol{\Theta}_n)$ be a univalent Segal sheaf. We must show the morphism between univalent Segal sheaves

(35) $$(25) \colon \mathcal{F} \longrightarrow \big|\mathcal{F}\big(- \times \mathsf{E}(\bullet)\big)\big| =: L(\mathcal{F})$$

is an equivalence. Let $S, T \in \boldsymbol{\Theta}_n$. Consider the projection morphism $T \times \mathsf{E}(S) \to T$ in the ∞-category $\mathsf{Shv}(\boldsymbol{\Theta}_n)$. In the case that $T = *$, this is the case nearly by definition of the univalent-condition. The case of general T follows from this $T = *$ case because the Bousfield localization $\mathsf{Shv}(\boldsymbol{\Theta}_n) \to \mathsf{Shv}^{\mathsf{unv}}(\boldsymbol{\Theta}_n)$ preserves products (as established in [**Re2**]). Using that $\mathcal{F}$ is univalent, the resulting map between spaces

$$\mathcal{F}(T) \longrightarrow \mathcal{F}\big(T \times \mathsf{E}(S)\big)$$

is an equivalence. It follows that the morphism (35) is an equivalence, as desired.

3. Using the formula to prove the main result

We use the formula of Proposition 2.6 for univalent-completion to prove Theorem 0.27. Specifically, we use that formula to prove Lemma 1.9, Lemma 1.14, and Lemma 1.16, then draw Theorem 0.27 as a consequence.

3.1. Proof of Lemma 1.9. Recall from Proposition 2.6 the explicit description of the functor $(-_{|\boldsymbol{\Theta}_\bullet^{\mathsf{op}}})^\frown_{\mathsf{unv}}\colon \mathsf{Shv}(\boldsymbol{\Theta}_n) \to \Gamma\big(\mathsf{Shv}^{\mathsf{unv}}(\boldsymbol{\Theta}_\bullet)\big)$. Through that description, the problem is to show that, for each Segal sheaf $\mathcal{F} \in \mathsf{Shv}(\boldsymbol{\Theta}_n)$, and for each $0 \leq i \leq j \leq n$, the canonical functor between (∞, j)-categories

$$(\mathcal{F}_{|\boldsymbol{\Theta}_i^{\mathsf{op}}})^\frown_{\mathsf{unv}} \longrightarrow (\mathcal{F}_{|\boldsymbol{\Theta}_j^{\mathsf{op}}})^\frown_{\mathsf{unv}}$$

is i-connective. So fix $\mathcal{F} \in \mathsf{Shv}(\boldsymbol{\Theta}_n)$ and $0 \leq i \leq j \leq n$. By Definition 0.9 of *i-connective*, we must show that, for each $0 \leq k \leq i$, the solid diagram among (∞, j)-categories

$$(36)\qquad \begin{array}{ccc} \partial c_k & \longrightarrow & (\mathcal{F}_{|\boldsymbol{\Theta}_i^{\mathsf{op}}})^\frown_{\mathsf{unv}} \\ \big\downarrow & \nearrow & \big\downarrow \\ c_k & \longrightarrow & (\mathcal{F}_{|\boldsymbol{\Theta}_j^{\mathsf{op}}})^\frown_{\mathsf{unv}} \end{array}$$

can be filled. So fix such a $0 \leq k \leq i$.

Consider the canonical diagram among spaces of functors:

$$(37)\qquad \begin{array}{ccc} \mathsf{Cat}_j\big(c_k, (\mathcal{F}_{|\boldsymbol{\Theta}_i^{\mathsf{op}}})^\frown_{\mathsf{unv}}\big) & \longrightarrow & \mathsf{Cat}_j\big(c_k, (\mathcal{F}_{|\boldsymbol{\Theta}_j^{\mathsf{op}}})^\frown_{\mathsf{unv}}\big) \\ \big\downarrow & & \big\downarrow \\ \mathsf{Cat}_j\big(\partial c_k, (\mathcal{F}_{|\boldsymbol{\Theta}_i^{\mathsf{op}}})^\frown_{\mathsf{unv}}\big) & \longrightarrow & \mathsf{Cat}_j\big(\partial c_k, (\mathcal{F}_{|\boldsymbol{\Theta}_j^{\mathsf{op}}})^\frown_{\mathsf{unv}}\big). \end{array}$$

Unwinding definitions, the problem of finding a filler in (36) is identical to showing the resulting map to the pullback

$$\mathsf{Cat}_j\big(c_k, (\mathcal{F}_{|\boldsymbol{\Theta}_i^{\mathsf{op}}})^\frown_{\mathsf{unv}}\big) \longrightarrow \mathsf{Cat}_j\big(\partial c_k, (\mathcal{F}_{|\boldsymbol{\Theta}_i^{\mathsf{op}}})^\frown_{\mathsf{unv}}\big) \underset{\mathsf{Cat}_j\left(\partial c_k, (\mathcal{F}_{|\boldsymbol{\Theta}_j^{\mathsf{op}}})^\frown_{\mathsf{unv}}\right)}{\times} \mathsf{Cat}_j\big(c_k, (\mathcal{F}_{|\boldsymbol{\Theta}_j^{\mathsf{op}}})^\frown_{\mathsf{unv}}\big)$$

is surjective (on path components).

Recall Definition 2.1 of the functor $\mathsf{E}\colon \boldsymbol{\Theta}_n \to \mathsf{PShv}(\boldsymbol{\Theta}_n)$. For each $\ell \leq n$, denote its restriction

$$\mathsf{E}_{|\ell}\colon \boldsymbol{\Theta}_\ell \hookrightarrow \boldsymbol{\Theta}_n \xrightarrow{\mathsf{E}} \mathsf{PShv}(\boldsymbol{\Theta}_n)\ .$$

The canonical morphism $\partial c_k \to c_k$ then determines, for each $\ell \leq n$, the morphism

$$\mathcal{F}(c_k \times \mathsf{E}_{|\ell}(\bullet)) \longrightarrow \mathcal{F}(\partial c_k \times \mathsf{E}_{|\ell}(\bullet))$$

between ℓ-groupoid objects in $\mathcal{S}\mathsf{paces}$. Taking colimits results in a commutative diagram among spaces:

$$\begin{array}{ccc} \left|\mathcal{F}(c_k \times \mathsf{E}_{|i}(\bullet))\right| & \longrightarrow & \left|\mathcal{F}(c_k \times \mathsf{E}_{|j}(\bullet))\right| \\ \downarrow & & \downarrow \\ \left|\mathcal{F}(\partial c_k \times \mathsf{E}_{|i}(\bullet))\right| & \longrightarrow & \left|\mathcal{F}(\partial c_k \times \mathsf{E}_{|j}(\bullet))\right|. \end{array}$$

Through Proposition 2.6, which identifies univalent-completion in terms of E, this commutative diagram among spaces is identified as the commutative diagram (37). So we must show that the canonical map between spaces

$$\left|\mathcal{F}(c_k \times \mathsf{E}_{|i}(\bullet))\right| \longrightarrow \left|\mathcal{F}(\partial c_k \times \mathsf{E}_{|i}(\bullet))\right| \underset{\left|\mathcal{F}(\partial c_k \times \mathsf{E}_{|j}(\bullet))\right|}{\times} \left|\mathcal{F}(c_k \times \mathsf{E}_{|j}(\bullet))\right| \tag{38}$$

is surjective (on path components). This map (38) factors as the string of maps,

$$\begin{aligned} \left|\mathcal{F}(c_k \times \mathsf{E}_{|i}(\bullet))\right| &\underset{(a)}{\simeq} \left|\iota^*\mathcal{F}(c_k \times \mathsf{E}_{|j}(\bullet))\right| \\ &\xrightarrow[(b)]{\simeq} \left|\iota^*\Big(\mathcal{F}(\partial c_k \times \mathsf{E}_{|j}(\bullet)) \underset{\mathcal{F}(\partial c_k \times \mathsf{E}_{|j}(\bullet))}{\times} \mathcal{F}(c_k \times \mathsf{E}_{|j}(\bullet))\Big)\right| \\ &\xleftarrow[(c)]{\simeq} \left|\iota^*\Big(\iota_!\iota^*\mathcal{F}(\partial c_k \times \mathsf{E}_{|j}(\bullet)) \underset{\mathcal{F}(\partial c_k \times \mathsf{E}_{|j}(\bullet))}{\times} \mathcal{F}(c_k \times \mathsf{E}_{|j}(\bullet))\Big)\right| \\ &\xrightarrow[(d)]{\text{surjective}} \left|\iota_!\iota^*\mathcal{F}(\partial c_k \times \mathsf{E}_{|j}(\bullet)) \underset{\mathcal{F}(\partial c_k \times \mathsf{E}_{|j}(\bullet))}{\times} \mathcal{F}(c_k \times \mathsf{E}_{|j}(\bullet))\right| \\ &\xrightarrow[(e)]{\simeq} \left|\iota_!\iota^*\mathcal{F}(\partial c_k \times \mathsf{E}_{|j}(\bullet))\right| \underset{\left|\mathcal{F}(\partial c_k \times \mathsf{E}_{|j}(\bullet))\right|}{\times} \left|\mathcal{F}(c_k \times \mathsf{E}_{|j}(\bullet))\right| \\ &\underset{(f)}{\simeq} \left|\mathcal{F}(\partial c_k \times \mathsf{E}_{|i}(\bullet))\right| \underset{\left|\mathcal{F}(\partial c_k \times \mathsf{E}_{|j}(\bullet))\right|}{\times} \left|\mathcal{F}(c_k \times \mathsf{E}_{|j}(\bullet))\right|, \end{aligned}$$

which we now explain. We use the notation $\iota\colon \mathbf{\Theta}_i \hookrightarrow \mathbf{\Theta}_j$ for the standard fully faithful right adjoint functor. This functor determines the adjunction $\iota_!\colon \mathsf{PShv}(\mathbf{\Theta}_i) \rightleftarrows \mathsf{PShv}(\mathbf{\Theta}_j)\colon \iota^*$, whose left adjoint is given by left Kan extension along ι^{op}, and whose right adjoint is given by restriction along ι^{op}. With this notation, the equivalence (a) is definitional, from the notation for $\mathsf{E}_{|\ell}$. The equivalence (b) is a trivial pullback. Because ι is fully faithful, the unit of this $(\iota_!, \iota^*)$-adjunction is an equivalence. Together with the fact that the right adjoint ι^* preserves pullbacks, this establishes the equivalence (c). The map (d) is the canonical one, induced by restriction along the functor ι. Because this functor ι is a right adjoint, for any presheaf $\mathcal{X}$ on $\mathbf{\Theta}_j$, the canonical map between colimits $|\iota^*\mathcal{X}| \to |\mathcal{X}|$ is surjective (on path components). In particular, the map (d) is surjective (on path components). The equivalence (f) is the fact that left Kan extensions compose, together with the definitional identification $\mathcal{F}(\partial c_k \times \mathsf{E}_{|i}(\bullet)) = \iota^*\mathcal{F}(\partial c_k \times \mathsf{E}_{|j}(\bullet))$ as that supporting the equivalence (a). The equivalence (e) follows through the same logic, which reduces from presheaves on $\mathbf{\Theta}_n$ to simplicial spaces, as in the final part of **Proof of (1)**, in §2.2. This completes the proof of Lemma 1.9.

3.2. Proof of Lemma 1.14. Let $\mathcal{F} \in \mathsf{Shv}(\mathbf{\Theta}_n)$ be a Segal sheaf. Following §1.3.3, we must show, for each $T \in \mathbf{\Theta}_n$, that the diagram (18) is a limit

diagram. Through Proposition 2.6, we can describe each instance of univalent-completion appearing in (18) by the expression involving $\mathsf{E}(\bullet)$:
(39)

$$\begin{array}{ccccc} \mathcal{F}(T) & \longrightarrow & & \longrightarrow & \big|\mathcal{F}(T\times\mathsf{E}(\bullet))\big| \\ \downarrow & & & & \downarrow \\ \mathcal{F}(T_{<n}) & \longrightarrow & \mathsf{fN}\Big(\big|\mathcal{F}(-\times\mathsf{E}(\bullet<n))\big|\Big)(T_{<n}) & \longrightarrow & \big|\mathcal{F}(T_{<n}\times\mathsf{E}(\bullet))\big| \end{array}$$

is a limit diagram. We do this by induction on n. For $n=0$, this assertion is vacuously true. So assume $n>0$. By induction on n, the bottom left horizontal morphism in (39) is an equivalence. Using that the ∞-category $\mathcal{S}\mathsf{paces}$ is an ∞-topos, Theorem 6.1.0.6 of [**Lu1**] grants that the square (39) is a pullback square provided, for each $S\in\mathbf{\Theta}_n$, the square

$$\begin{array}{ccc} \mathcal{F}(T) & \longrightarrow & \mathcal{F}\big(T\times\mathsf{E}(S)\big) \\ \downarrow & & \downarrow \\ \mathcal{F}(T_{<n}) & \longrightarrow & \mathcal{F}\big(T_{<n}\times\mathsf{E}(S)\big) \end{array}$$

is a pullback. To show this square is pullback it is sufficient to show that the square among strict n-categories

$$\begin{array}{ccc} T_{<n}\times\mathsf{E}(S) & \longrightarrow & T\times\mathsf{E}(S) \\ \downarrow & & \downarrow \\ T_{<n} & \longrightarrow & T \end{array} \tag{40}$$

is a pushout in $\mathsf{Shv}(\mathbf{\Theta}_n)$. Because $S\to\mathsf{E}(S)$ is an epimorphism in $\mathsf{Shv}(\mathbf{\Theta}_n)$, it follows through the Cartesian Bousfield localization of univalent-completion (established in [**Re2**]) that the top horizontal morphism in (40) is an epimorphism in $\mathsf{Shv}(\mathbf{\Theta}_n)$. Thus, the square (40) among strict n-categories is a pushout in $\mathsf{Shv}(\mathbf{\Theta}_n)$ provided the square among strict n-categories

$$\begin{array}{ccc} T_{<n}\times S & \longrightarrow & T\times S \\ \downarrow & & \downarrow \\ T_{<n} & \longrightarrow & T \end{array}$$

is a pushout in $\mathsf{Shv}(\mathbf{\Theta}_n)$. This latter square is trivially a pushout, after the Cartesian Bousfield localization of Segal completion (established in [**Re2**]).

3.3. Proof of Lemma 1.16. From the description of the counit in §1.3.4, we must show, for each $0\le i\le n$, that the functor between (∞,i)-categories

$$\mathrm{counit}\colon \big(\mathsf{fN}(\underline{\mathcal{C}})_{|\mathbf{\Theta}_i^{\mathrm{op}}}\big)^{\widehat{}}_{\mathsf{unv}} \longrightarrow \mathcal{C}_i$$

is an equivalence. So fix $0\le i\le n$. Through the nearly definitional Observation 0.8, it is enough to show, for each $T\in\mathbf{\Theta}_i$, that the map between spaces

$$\big(\mathsf{fN}(\underline{\mathcal{C}})_{|\mathbf{\Theta}_i^{\mathrm{op}}}\big)^{\widehat{}}_{\mathsf{unv}}(T) \longrightarrow \mathcal{C}_i(T) \tag{41}$$

is an equivalence. So let $T \in \mathbf{\Theta}_i$. Through Proposition 2.6, this map (41) is identified as the map

$$\left|\mathsf{fN}(\underline{\mathcal{C}})\big(T \times \mathsf{E}(\bullet_{\leq i})\big)\right| \longrightarrow \left|\mathcal{C}_i\big(T \times \mathsf{E}(\bullet_{\leq i})\big)\right| \xleftarrow{\simeq} \mathcal{C}_i(T) \tag{42}$$

in which the leftward arrow is an equivalence because $\mathcal{C}_i$ is, by definition, univalent.

We establish that the map (42) is an equivalence using, for each $S \in \mathbf{\Theta}_i$, the following diagram among spaces

$$\begin{array}{ccccc}
\mathsf{fN}(\underline{\mathcal{C}})\big(T \times \mathsf{E}(S)\big) & & & & \\
\downarrow & \searrow & & & \\
\mathsf{fN}(\underline{\mathcal{C}})\big(T_{<i} \times \mathsf{E}(S)\big) & & \left|\mathsf{fN}(\underline{\mathcal{C}})\big(T \times \mathsf{E}(\bullet_{\leq i})\big)\right| & \xrightarrow[(a)]{} & \mathcal{C}_i(T) \\
\downarrow & \searrow & \downarrow & & \downarrow \\
\mathsf{fN}(\underline{\mathcal{C}})\big(T_{<i} \times \mathsf{E}(S_{<i})\big) & & \left|\mathsf{fN}(\underline{\mathcal{C}})\big(T_{<i} \times \mathsf{E}(\bullet_{\leq i})\big)\right| & \xrightarrow[(b)]{} & \mathcal{C}_i(T_{<i}) \\
& \searrow & \downarrow & & \downarrow \\
& & \left|\mathsf{fN}(\underline{\mathcal{C}})\big(T_{<i} \times \mathsf{E}(\bullet_{\leq i})\big)\right|^{S_{<i}} & \longrightarrow & \mathcal{C}_i(T_{<i})^{S_{<i}},
\end{array}$$

(with diagonal arrows $\mathsf{fN}(\underline{\mathcal{C}})\big(T \times \mathsf{E}(S)\big) \to \mathcal{C}_i(T)$, $\mathsf{fN}(\underline{\mathcal{C}})\big(T_{<i} \times \mathsf{E}(S)\big) \to \mathcal{C}_i(T_{<i})$, $\mathsf{fN}(\underline{\mathcal{C}})\big(T_{<i} \times \mathsf{E}(S_{<i})\big) \to \mathcal{C}_i(T_{<i})^{S_{<i}}$)

which we now explain. This diagram is a representation of the ∞-category $[2] \times [2]$ in the ∞-category $\mathcal{S}\mathsf{paces}$, where the second factor selects the vertical sequences.

- For each $u \in [2]$, the representation $\{u\} \times \{0 < 1\} \to [2] \times [2] \to \mathcal{S}\mathsf{paces}$ selects the map that is induced by the canonical functor $T_{<i} \to T$ between i-categories.
- For each $u \in [2]$, the representation $\{u\} \times \{1 < 2\} \to [2] \times [2] \to \mathcal{S}\mathsf{paces}$ selects the map that is induced by the canonical functor $S_{<i} \to S$ between i-categories.
- For each $v \in [2]$, the representation $\{0 < 1\} \to \{v\} \times [2] \times [2] \to \mathcal{S}\mathsf{paces}$ selects the map that is induced by the canonical map to a colimit.
- For each $v \in [2]$, the representation $\{1 < 2\} \times \{v\} \to [2] \times [2] \to \mathcal{S}\mathsf{paces}$ selects the map that is that determined by (42). In particular, the representation (a): $\{1 < 2\} \times \{0\} \to [2] \times [2] \to \mathcal{S}\mathsf{paces}$ is precisely (42).
- This entire diagram commutes because $\mathcal{C}_i$ is univalent.

So we must show that the map (a) is an equivalence.

We isolate some observations concerning surjectivity.

(Surj 1): Observation 1.15 grants that, for $v = 0, 1 \in [2]$, the representation $\{0 < 2\} \times \{v\} \to [2] \times [2] \to \mathcal{S}\mathsf{paces}$ selects a map between spaces that is surjective (on path components).

(Surj 2): Because this functor $\iota\colon \mathbf{\Theta}_0 \hookrightarrow \mathbf{\Theta}_i$ is a right adjoint, for any presheaf $\mathcal{F}$ on $\mathbf{\Theta}_i$, the canonical map between colimits $\mathcal{F}[0] = |\iota^*\mathcal{F}| \to |\mathcal{F}|$ is surjective (on path components). Applying this observation to $\mathcal{F} = \mathsf{fN}(\underline{\mathcal{C}})\big(T \times \mathsf{E}(\bullet)\big)$ and to $\mathcal{F} = \mathsf{fN}(\underline{\mathcal{C}})\big(T_{<i} \times \mathsf{E}(\bullet)\big)$ gives that, for $v = 0, 1 \in [2]$, the representation $\{0 < 1\} \times \{v\} \to [2] \times [2] \to \mathcal{S}\mathsf{paces}$ selects a map between spaces that is surjective (on path components) in the case that $S = [0]$.

The general case for $S \in \mathbf{\Theta}_i$ follows from the 2-out-of-3 property for surjections, again using that $[0] \in \mathbf{\Theta}_i$ is a final object.

(Surj 3): After the previous point **(Surj 2)**, commutativity of the square selected by the representation $\{0<1\} \times \{1<2\} \to [2] \times [2] \to \mathcal{S}\mathsf{paces}$ implies the map selected by the representation $\{0<1\} \times \{2\} \to [2] \times [2] \to \mathcal{S}\mathsf{paces}$ is surjective over the image of the diagonal map, which is selected by the representation $\{1\} \times \{1<2\} \to [2] \times [2] \to \mathcal{S}\mathsf{paces}$.

(Surj 4): By the 2-out-of-3 property for surjections, we conclude that, for $v = 0, 1 \in [2]$, the representation $\{1<2\} \times \{v\} \to [2] \times [2] \to \mathcal{S}\mathsf{paces}$ selects a map that is surjective (on path components).

Now, it follows from Corollary 1.12 that the representation $\{0<2\} \times \{1<2\} \to [2] \times [2] \to \mathcal{S}\mathsf{paces}$ selects a pullback diagram. As established in **Proof of (1)** of §2.2, the functor

$$\mathbf{\Theta}_i^{\mathsf{op}} \xrightarrow{\mathsf{fN}(\mathcal{C})\left(T_{<i} \times \mathsf{E}(\bullet_{\leq i})\right)} \mathcal{S}\mathsf{paces}$$

is an i-groupoid object. Precisely because i-groupoid objects in $\mathcal{S}\mathsf{paces}$ are effective, the representation $\{0<1\} \times \{1<2\} \to [2] \times [2] \to \mathcal{S}\mathsf{paces}$ selects a pullback diagram. Taking the case that $S = [1]$, the two squares $\{0<u\} \times \{0<1\} \to [2] \times [2] \to \mathcal{S}\mathsf{paces}$, for $u = 1, 2$, being pullback implies, for each point $x \in \mathsf{fN}(\underline{\mathcal{C}})\left(T_{<i} \times \mathsf{E}([1])\right)$, that the induced map between based loop spaces

$$\Omega_{|x|}\Big|\mathsf{fN}(\underline{\mathcal{C}})\left(T_{<i} \times \mathsf{E}(\bullet_{\leq i})\right)\Big| \xrightarrow{\Omega(b)} \Omega_{(b)|x|}\mathcal{C}_i(T_{<i})$$

is an equivalence. Using that the ∞-topos $\mathcal{S}\mathsf{paces}$ is hypercomplete, we conclude from this, together with the above **(Surj)** observations concerning the bottom horizontal arrow, that the map (b) is an equivalence between spaces.

The **(Surj)** observations give that each of the representations $\{0<1\} \times \{1\} \to [2] \times [2] \to \mathcal{S}\mathsf{paces}$ and $\{0<2\} \times \{1\} \to [2] \times [2] \to \mathcal{S}\mathsf{paces}$ is surjective (on path components). It follows, again, from Corollary 1.12 that the representation $\{0<2\} \times \{0<1\} \to [2] \times [2] \to \mathcal{S}\mathsf{paces}$ selects a pullback diagram. Established above is that the map (b) is an equivalence. Using, again, that the ∞-topos $\mathcal{S}\mathsf{paces}$ is hypercomplete, we conclude from (b) being an equivalence, that the map (a) is an equivalence, as desired, provided the representation $\{0<1\} \times \{0<1\} \to [2] \times [2] \to \mathcal{S}\mathsf{paces}$ selects a pullback square. So it remains to show just that.

Let $S \to S'$ be a morphism in $\mathbf{\Theta}_i$. Consider the evident square among strict i-categories:

$$\begin{array}{ccc} T_{<i} \times \mathsf{E}(S') & \longrightarrow & T \times \mathsf{E}(S') \\ \downarrow & & \downarrow \\ T_{<i} \times \mathsf{E}(S) & \longrightarrow & T \times \mathsf{E}(S). \end{array} \tag{43}$$

Applying the left adjoint functor $(-_{|\mathbf{\Theta}_\bullet^{\mathsf{op}}})^{\widehat{}}_{\mathsf{unv}} \colon \mathsf{Shv}(\mathbf{\Theta}_i) \to \mathsf{fCat}_i$ of (17) to this square (43) results in square among flagged (∞, i)-categories. Using the fact that the Bousfield localization implementing Segal completion is Cartesian (as established in [**Re2**]), this square in flagged (∞, i)-categories is a pushout. From the definition of fN as a restricted Yoneda functor, it follows that the square among spaces

$$
\begin{array}{ccc}
\mathsf{fN}(\underline{\mathcal{C}})\big(T\times\mathsf{E}(S)\big) & \longrightarrow & \mathsf{fN}(\underline{\mathcal{C}})\big(T\times\mathsf{E}(S')\big)\\
\big\downarrow & & \big\downarrow\\
\mathsf{fN}(\underline{\mathcal{C}})\big(T_{<i}\times\mathsf{E}(S)\big) & \longrightarrow & \mathsf{fN}(\underline{\mathcal{C}})\big(T_{<i}\times\mathsf{E}(S')\big)
\end{array}
\tag{44}
$$

is a pullback. Using that the ∞-category $\mathcal{S}\mathsf{paces}$ is an ∞-topos, Theorem 6.1.0.6 of [**Lu1**] gives that the resulting square involving colimits

$$
\begin{array}{ccc}
\mathsf{fN}(\underline{\mathcal{C}})\big(T\times\mathsf{E}(S)\big) & \longrightarrow & \Big|\mathsf{fN}(\underline{\mathcal{C}})\big(T\times\mathsf{E}(\bullet)\big)\Big|\\
\big\downarrow & & \big\downarrow\\
\mathsf{fN}(\underline{\mathcal{C}})\big(T_{<i}\times\mathsf{E}(S)\big) & \longrightarrow & \Big|\mathsf{fN}(\underline{\mathcal{C}})\big(T_{<i}\times\mathsf{E}(\bullet)\big)\Big|.
\end{array}
$$

is a pullback. As this square is the representation $\{0<1\}\times\{0<1\}\to[2]\times[2]\to\mathcal{S}\mathsf{paces}$, this completes this proof.

3.4. Proof of the main result (Theorem 0.27). To prove Theorem 0.27, it is enough to show that the unit and the counit of the adjunction (13) are equivalences. Through the discussion in §1.3.3, that the unit of the adjunction (15) is an equivalence is exactly the statement of Lemma 1.14. Through the discussion in §1.3.4, that the counit of the adjunction (15) is an equivalence is exactly the statement of Lemma 1.16.

3.5. Proof of Corollary 0.33. We prove Corollary 0.33. Recall Definition 0.29 of n-groupoid objects, and Definition 0.31 of n-flagged ∞-groupoids.

Let $0\leq i\leq n$. Using that the standard fully faithful functor $*=\mathbf{\Theta}_0^{\mathsf{op}}\hookrightarrow\mathbf{\Theta}_i^{\mathsf{op}}$ is a fully faithful left adjoint, restriction and left Kan extension define a localization

$$
\mathcal{S}\mathsf{paces}=\mathsf{PShv}(\mathbf{\Theta}_0)\ \rightleftarrows\ \mathsf{PShv}(\mathbf{\Theta}_i)
$$

in which the left adjoint is fully faithful. Evidently, this left adjoint factors as the commutative diagram among ∞-categories and fully faithful functors there among:

$$
\begin{array}{ccc}
\mathcal{S}\mathsf{paces} & \dashrightarrow & i\mathsf{Gpd}[\mathcal{S}]\\
\vdots & \searrow & \big\downarrow{\scriptstyle \text{Obs } 0.32}\\
\mathsf{Shv}^{\mathsf{unv}}(\mathbf{\Theta}_i) & \longrightarrow & \mathsf{Shv}(\mathbf{\Theta}_i).
\end{array}
\tag{45}
$$

Inspecting the definitions of these full ∞-subcategories of $\mathsf{PShv}(\mathbf{\Theta}_i)$ (as Bousfield localizations thereof) reveals that this is a limit diagram. In other words, if a Segal sheaf on $\mathbf{\Theta}_i$ is both an n-groupoid object in $\mathcal{S}\mathsf{paces}$ and univalent-complete, then it is a constant presheaf.

LEMMA 3.1. *For each $0 \leq i \leq n$, the adjunction (13) restricts through Observation 0.32 as an adjunction:*

$$
\begin{array}{ccc}
& \overset{|-|}{\dashrightarrow} & \\
i\mathsf{Gpd}[\mathcal{S}] & \xleftarrow{\quad(45)\quad} & \mathcal{S}\mathsf{paces} \\
{\scriptstyle \mathrm{Obs}\ 0.32}\downarrow & \overset{(-)^{\frown}_{\mathsf{unv}}}{\longrightarrow} & \downarrow{\scriptstyle \mathrm{constant}} \\
\mathsf{Shv}(\mathbf{\Theta}_i) & \xleftarrow[\text{inclusion}]{} & \mathsf{Shv}^{\mathsf{unv}}(\mathbf{\Theta}_i)
\end{array}
$$

PROOF. The top horizontal arrow $\mathcal{S}\mathsf{paces} \hookrightarrow i\mathsf{Gpd}[\mathcal{S}]$ is a right adjoint; its left adjoint is the functor $|-|\colon i\mathsf{Gpd}[\mathcal{S}] \to \mathcal{S}\mathsf{paces}$ given by taking the colimit of an i-groupoid object $\mathbf{\Theta}_i^{\mathsf{op}} \to \mathcal{S}\mathsf{paces}$. Because the straight diagram in the statement of the lemma commutes, there is a canonical lax-commutative diagram:

$$
\begin{array}{ccc}
& \overset{|-|}{\longrightarrow} & \\
i\mathsf{Gpd}[\mathcal{S}] & \Uparrow & \mathcal{S}\mathsf{paces} \\
{\scriptstyle \mathrm{Obs}\ 0.32}\downarrow & & \downarrow{\scriptstyle \mathrm{constant}} \\
\mathsf{Shv}(\mathbf{\Theta}_i) & \xrightarrow{(-)^{\frown}_{\mathsf{unv}}} & \mathsf{Shv}^{\mathsf{unv}}(\mathbf{\Theta}_i).
\end{array}
$$

It remains, then, to show that this lax-commutative diagram is, in fact, a commutative diagram. So let $\mathcal{G} \in i\mathsf{Gpd}[\mathcal{S}]$ be an i-groupoid object. We must show that the canonical morphism between univalent Segal sheaves on $\mathbf{\Theta}_i$ under $\mathcal{G}$,

$$
\mathcal{G}^{\frown}_{\mathsf{unv}} \longrightarrow |\mathcal{G}| ,
$$

is an equivalence. Through the universal property of the univalent-completion of $\mathcal{G}$, this is to show that, for each univalent Segal sheaf $\mathcal{C}$ on $\mathbf{\Theta}_i$, each horizontal morphism in $\mathsf{Shv}(\mathbf{\Theta}_i)$,

$$
\begin{array}{ccc}
\mathcal{G} & \longrightarrow & \mathcal{C} \\
\downarrow & \nearrow\!\!\!\searrow & \uparrow \\
|\mathcal{G}| & \dashrightarrow & \mathcal{C}^{\sim},
\end{array}
$$

admits a unique factorization as in the upward diagonal arrow, which we now explain. The fully faithful functor $i\mathsf{Gpd}[\mathcal{S}] \hookrightarrow \mathsf{Shv}(\mathbf{\Theta}_i)$ is a left adjoint; we denote its right adjoint as $(-)^{\sim}\colon \mathsf{Shv}(\mathbf{\Theta}_i) \to i\mathsf{Gpd}[\mathcal{S}]$. The definition of *univalent* Segal sheaves on $\mathbf{\Theta}_i$ is just so that this adjunction restricts as an adjunction

$$
\mathcal{S}\mathsf{paces} \rightleftarrows \mathsf{Shv}^{\mathsf{unv}}(\mathbf{\Theta}_i)\colon (-)^{\sim}.
$$

This adjunction determines a unique downward diagonal factorization, and also ensures that the bottom right presheaf $\mathcal{C}^{\sim}$ on $\mathbf{\Theta}_i$ is, in fact, constant. Because $\mathcal{C}^{\sim}$ is constant, there is a unique factorization as in the bottom horizontal arrow. This establishes a unique factorization as in the upward diagonal arrow, as desired. □

REMARK 3.2. Lemma 3.1 articulates that the univalent-completion of an i-groupoid object $\mathbf{\Theta}_i^{\mathsf{op}} \xrightarrow{\mathcal{G}} \mathcal{S}\mathsf{paces}$ is nothing other than its colimit $|\mathcal{G}|$, regarded as a constant presheaf on $\mathbf{\Theta}_i$ which is thusly univalent and Segal.

LEMMA 3.3. *The adjunction (17) restricts through Observation 0.32 as an adjunction:*

$$\big|(-)_{|\boldsymbol{\Theta}_\bullet^{\mathrm{op}}}\big|\colon n\,\mathsf{Gpd}[\mathcal{S}] \rightleftarrows \mathsf{fGpd}_n\colon \mathsf{fN}\ .$$

PROOF. Lemma 3.1 implies, and identifies how, the restriction of $\widehat{(-_{|\boldsymbol{\Theta}_\bullet^{\mathrm{op}}})}_{\mathsf{unv}}$ to $n\,\mathsf{Gpd}[\mathcal{S}]$ factors through fGpd_n. It remains to show that the restriction of fN to fGpd_n factors through $n\,\mathsf{Gpd}[\mathcal{S}]$.

By direct inspection, there is a canonical lax-commutative diagram among ∞-categories,

$$\begin{array}{ccccccc}
\boldsymbol{\Theta}_n & \longrightarrow & \mathsf{Cat}_n & \longrightarrow & \mathsf{fCat}_n & \longrightarrow & \Gamma\big(\mathsf{Shv}^{\mathsf{unv}}(\boldsymbol{\Theta}_\bullet)\to[n]\big) \\
{\scriptstyle \mathsf{E}}\big\downarrow & & & & \big\uparrow & & \big\uparrow \\
n\,\mathsf{Gpd}[\mathcal{S}] & & \Uparrow & & \mathsf{fGpd}_n & \longrightarrow & \mathsf{Fun}([n],\mathcal{S}\mathsf{paces}),
\end{array}$$

(with a curved arrow $n\,\mathsf{Gpd}[\mathcal{S}] \to \mathsf{fGpd}_n$ along the bottom)

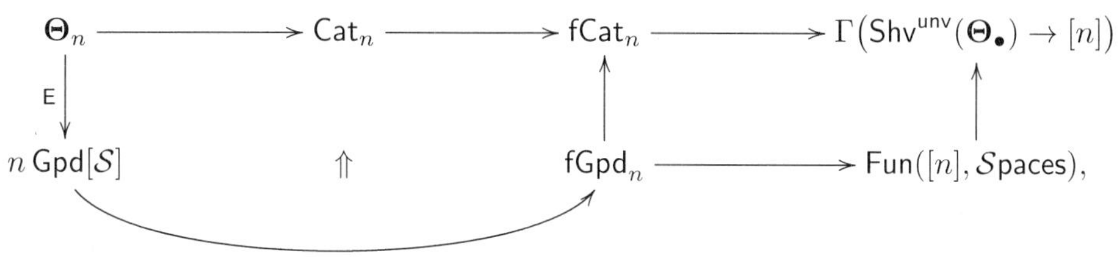

where the bottom curved arrow has been established in the first line of this proof, and the right horizontal arrows are the evident fully faithful functors. For each $0\leq i\leq n$, the standard fully faithful functor $\mathcal{S}\mathsf{paces}\hookrightarrow\mathsf{Cat}_i$ is a right adjoint. Their left adjoints assemble as a left adjoint to the standard fully faithful functor $\mathsf{Fun}([n],\mathcal{S}\mathsf{paces})\hookrightarrow\Gamma\big(\mathsf{Shv}^{\mathsf{unv}}(\boldsymbol{\Theta}_\bullet)\to[n]\big)$ whose image consists of those sections that take values in ∞-groupoids. Replacing the right upward arrow in the above diagram by this left adjoint determines another lax-commutative diagram among ∞-categories:

$$\begin{array}{ccccccc}
\boldsymbol{\Theta}_n & \longrightarrow & \mathsf{Cat}_n & \longrightarrow & \mathsf{fCat}_n & \longrightarrow & \Gamma\big(\mathsf{Shv}^{\mathsf{unv}}(\boldsymbol{\Theta}_\bullet)\to[n]\big) \\
{\scriptstyle \mathsf{E}}\big\downarrow & & & & & & \big\downarrow \\
n\,\mathsf{Gpd}[\mathcal{S}] & & \Uparrow & & \mathsf{fGpd}_n & \longrightarrow & \mathsf{Fun}([n],\mathcal{S}\mathsf{paces}).
\end{array}$$

(with a curved arrow $n\,\mathsf{Gpd}[\mathcal{S}] \to \mathsf{fGpd}_n$ along the bottom)

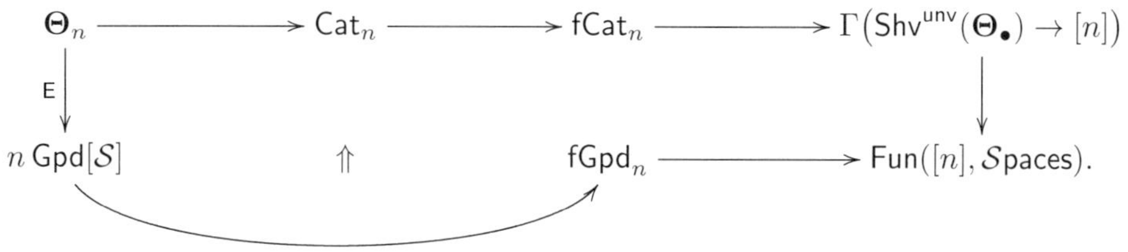

Observation 2.3 implies that this lax-commutative diagram is, in fact, a commutative diagram. Now, the functor fN is defined as the restricted Yoneda functor along the composite functor $\boldsymbol{\Theta}_n\to\mathsf{fCat}_n$ appearing in the above diagram. Commutativity of the above diagram thusly gives that, for each n-flagged ∞-groupoid $\underline{X}=(X_0\to\cdots\to X_n)$, and each $T\in\boldsymbol{\Theta}_n$, that the canonical monomorphism between spaces

$$\mathsf{fN}(\underline{X})(\mathsf{E}(T))\longrightarrow\mathsf{fN}(\underline{X})(T)$$

is an equivalence. We conclude that flagged (∞,n)-category $\mathsf{fN}(\underline{X})$ is, in fact, an n-flagged ∞-groupoid, as desired. $\square$

REMARK 3.4. In Lemma 3.3, the left adjoint carries an n-groupoid object $\boldsymbol{\Theta}_n^{\mathrm{op}}\xrightarrow{\mathcal{G}}\mathcal{S}\mathsf{paces}$ to the sequence of spaces $|\mathcal{G}_{|\boldsymbol{\Theta}_0^{\mathrm{op}}}|\to|\mathcal{G}_{|\boldsymbol{\Theta}_1^{\mathrm{op}}}|\to\cdots\to|\mathcal{G}_{|\boldsymbol{\Theta}_n^{\mathrm{op}}}|$. Lemma 1.9 ensures that this sequence of spaces indeed satisfies the requisite connectivity conditions to be an n-flagged ∞-groupoid. The right adjoint in Lemma 3.3 carries an n-flagged ∞-groupoid $X_0\to\cdots\to X_n$ to its n-Cech nerve, elaborated in Remark 0.34.

REMARK 3.5. Note that the right adjoint in Lemma 3.3 is defined on the full ∞-subcategory $\mathsf{Fun}([n], \mathcal{S}\mathsf{paces}) \subset \Gamma\big(\mathsf{Shv}^{\mathsf{unv}}(\boldsymbol{\Theta}_\bullet) \to [n]\big)$. As so, for $X_\bullet \in \mathsf{Fun}([n], \mathcal{S}\mathsf{paces})$ a sequence of spaces, the counit

$$\big|\mathsf{fN}(X_\bullet)_{|\boldsymbol{\Theta}_\bullet^{\mathrm{op}}}\big| \longrightarrow X_\bullet$$

is the terminal n-flagged ∞-groupoid over $X_\bullet$. In this way, this is counit is a relative connective cover of $X_\bullet$. Indeed, for $n = 1$, and should $X_0 \simeq *$, then this morphism is $(* \to X_1') \to (* \to X_1)$ where $X_1' \to X_1$ is the canonical map from the connected component of X_1 through which $* \to X_1$ factors. And for general n, yet $X_0 \xrightarrow{\simeq} \dots \xrightarrow{\simeq} X_{n-1} \simeq *$, this morphism is $(* \to \cdots \to * \to X_n') \to (* \to \cdots \to * \to X_n)$ where $X_n' \to X_n$ is the n-connective cover of the based space $* \to X_n$.

PROOF OF COROLLARY 0.33. It remains to show that the unit and the counit of the adjunction of Lemma 3.3 are equivalences. Through Lemma 3.3, it is enough to observe that the unit and the counit of the extended adjunction (17) are equivalences. That this is so is precisely Lemma 1.14 and Lemma 1.16, respectively. □

Acknowledgments

We are grateful to Jacob Lurie, for his contributions to foundational topos theory, and to Charles Rezk, for his careful and operationally practical exposition of (∞, n)-categories.

References

[AF1] David Ayala and John Francis, *Fibrations of ∞-categories*. Preprint, 2017.

[AF2] David Ayala and John Francis, *Poincaré/Koszul duality*. Preprint, 2015.

[AFR] David Ayala, John Francis, and Nick Rozenblyum, *Factorization homology I: Higher categories*, Adv. Math. **333** (2018), 1042–1177, DOI 10.1016/j.aim.2018.05.031. MR3818096

[BS] Clark Barwick and Christopher Schommer-Pries, *On the unicity of the homotopy theory of higher categories*. Preprint, 2013.

[Be] Clemens Berger, *Iterated wreath product of the simplex category and iterated loop spaces*, Adv. Math. **213** (2007), no. 1, 230–270, DOI 10.1016/j.aim.2006.12.006. MR2331244

[GH] David Gepner and Rune Haugseng, *Enriched ∞-categories via non-symmetric ∞-operads*, Adv. Math. **279** (2015), 575–716, DOI 10.1016/j.aim.2015.02.007. MR3345192

[Jo1] André Joyal, *Quasi-categories and Kan complexes*, J. Pure Appl. Algebra **175** (2002), no. 1-3, 207–222, DOI 10.1016/S0022-4049(02)00135-4. MR1935979

[Jo2] André Joyal,*Disks, duality and Theta-categories*. Unpublished work, 1997.

[JT] André Joyal and Myles Tierney, *Quasi-categories vs Segal spaces*, Categories in algebra, geometry and mathematical physics, Contemp. Math., vol. 431, Amer. Math. Soc., Providence, RI, 2007, pp. 277–326, DOI 10.1090/conm/431/08278. MR2342834

[Lu1] Jacob Lurie, *Higher topos theory*, Annals of Mathematics Studies, vol. 170, Princeton University Press, Princeton, NJ, 2009. MR2522659

[Lu2] Jacob Lurie, *Higher algebra*. Preprint, 2016.

[Lu3] Jacob Lurie, *On the classification of topological field theories*, Current developments in mathematics, 2008, Int. Press, Somerville, MA, 2009, pp. 129–280. MR2555928

[PTVV] Tony Pantev, Bertrand Toën, Michel Vaquié, and Gabriele Vezzosi, *Shifted symplectic structures*, Publ. Math. Inst. Hautes Études Sci. **117** (2013), 271–328, DOI 10.1007/s10240-013-0054-1. MR3090262

[Re1] Charles Rezk, *A model for the homotopy theory of homotopy theory*, Trans. Amer. Math. Soc. **353** (2001), no. 3, 973–1007, DOI 10.1090/S0002-9947-00-02653-2. MR1804411

[Re2] Charles Rezk, *A Cartesian presentation of weak n-categories*, Geom. Topol. **14** (2010), no. 1, 521–571, DOI 10.2140/gt.2010.14.521. MR2578310

Department of Mathematics, Montana State University, Bozeman, Montana 59717
Email address: david.ayala@montana.edu

Department of Mathematics, Northwestern University, Evanston, Illinois 60208
Email address: jnkf@northwestern.edu

Contemporary Mathematics
Volume **718**, 2018
https://doi.org/10.1090/conm/718/14479

How to derive Feynman diagrams for finite-dimensional integrals directly from the BV formalism

Owen Gwilliam and Theo Johnson-Freyd

ABSTRACT. The Batalin-Vilkovisky formalism in quantum field theory was originally invented to address the difficult problem of finding diagrammatic descriptions of oscillating integrals with degenerate critical points. But since then, BV algebras have become interesting objects of study in their own right, and mathematicians sometimes have good understanding of the homological aspects of the story without any access to the diagrammatics. In this note we reverse the usual direction of argument: we begin by asking for an explicit calculation of the homology of a BV algebra, and from it derive Wick's Theorem and the other Feynman rules for finite-dimensional integrals.

PREFATORY REMARK. We wrote this note in the spring of 2011 as graduate students, when we were seeking to understand how one might invent such an elaborate procedure and found this elementary motivating example. In the following years, our understanding of the Batalin–Vilkovisky formalism continued to evolve [**Gwi12**, **JF15**, **GH18**]. To get other perspectives and to find recent applications of the formalism, we recommend [**Li17**, **Mne17**].

0. Introduction

Anyone with a glancing interest in physics eventually sees Feynman diagrams and is told that they play a key role in quantum field theory, notably in the process of renormalization of path integrals. If a mathematician digs a bit deeper, she learns that Feynman diagrams fit into a method for constructing asymptotic series associated to oscillating integrals. Recall that an oscillating integral is an expression of the form

$$\int_{\mathbb{R}} f(x) e^{\sqrt{-1}g(x)}\, \mathrm{d}x,$$

with f and g real functions. The adjective "oscillating" refers to the variation of the exponential term, especially when $|g'|$ is large. In section 4, we will return to this relationship between Feynman diagrams and integration, but our main goal is to provide an alternative story about how to invent Feynman diagrams, rooted in homological algebra. Again, the basic problem has its source in physics. In the process of trying to better understand quantum field theory and string theory, physicists rediscovered many homological tools, notably the Chevalley–Eilenberg description

2010 *Mathematics Subject Classification.* Primary 81S40. Secondary 18G40.
This work was supported by NSF grants DMS-0636646 and DMS-0901431.

of equivariant cohomology, a version of which is called the "BRST formalism" in the physics literature. Building on this, the Russian physicists Batalin and Vilkovisky introduced a formalism to tackle supergravity [**BV81, BV83, BV84, BV85**]; the BV formalism is now widely considered the most powerful approach to quantizing gauge theories. We focus this article on the simplest examples of the homological problem that appears in the BV formalism. The problem is as follows.

Fix a field $\mathbb{K}$ of characteristic 0 and a positive integer N. Consider the graded-commutative algebra

$$V_\bullet = \mathbb{K}[\![x_1, \ldots, x_N, \xi_1, \ldots, \xi_N, \hbar]\!]$$

with its usual topology as a power-series algebra, where $x_1, \ldots, x_N, \hbar$ are in degree 0 (and hence commute with everything), and $\xi_1, \ldots, \xi_N$ are in homological degree 1 (and hence anticommute with each other). Observe that $V_\bullet$ is concentrated between degrees 0 and N, because $\xi_i^2 = 0$. We make $V_\bullet$ into a chain complex, but *not* a differential graded algebra, as follows. Pick a symmetric invertible $N \times N$ matrix $a = (a_{i,j})_{i,j=1}^N$. Pick a power series $b(x) \in \mathbb{K}[\![x_1, \ldots, x_N]\!] \subset V_\bullet$ with only cubic and higher terms. Equip $V_\bullet$ with the degree-(-1) differential:

$$Q = \sum_{i,j=1}^N a_{i,j} x_i \frac{\partial}{\partial \xi_j} - \sum_{i=1}^N \frac{\partial b(x)}{\partial x_i} \frac{\partial}{\partial \xi_i} - \hbar \sum_{i=1}^N \frac{\partial^2}{\partial x_i \partial \xi_i}.$$

These partial derivatives behave as usual so long as one takes into account the appropriate signs. For instance,

$$\frac{\partial}{\partial \xi_i} \frac{\partial}{\partial \xi_j} = - \frac{\partial}{\partial \xi_j} \frac{\partial}{\partial \xi_i}$$

because the partial derivative $\frac{\partial}{\partial \xi_i}$ has odd degree. After all, it removes a copy of ξ_i from any monomial $x_1^{m_1} \cdots x_N^{m_N} \xi_1^{n_1} \cdots \xi_N^{n_N}$ and hence lowers the homological degree of the monomial by 1.

The fundamental goal is to understand the homology of this chain complex and, even more, to know how to explicitly describe the image of closed elements inside the homology. Let us take as an ansatz that the homology of $(V_\bullet, Q)$ in degree 0 is isomorphic to $\mathbb{K}[\![\hbar]\!]$ as a $\mathbb{K}[\![\hbar]\!]$-module. As $V_\bullet$ is zero in positive degrees, every element in $V_0 = \mathbb{K}[\![x_1, \ldots, x_N, \hbar]\!]$ is a cycle. The image of Q in V_0 is contained in the ideal generated by $x_1, \ldots, x_N, \hbar$, since every term in Q either adds a power of x_i or $\hbar$. Thus the number $1 \in V_0$ is not a boundary. For $f \in V_0$ we write $[f]$ for its image in homology $H_0(V)$. According to the ansatz, for every $f \in V_0$ we should be able to write

$$[f] = \langle f \rangle [1]$$

for a unique number $\langle f \rangle \in \mathbb{K}[\![\hbar]\!]$. By succeeding at this, we will verify the ansatz.

The $\langle f \rangle$ are computed by Feynman diagrams. We will warm up with two examples, and then explain in general how the diagrammatic description arises naturally from the homology calculation. The chain complex $(V_\bullet, Q)$ is an example of a *BV algebra*, and we hope that this example can provide the reader with more intuition for the general BV formalism in perturbative quantum field theory. We end this note with some general discussion motivating the BV approach in finite-dimensional non-gauged integrals. We expect that this note contains no results that are not known to the experts, but we hope that it gives an entry point for non-experts trying to learn BV theory. For other parts of this story, see for example [**ABF10, Fio03, Sta97, Wit90**].

1. Example: Wick's lemma

As our first warm-up, we set $N = 1$ and $b = 0$. Then our complex is

$$\begin{array}{ccc} V_1 & \xrightarrow{Q} & V_0 \\ \shortparallel & & \shortparallel \\ \mathbb{K}[\![x,\hbar]\!]\,\xi & & \mathbb{K}[\![x,\hbar]\!] \end{array}$$

where

$$Q = ax\frac{\partial}{\partial\xi} - \hbar\frac{\partial^2}{\partial x\partial\xi}.$$

Given $\xi\, f(x) \in V_1$, $Q(\xi f) = axf(x) - \hbar f'(x)$. A little formal calculus tells us that $Q(\xi f) = 0$ only when $f(x) = \exp(ax^2/2\hbar)$, but as this f is not in $\mathbb{K}[\![x,\hbar]\!]$, the homology in degree 1 vanishes.

The boundaries comprise the closed $\mathbb{K}[\![\hbar]\!]$-span of the image of Q on the monomials ξx^n, where "closed" means that we take the closure in the power series topology. The base case is $Q(\xi) = ax$. For $n > 0$, we have the elements $Q(\xi x^n) = ax^{n+1} - \hbar n x^{n-1}$. Hence in homology we have $[x^{n+1}] = \frac{\hbar}{a}n[x^{n-1}]$. By recursion $\langle x^{2n+1}\rangle = 0$ and

$$\langle x^{2n}\rangle = \left(\frac{\hbar}{a}\right)^n (2n-1)!!,$$

where

$$(2n-1)!! = (2n-1)(2n-3)\cdots(3)(1) = \frac{(2n)!}{2^n n!}.$$

This classic result about the moments of a Gaussian distribution is known as *Wick's formula* by physicists [**Wic50**]. The combinatorial interpretation is that $(2n-1)!!$ counts the number of pairings of $2n$ objects; each of the n connections in such a pairing should then be understood as contributing a factor of $\hbar/a$. (Note that we have confirmed the ansatz for $H_0(V_\bullet, Q)$ in this case.)

EXERCISE 1.1. We encourage the reader to pause from this note long enough to consider: what happens when $N > 1$ and $b = 0$?

2. Example: Counting trivalent graphs

We consider now the situation that $N = 1$, $a = 1$, and $b(x) = x^3/6$. The boundaries are spanned by the images of monomials

$$Q(x^n\xi) = x^{n+1} - x^{n+2}/2 - \hbar n x^{n-1},$$

where we interpret x^{-1} as 0 to include the case $n = 0$ conveniently. Hence

$$[x^{n+1}] = \frac{1}{2}[x^{n+2}] + \hbar\, n\,[x^{n-1}].$$

One can thus begin substituting $[x] = \frac{1}{2}[x^2] = \frac{1}{2}\big(\frac{1}{2}[x^3] + \hbar[1]\big) = \frac{1}{2}\big(\frac{1}{2}\big(\frac{1}{2}[x^4] + 2\hbar[x]\big) + \hbar[1]\big) = \ldots.$ At each stage one adds terms in either high degree in x or high degree in $\hbar$, and so the subsequent infinite sum converges to something in $\mathbb{K}[\![\hbar]\!]\cdot[1]$. But if we use this *ad hoc* approach to the substitutions, it becomes combinatorially difficult to compute the coefficients in $\langle x^n\rangle = [x^n]/[1] \in \mathbb{K}[\![\hbar]\!]$.

Instead, we posit the following answer, and check that it satisfies the necessary conditions. A *Feynman diagram for* $\langle x^n \rangle$ is a connected finite graph with all vertices trivalent, except for one marked vertex that is n-valent with totally-ordered incident half-edges (self-loops and parallel edges are allowed). An *automorphism* of a Feynman diagram for $\langle x^n \rangle$ is a permutation of the half-edges of the graph which does not change the data of which half-edges are part of the same edge nor the data of which half edges are incident on a vertex; the permutation also should act trivially on the half edges incident to the marked vertex. If Γ is a Feynman diagram for $\langle x^n \rangle$, its *first Betti number* $\beta(\Gamma)$ is the number of (full) edges minus the number of non-marked vertices — an easy calculation shows that $\beta(\Gamma) = (v(\Gamma) + n)/2$, where $v(\Gamma)$ is the number of unmarked vertices in Γ.

We list the Feynman diagrams for $\langle x^2 \rangle$ with Betti number 1 or 2, along with their numbers of automorphisms:

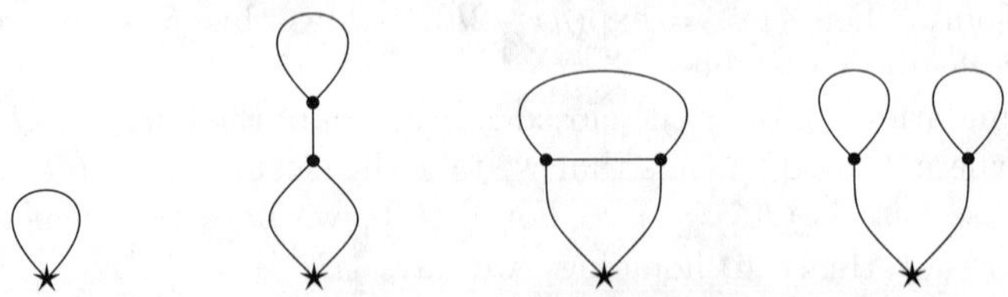

$|\mathrm{Aut}| = 1$ $\quad$ $|\mathrm{Aut}| = 2$ $\quad$ $|\mathrm{Aut}| = 2$ $\quad$ $|\mathrm{Aut}| = 4$

We claim that $\langle x^n \rangle = c_n$, where c_n is the following Feynman-style sum:

$$c_n = \sum_{\substack{\Gamma \text{ a Feynman} \\ \text{diagram for } \langle x^n \rangle}} \frac{\hbar^{\beta(\Gamma)}}{|\mathrm{Aut}\,\Gamma|}.$$

In particular, $c_0 = 1 = \langle 1 \rangle$ as it should be, and $c_n \in \hbar^{n-1}\mathbb{K}^n$. It therefore suffices to verify that the c_ns satisfy the recursion relation

$$c_{n+1} = \frac{1}{2} c_{n+2} + \hbar n c_{n-1}, \quad n \geq 0;$$

we set $c_{-1} = 0$ to get things started. Indeed, the *ad hoc* substitutions above imply that this recursion relation has at most one solution with $c_0 = 1$ such that $c_n \to 0$ in the power series topology on $\mathbb{K}[\![\hbar]\!]$.

We verify the recursion relation as follows. Let Γ be a Feynman diagram for $\langle x^{n+1} \rangle$. The half-edges ending on the marked vertex are totally ordered by $1, 2, \ldots, n+1$. Start walking along the last half-edge. You will either arrive back at the marked vertex along half-edge number j, or you will arrive at a trivalent vertex. In the first case, delete this loop, thereby creating a Feynman diagram for $\langle x^{n-1} \rangle$. In the second case, you can "unzip" Γ along half-edge number $n + 1$, between the marked vertex and the trivalent vertex. (That is, you slide the point down the half-edge to the starred vertex, much as you slide a zipper down a jacket.) You thereby create a Feynman diagram for $\langle x^{n+2} \rangle$.

Delete $\qquad$ Unzip

Every diagram for $\langle x^{n+2}\rangle$ or $\langle x^{n-1}\rangle$ can be created this way from some diagram for $\langle x^{n+1}\rangle$. To reverse the process, either zip up the last two edges in a diagram for $\langle x^{n+2}\rangle$, or add a self-loop to the marked vertex in a diagram for $\langle x^{n-1}\rangle$.

We must now count, with symmetry, how many times each diagram is created. First, we consider the unzipping operation. If our initial diagram Γ for $\langle x^{n+1}\rangle$ had a symmetry switching the two half-edges on that first trivalent vertex, then unzipping broke this symmetry, and so divided the automorphism group by 2. If it did not have such a symmetry, then in fact there were two distinct ways to unzip Γ to get the same diagram for $\langle x^{n+2}\rangle$, because we did not say which edge should become number $n+1$ and which should become number $n+2$. Together, these are the factor of $\frac{1}{2}$ in $c_{n+1} = \frac{1}{2}c_{n+2} + \dots$. Second, we have the case of deleting a self-loop: for a given diagram for $\langle x^{n-1}\rangle$, there are precisely n ways to add a self-loop to the marked vertex such that one end of the self-loop is half-edge number $n+1$, hence the factor of n in $c_{n+1} = \dots + \hbar n c_{n-1}$. The unzipping operation does not change the Betti number of a diagram, but deleting a self-loop does, hence the factors of $\hbar$. This completes the proof.

EXERCISE 2.1. What happens when $N = 1$, $a = 1$, and $b(x) = \frac{x^4}{4!}$? What about when $b(x) = \frac{x^3}{6} + \frac{x^4}{4!}$?

REMARK 2.2. There is a different solution to the recursion relation, which highlights some interesting issues about convergence. Define the following sequence in $\mathbb{K}[\![\hbar]\!]$:

$$d_n = 12^n \sum_{k=\lceil \frac{n}{2}\rceil}^{\infty} \left(\frac{\hbar}{288}\right)^k \frac{(6k-2n)!}{(3k-n)!\,(2k-n)!}$$

These d_n also satisfy the recursion relation, and $d_0 = 1 + \frac{5}{24}\hbar + \dots$ is invertible, so $c_n = d_n/d_0$. When $\mathbb{K} = \mathbb{R}$, one can estimate the growth rates of the coefficients of d_n: $\frac{(6k-2n)!}{(3k-n)!\,(2k-n)!} \sim 6^k k!$, and so $d_n \in \mathbb{R}[\![\hbar]\!]$ has zero radius of convergence. In fact, the power series c_n also have zero radius of convergence for $n \geq 1$, because their coefficients in $\hbar$ grow roughly as the coefficients of $\log d_0$. For example, one can show that $d_1 = 3\frac{\partial d_0}{\partial \hbar}$, and so $c_1 = 3\frac{\partial}{\partial \hbar}\log d_0$, so if c_1 were the Taylor expansion at 0 of something analytic in $\hbar$, then d_0 would also have positive radius of convergence. (On the other hand, the series c_n do have positive radius of convergence over $\mathbb{Q}_p$ for $p > 0$.)

3. The general case: From homological algebra to diagrammatics

We now consider the general case. Recall that we fix N a positive integer and $\mathbb{K}$ a field of characteristic 0, and we build the graded commutative algebra $V_\bullet = \mathbb{K}[\![x_1, \dots, x_N, \xi_1, \dots, \xi_N, \hbar]\!]$ as a completed symmetric algebra, with the generators are $x_1, \dots, x_N, \hbar$ in grading zero and $\xi_1, \dots, \xi_N$ in grading 1. We then choose an invertible symmetric $N \times N$ matrix (a_{ij}) and a power series $b(x) \in \mathbb{K}[\![x_1, \dots, x_N]\!]$ that vanishes at least to order 3. With this data, we make $V_\bullet$ into a chain complex by choosing the differential:

$$Q = \sum_{i,j=1}^{N} a_{ij} x_i \frac{\partial}{\partial \xi_j} - \sum_{i=1}^{N} \frac{\partial b(x)}{\partial x_i}\frac{\partial}{\partial \xi_i} - \hbar \sum_{i=1}^{N} \frac{\partial^2}{\partial x_i \partial \xi_i}.$$

For $f \in \mathbb{K}[\![x_1, \dots, x_N, \hbar]\!]$, we write $[f]$ for its image in the Q-homology of $V_\bullet$. In light of the ansatz that $H_0(V_\bullet, Q) = \mathbb{K}[\![\hbar]\!]$, we are interested in computing

$\langle f\rangle = [f]/[1] \in \mathbb{K}[\![\hbar]\!]$. It suffices to compute $\langle \sum_{i_1,\dots,i_n=1}^N f_{i_1,\dots,i_n} x_{i_1}\cdots x_{i_n}\rangle$ for each n-tensor $(f_{i_1,\dots,i_n})_{\vec\imath\in\{1,\dots,N\}^n} \in \mathbb{K}^{N^n}$.

One approach is to tackle the algebra directly. We (temporarily) adopt the following index-full notation. For $\vec\imath \in \{1,\dots,N\}^m$, we write $x_{\vec\imath}$ for $x_{i_1}\cdots x_{i_m} \in \mathbb{K}[x_1,\dots,x_N]$, so that for example the $m$fold power of x_1 is written $(x_1)^m = x_{1,\dots,1}$. We define the Taylor coefficients of b via $b^{(m)}_{\vec\imath} = \frac{\partial^m b}{\partial x_{i_1}\cdots\partial x_{i_m}}\big|_0$. In particular, each $b^{(m)}$ is a symmetric m-tensor, and $\frac{\partial b(x)}{\partial x_i} = \sum_{m=2}^\infty \frac{1}{m!}\sum_{\vec\jmath\in\{1,\dots,N\}^m} b^{(m+1)}_{i,\vec\jmath}\, x_{\vec\jmath}$.

For each $(n+1)$-tensor $(f_{i,\vec\imath})_{i\in\{1,\dots,N\},\,\vec\imath\in\{1,\dots,N\}^n}$, consider the element

$$\sum_{i,\vec\imath,j} f_{i,\vec\imath}\, x_{\vec\imath}\, (a^{-1})_{i,j}\, \xi_j \in V_1,$$

where a^{-1} is the matrix inverse to $(a_{i,j})$. These elements, running over all such tensors, span V_1 (rather, for the power-series topology, their span is dense in V_1). Thus, the image under Q of this basis spans the boundaries in V_0 (in the power-series topology):

$$\begin{aligned} Q\left(\sum_{i,\vec\imath} f_{i,\vec\imath}\, x_{\vec\imath}\, (a^{-1})_{i,j}\, \xi_j\right) \\ = \sum_{i,\vec\imath} f_{i,\vec\imath}\, x_i x_{\vec\imath} - \sum_{m=2}^{\infty}\sum_{i,\vec\imath,j,\vec\jmath} \frac{1}{m!} b^{(m+1)}_{j,\vec\jmath}\, x_{\vec\jmath}\, (a^{-1})_{i,j}\, f_{i,\vec\imath}\, x_{\vec\imath} \\ - \hbar \sum_{i,\vec\jmath}\sum_{k=1}^{n} f_{i,\vec\jmath}\, (a^{-1})_{i,j_k}\, x_{j_1,\dots,\hat\jmath_k,\dots,j_n}. \end{aligned}$$

By "$\hat\jmath_k$" we mean "remove this term from the list." Thus we can write the class $\left[\sum_{i,\vec\imath} f_{i,\vec\imath}\, x_i x_{\vec\imath}\right]$ as a sum of various other terms, each of which has either more xs or more $\hbar$s.

This index-full notation is, of course, a mess. Much better is a Feynman-diagrammatic notation generalized by Penrose to handle contractions of tensors [**Pen71**]. We define a *Feynman diagram* to be a finite connected graph (self-loops and parallel edges are allowed) built from the following pieces:

- Precisely one *marked* vertex, with valence n, which is labeled by an n-tensor $f \in \mathbb{K}^{N^n}$, and whose incident half-edges are totally ordered; we will draw the marked vertex with a star ★, and leave the tensor and the total ordering implicit.
- Some number of *internal* vertices, which are required to have valence 3 or more; we will draw internal vertices as solid bullets •.
- Some number of univalent *external* vertices; we will draw external vertices as open circles ∘.

An *automorphism* of a Feynman diagram is a permutation of its half-edges that does not change the combinatorial type of the diagram — it may separately permute both the internal and external vertices, but it should not permute the half-edges incident to the marked vertex. Given a Feynman diagram Γ, its *first Betti number* $\beta(\Gamma)$ is its total number of edges minus its number of un-marked vertices. We say that an edge is *internal* if it connects internal and marked vertices, and *external* if one of its ends is an external vertex.

We now define the *evaluation* $\mathrm{ev}(\Gamma)$ of a Feynman diagram Γ as follows. First, suppose we are given a labeling of the half-edges by numbers $\{1, \ldots, N\}$. To such a labeled Feynman diagram we associate a product of matrix coefficients:

- The marked vertex contributes $f_{\vec{\imath}}$, where $\vec{\imath}$ is the vector of labels formed by reading the labels on the incident half-edges in the prescribed order (recall that part of the data of Γ was a total ordering of these vertices).
- Each internal vertex with valence m contributes $b^{(m)}_{\vec{\imath}}$, where $\vec{\imath}$ is the vector of labels formed by reading the incident half-edges in any order (recall that the tensors $b^{(m)}$ are symmetric).
- Each external vertex with incident half-edge labeled by $i \in \{1, \ldots, N\}$ contributes the variable $x_i \in V_0 = \mathbb{K}[\![x_1, \ldots, x_N, \hbar]\!]$.
- Each internal edge with half-edges labeled i, j contributes $(a^{-1})_{i,j} = (a^{-1})_{j,i}$.
- Each external edge with half-edges labeled i, j contributes $\delta_{i,j}$.

Thus a labeled Feynman diagram evaluates to some monomial in V_0. The evaluation $\mathrm{ev}(\Gamma)$ of an unlabeled Feynman diagram Γ is defined to be the sum over all possible labelings of its evaluation as a labeled Feynman diagram. Finally, we give a map $\{\text{Feynman diagrams}\} \to V_0$ by:

$$\Gamma \mapsto \frac{\mathrm{ev}(\Gamma)\hbar^{\beta(\Gamma)}}{|\mathrm{Aut}(\Gamma)|}.$$

In this notation, our index-full calculation above was the statement that, for a fixed tensor f labeling the marked vertices:

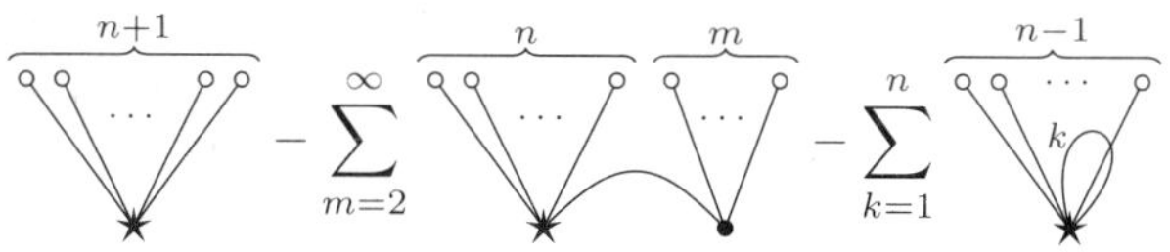

is a boundary in V_0.

In the final diagram, the self-loop connects the kth and $(n+1)$th half-edges on the marked vertex.

We can therefore evaluate any $\langle \sum_{\vec{\imath}} f_{\vec{\imath}} x_{\vec{\imath}} \rangle$, and indeed $\langle \frac{\mathrm{ev}(\Gamma)\hbar^{\beta(\Gamma)}}{|\mathrm{Aut}(\Gamma)|} \rangle$ for any Feynman diagram Γ, by playing Hercules' game of the many-headed Hydra, as follows. Pick some external vertex. Either we try to chop it off, forming a new internal vertex, at which point our Hydra grows at least two new external vertices ("heads"), or we attach it to some other external vertex, thus forming a loop and increasing the Betti number of the graph. (This description holds modulo boundaries in $V_\bullet$.) In the power-series topology, any sequence of Hydra with strictly-increasing head number converges to 0, and for any given β the game only produces finitely many graphs with Betti number $\leq \beta$. Thus the whole game converges in the power-series topology.

What does it converge to? The only Feynman diagrams left at the end of the game are those with no external vertices at all, since these are the only Hydra that do not have a head that Hercules can chop off. All together, we have proved that for any tensor $f_{\vec{\imath}}$,

$$\left\langle \sum_{\vec{\imath}} f_{\vec{\imath}} x_{\vec{\imath}} \right\rangle = \sum_{\substack{\text{Feynman diagrams } \Gamma \\ \text{with no external vertices} \\ \text{and marked vertex labeled by } f}} \frac{\mathrm{ev}(\Gamma)\, \hbar^{\beta(\Gamma)}}{|\mathrm{Aut}(\Gamma)|}$$

as a formal power series in $\mathbb{K}[\![\hbar]\!]$.

REMARK 3.1. These techniques and this result are examples of *homological perturbation theory*, which allows one to understand a chain complex as a "perturbation" of a simpler chain complex. On the same graded vector space $V_\bullet = \mathbb{K}[\![x_1, \ldots, x_N, \xi_1, \ldots, \xi_N, \hbar]\!]$ we could consider the very simple differential $Q_0 = \sum_{i,j=1}^N a_{ij} x_i \frac{\partial}{\partial \xi_j}$. An easy exercise shows that the homology of $V_\bullet$ consists of a copy of $\mathbb{K}[\![\hbar]\!]$ in degree 0, and that for Q_0 the map $\langle \cdot \rangle : V_\bullet \to \mathbb{K}[\![\hbar]\!]$ is the algebra homomorphism that sets all the x_i and ξ_j to 0.

The differential Q that we actually care about is a "perturbation" of Q_0 in the sense that $Q - Q_0$ is "small", and, in fact, "much smaller than Q_0" in the power-series topology. There is a general theory that allows one to analyze (and, indeed, write formulas for) chain complexes which are perturbations of understood complexes. In particular, whenever one has a perturbation of a chain complex, one can construct a differential δ on $\mathrm{H}_\bullet(V_\bullet, Q_0)$, such that the homology of $(\mathrm{H}_\bullet(V_\bullet, Q_0), \delta)$ is identified with the homology of $(V_\bullet, Q)$. But in our case $\mathrm{H}_\bullet(V_\bullet, Q_0) = \mathbb{K}[\![\hbar]\!]$, and so the differential δ is necessarily zero. Thus $\mathrm{H}_\bullet(V_\bullet, Q) = \mathbb{K}[\![\hbar]\!]$ as well, justifying our original ansatz from the introduction. Perturbing the differential does, however, change the map $\langle \cdot \rangle$, and although the perturbed map can be fully described with general homological perturbation theory, in our situation of interest the direct analysis is shorter. For more details on homological perturbation theory, see [**Cra04**].

4. Motivation: Finite dimensional integrals

A different explanation is available for why $\langle f \rangle$ is computable as a sum of Feynman diagrams. To explain it, we move to a somewhat more general problem, explained in detail in [**Wit90**]. In brief, we relate the usual way of encoding integration using homological algebra — namely, the de Rham complex — to the homological constructions we've discussed so far.

Let X be an N-dimensional smooth manifold over $\mathbb{R}$. In analogy with the de Rham complex $\Omega^\bullet = \Gamma(\bigwedge^\bullet \mathrm{T}^* X)$, the manifold X determines a graded commutative algebra $V_\bullet = \Gamma(\bigwedge^\bullet \mathrm{T}X)$ of antisymmetric multivector fields. Unlike the de Rham complex, $V_\bullet$ is not canonically a chain complex. But we can make it into one: when X is oriented, any choice of nowhere-vanishing smooth measure μ on X determines an isomorphism of graded vector spaces $i_\mu : V_\bullet \xrightarrow{\sim} \Omega^{N-\bullet}$, sending a multivector field $Y_1 \wedge \cdots \wedge Y_k$ to its contraction with μ, which will be an $(N-k)$-form. For instance, when $k = 0$, we send a function $f \in V_0$ to the top form $f\mu \in \Omega^N$. Under this isomorphism of vector spaces, we can transfer the exterior derivative $\mathrm{d} : \Omega^\bullet \to \Omega^{\bullet+1}$ to define the *divergence* operator $\Delta_\mu : V_\bullet \to V_{\bullet-1}$. We remark that Δ_μ only depends on μ up to global rescalings: $\Delta_\mu = \Delta_{r\mu}$ for $r \in \mathbb{R}^\times$.

We call the resulting differential "Δ" because it is a second-order differential operator for the algebra structure on $V_\bullet$. Pick local coordinates $x_1, \ldots, x_N$ on X such that $\mu = |\mathrm{d}x_1 \cdots \mathrm{d}x_N|$. Then locally $V_\bullet$ has generators $x_1, \ldots, x_N$ in degree 0 and $\xi_1, \ldots, \xi_N$ in degree 1, where the generator $\xi_i \in V_1 = \Gamma(\mathrm{T}X)$ is nothing but the vector field $\frac{\partial}{\partial x_i}$. Then $\Delta_\mu = \sum_i \frac{\partial^2}{\partial x_i \partial \xi_i}$. Observe that $(V_\bullet, \Delta_\mu)$ is *not* a dg algebra. Rather, it is an example of a BV algebra. The notion of a BV algebra axiomatizes the structure implicit in this example of multivector fields, and it allows one to apply the diagrammatics we've developed to other circumstances. The origins of BV algebras lie in quantum field theory. In that case, the "fields" form an infinite-dimensional manifold. The ordinary de Rham complex is not the right way to obtain volume forms or measures since there are no "top forms" for an infinite-dimensional manifold. By contrast, the multivector fields still work and thus provide a homological approach to defining volume forms.

A *BV algebra* is a graded-commutative algebra $V_\bullet$ with a degree-(-1) square-zero second-order differential operator Δ. The failure of Δ to be a derivation equips $V_\bullet$ with a Poisson bracket $\{-,-\}$ of degree-(-1): we set

$$\{f,g\} := \Delta(fg) - \Big((\Delta f)g + (-1)^{|f|} f(\Delta g)\Big),$$

and see that this is a derivation in each slot because Δ is a second-order differential operator. In our case of the multivector fields, the failure of Δ_μ to be a derivation is measured by the *Schouten-Nijenhuis bracket*, the usual Lie bracket of vector fields extended to the exterior algebra in the natural way.

Provided X is compact and connected, the Nth cohomology of $\Omega^\bullet$, and hence the 0th homology of $V_\bullet$, is one-dimensional. The map

$$[\cdot] : V_0 = \mathscr{C}^\infty(X) \to H_0(V_\bullet) \cong \mathbb{R}$$

is, up to rescaling, precisely the usual integration map $f \mapsto \int_X f\mu$. In particular, the boundaries in V_0 are the functions with total integral 0, and, defining $\langle f \rangle = [f]/[1]$ as before, we see that $\langle f \rangle$ is precisely the *expectation value* $\langle f \rangle = \frac{\int f\mu}{\int \mu}$.

Pick a function $s \in \mathscr{C}^\infty(X)$, and let $\hbar$ range over $\mathbb{R}_+$. Fix a volume form μ on X, and consider the family of measures $\exp(-s/\hbar)\mu$. This produces a family of divergence operators and so of chain complexes, but all the complexes are isomorphic to the de Rham complex on X, by construction. It's not difficult to compute the corresponding BV structures:

$$\Delta_{\exp(-s/\hbar)\mu} = -\frac{1}{\hbar}\sum_{i=1}^{N} \frac{\partial s}{\partial x_i}\frac{\partial}{\partial \xi_i} + \Delta_\mu,$$

where $x_1, \dots, x_N$ are any system of local coordinates on X, and $\xi_1, \dots, \xi_N$ are the corresponding degree-1 generators of $V_\bullet$. (In coordinate-free language, the operator $\sum_{i=1}^N \frac{\partial s}{\partial x_i}\frac{\partial}{\partial \xi_i}$ on $V_\bullet = \Gamma(\bigwedge^\bullet \mathrm{T}X)$ is the operator "contract with $\mathrm{d}s$.") As all the complexes are isomorphic, the homology of $\Delta_{\exp(-s/\hbar)\mu}$ is the same as the homology of $-\hbar\Delta_{\exp(-s/\hbar)\mu} = \sum \frac{\partial s}{\partial x_i}\frac{\partial}{\partial \xi_i} - \hbar\Delta_\mu$.

Let us now consider the case when s has a unique minimum (with invertible Hessian), in the limit as $\hbar \to 0$. Then the measure $\exp(-s/\hbar)\mu$ is asymptotically supported in an infinitesimal neighborhood of this critical point. Pick local coordinates $x_1, \dots, x_N$ with the critical point at the origin and satisfying $\mu = |\mathrm{d}x_1 \cdots \mathrm{d}x_N|$. Then:

$$-\hbar\Delta_{\exp(-s/\hbar)\mu} = \sum_{i=1}^{N} \frac{\partial s}{\partial x_i}\frac{\partial}{\partial \xi_i} - \hbar \sum_{i=1}^{N} \frac{\partial^2}{\partial x_i \partial \xi_i}.$$

The choice of coordinates determines a splitting $s(x) = s(0) + \frac{1}{2}\sum_{i,j} a_{i,j}x_i x_j - b(x)$ for $(a_{i,j})_{i,j=1}^N$ a symmetric positive-definite matrix, and $b(x) \in \mathscr{C}^\infty(X)$ vanishing at least cubicly. Then in terms of Taylor series, $-\hbar\Delta_{\exp(-s/\hbar)\mu}$ is precisely the differential Q that we considered above. The case when the matrix $(a_{i,j})$ is symmetric and invertible but not positive-definite can be treated similarly, by considering oscillating measures of the form $\exp(\sqrt{-1}s/\hbar)\mu$ rather than the exponentially suppressed measures $\exp(-s/\hbar)\mu$.

Then to calculate the $\hbar \to 0$ asymptotics of expectation values $\langle f \rangle$, we can on the one hand proceed as above with homological calculations, and on the other hand we can study asymptotics of finite-dimensional integrals. But these asymptotics are well-known to be computed by Feynman diagrams: one simply uses repeated integration by parts to recover the kind of diagrams we have drawn [**Dys48**, **Fey50**, **Pol05**]. This is the secret reason that the homological problem we began with had a diagrammatic answer.

REMARK 4.1. When $\pi_1(X) \neq 1$, not all BV algebra structures compatible with the Schouten-Nijenhuis bracket arise from measures. When X is connected, the data of a nowhere-vanishing measure up to global rescaling is the same as a flat, holonomy-free connection on the density line bundle over X. In general, BV algebra structures on $V_\bullet$ are in bijection with flat connections on the density line bundle, possibly with nontrivial holonomy [**Kos85**]. Among other lessons of the BV formalism is that it makes sense to talk about "expectation values" for this more general kind of "measure."

Acknowledgments

We had many rich and helpful discussions with Dan Berwick-Evans, Kevin Costello, Nicolai Reshetikhin, Thel Seraphim, and Yuan Shen about this material and many related questions. We learned the p-adic remark at the end of Section 2 from Boris Hanin. Philsang Yoo provided detailed comments on a draft of this paper. The original work was completed while the second author was a visitor at Northwestern University, which he thanks for its hospitality. We wish to thank an attentive, thoughtful referee for advice leading to this improved version.

References

[ABF10] C. Albert, B. Bleile, and J. Fröhlich, *Batalin-Vilkovisky integrals in finite dimensions*, J. Math. Phys. **51** (2010), no. 1, 015213, 31, DOI 10.1063/1.3278524. MR2605846

[BV81] I. A. Batalin and G. A. Vilkovisky, *Gauge algebra and quantization*, Phys. Lett. B **102** (1981), no. 1, 27–31, DOI 10.1016/0370-2693(81)90205-7. MR616572

[BV83] I. A. Batalin and G. A. Vilkovisky, *Quantization of gauge theories with linearly dependent generators*, Phys. Rev. D (3) **28** (1983), no. 10, 2567–2582, DOI 10.1103/PhysRevD.28.2567. MR726170

[BV84] I. A. Batalin and G. A. Vilkovisky, *Closure of the gauge algebra, generalized Lie equations and Feynman rules*, Nuclear Phys. B **234** (1984), no. 1, 106–124, DOI 10.1016/0550-3213(84)90227-X. MR736479

[BV85] I. A. Batalin and G. A. Vilkovisky, *Existence theorem for gauge algebra*, J. Math. Phys. **26** (1985), no. 1, 172–184, DOI 10.1063/1.526780. MR776145

[Cra04] M. Crainic, *On the perturbation lemma, and deformations*, arXiv:math/0403266.

[Dys48] F. J. Dyson, *The radiation theories of Tomonaga, Schwinger, and Feynman*, Physical Rev. (2) **75** (1949), 486–502. MR0028203

[Fey50] R. P. Feynman, *Mathematical formulation of the quantum theory of electromagnetic interaction*, Physical Rev. (2) **80** (1950), 440–457. MR0041726

[Fio03] D. Fiorenza, *An introduction to the Batalin-Vilkovisky formalism*, Comptes Rendus des Rencontres Mathematiques de Glanon (2003).

[GH18] Owen Gwilliam and Rune Haugseng, *Linear Batalin–Vilkovisky quantization as a functor of ∞-categories*, Selecta Math. (N.S.) **24** (2018), no. 2, 1247–1313, DOI 10.1007/s00029-018-0396-0. MR3782422

[Gwi12] Owen Gwilliam, *Factorization Algebras and Free Field Theories*, ProQuest LLC, Ann Arbor, MI, 2012. Thesis (Ph.D.)–Northwestern University. MR3034665

[JF15] Theo Johnson-Freyd, *Homological perturbation theory for nonperturbative integrals*, Lett. Math. Phys. **105** (2015), no. 11, 1605–1632, DOI 10.1007/s11005-015-0791-9. MR3406714

[Kos85] Jean-Louis Koszul, *Crochet de Schouten-Nijenhuis et cohomologie* (French), Astérisque **Numéro Hors Série** (1985), 257–271. The mathematical heritage of Élie Cartan (Lyon, 1984). MR837203

[Li17] Si Li, *Effective Batalin–Vilkovisky quantization and geometric applications*, arXiv:1709.00669.

[Mne17] Pavel Mnev, *Lectures on Batalin–Vilkovisky formalism and its applications in topological quantum field theory*, arXiv:1707.08096.

[Pen71] Roger Penrose, *Applications of negative dimensional tensors*, Combinatorial Mathematics and its Applications (Proc. Conf., Oxford, 1969), Academic Press, London, 1971, pp. 221–244. MR0281657

[Pol05] Michael Polyak, *Feynman diagrams for pedestrians and mathematicians*, Graphs and patterns in mathematics and theoretical physics, Proc. Sympos. Pure Math., vol. 73, Amer. Math. Soc., Providence, RI, 2005, pp. 15–42, DOI 10.1090/pspum/073/2131010. MR2131010

[Sta97] J. Stasheff, *The (secret?) homological algebra of the Batalin-Vilkovisky approach*, Talk given at Conference on Secondary Calculus and Cohomological Physics, Moscow, Russia, 24-31 Aug, 1997.

[Wic50] G. C. Wick, *The evaluation of the collision matrix*, Physical Rev. (2) **80** (1950), 268–272. MR0038281

[Wit90] Edward Witten, *A note on the antibracket formalism*, Modern Phys. Lett. A **5** (1990), no. 7, 487–494, DOI 10.1142/S0217732390000561. MR1049114

Department of Mathematics, Northwestern University, Evanston, Illinois 60208
Current address: Max Planck Institute for Mathematics, Bonn, Germany
Email address: gwilliam@mpim-bonn.mpg.de

Department of Mathematics, University of California, Berkeley, California 94720
Current address: Perimeter Institute for Mathematical Physics, Waterloo, Canada
Email address: tjohnsonfreyd@perimeterinsitute.ca

Contemporary Mathematics
Volume **718**, 2018
https://doi.org/10.1090/conm/718/14477

Homotopy RG flow and the non-linear σ-model

Ryan Grady and Brian Williams

Abstract. The purpose of this note is to give a mathematical treatment to the low energy effective theory of the two-dimensional σ-model. Perhaps surprisingly, this low energy effective theory encodes much of the topology and geometry of the target manifold. In particular, we relate the β-function of our theory to the Ricci curvature of the target, recovering the physical result of Friedan.

Contents

1. Introduction

In this note we give a mathematical treatment to the low energy effective theory of the two-dimensional σ-model. This σ-model involves maps from a Riemann surface Σ to a Riemannian manifold X with metric h. Physically, a field is a smooth map $\varphi : \Sigma \to X$ and the action functional is given by

$$S(\varphi) = \int_\Sigma h(\partial\varphi, \overline{\partial}\varphi),$$

where $\partial, \overline{\partial}$ are the holomorphic and anti-holomorphic pieces of the de Rham differential on Σ. The map φ satisfies the classical equations of motion if and only if it is harmonic. Classically, of course, the theory is conformal. In particular, when we work locally on Σ the theory is scale invariant. We study the failure of the classical

2010 *Mathematics Subject Classification.* Primary 81T40, 81T17, 53C44; Secondary 81T15, 17B55.

The first author was partially supported by the National Science Foundation under Award DMS-1309118.

The second author enjoyed support of the National Science Foundation as a graduate student research fellow under Award DGE-1324585.

theory to be scale invariant at the quantum level. The quantity that measures this is precisely the *β-function* of the quantum field theory.

We do not study the full theory, but rather only the theory in perturbation around the space of constant maps in the space of all harmonic maps. Our low energy effective quantization encodes much of the topology and geometry of the target manifold. Our main result is to relate the one-loop β-function to the Ricci curvature of X.

THEOREM A. *The one-loop β-function for the two-dimensional σ-model with target the Riemannian manifold (X, h) satisfies*

$$\beta^{(1)}(h) = -\frac{1}{12\pi}\mathrm{Ric}(h).$$

In 1985, Dan Friedan [**7**] gave a physical argument for the validity of Theorem A; see also the more recent geometric overview of Carfora [**1**]. Recently, using Costello's approach to the Batalin–Vilkovisky (BV) formalism [**3**], Nguyen [**14**] has sketched an approach towards this result in the case that the target manifold X is a homogeneous space. Below, we give a mathematically rigorous treatment for a general target Riemannian manifold (X, h). Our approach also uses Costello's paradigm for BV theory, as well as the subsequent (rigorous) development of the β-function in this setting by Elliott, Williams, and Yoo [**6**].

The theorem identifies the one-loop contribution to the β-function, so to compute it we only consider the quantization to first-order in the quantum parameter $\hbar$. Key to this process, in the BV formalism, is to study the moduli of all such quantizations. Perturbatively, the object which controls this moduli space is the deformation complex. Its cohomology represents the space of all functionals by which we can deform the classical action functional.

THEOREM B. *Let* Def *denote the obstruction deformation complex for the σ-model of maps $\mathbb{C} \to (X, h)$. There is a quasi-isomorphism*

$$(\mathrm{Def})^{\mathrm{Aff}(\mathbb{C})} \simeq \mathrm{Riem}(X,h)[1] \oplus \Omega^3_{cl,X},$$

where $\Omega^3_{cl,X}$ is the sheaf of closed three-forms and

$$\mathrm{Riem}(X,h) = T_X \xrightarrow{\mathrm{d}} \mathrm{Sym}^2(T_X^\vee)[-1].$$

The first summand of Def deforms the metric on the target manifold, while the second summand introduces an H-flux. In this note, we don't discuss how the addition of an H-flux modifies the β-function. Though we expect that one could study this through the exact Courant algebroid determined by the H-flux 3-form; this would provide an interpretation of recent work of Ševera and Valach [**15**] in the BV formalism.

There are many remaining mathematical questions, especially regarding the observable theory for the σ-model, which are beyond the scope of this paper. For instance, the algebra of quantum observables [**4**] should be quite large and include the *bc*-$\beta\gamma$-vertex algebra. There is also the question of studying the non-perturbative behavior of the β-function. The full β-function should be a section of some line bundle on the full space of classical solutions (that is, harmonic maps) obtained by quantizing the theory in families.

In Section 2, we construct the classical BV theory using Lie theoretic techniques as in [**2**] and [**10**]. We also compute the obstruction deformation complex of the theory, culminating in Proposition 2.8. Section 3 is about quantizing the σ-model to one-loop, i.e., modulo $\hbar^2$; a cohomological argument shows there is no obstruction. In Section 4, we summarize the relevant results from **[3]** and **[6]** regarding the mathematical definition of the β-function. In particular, we have aimed to point out connections to and differences from the physics literature, e.g., as in Remark 4.5. Finally, in Section 5, we compute the one-loop β-function in terms of the Ricci tensor of the target manifold.

2. The classical σ-model

In this section we set up the mathematical model that we use to study the the σ-model of maps from a Riemann surface Σ to a Riemannian manifold (X, h) where h is the metric. Physically, a field is a map $\varphi : \Sigma \to X$ and the action functional is of the form

$$S(\varphi) = \int_\Sigma h(\partial\varphi, \overline{\partial}\varphi),$$

where $\partial, \overline{\partial}$ are the holomorphic and anti-holomorphic pieces of the de Rham differential on Σ. The classical solutions to the equations of motion are harmonic maps $\Sigma \to X$. This whole mapping space is quite complicated and difficult to study, and the approach we take is to work perturbatively around a fixed harmonic map $\varphi_0 : \Sigma \to X$ (in fact, we take φ_0 to be a constant map).

2.1. Smooth geometry and L_∞ algebras. The language we will use to describe the perturbative non-linear σ-model uses the formalism of L_∞ spaces. For examples on how this formalism has been used to study related theories see **[2, 9, 10, 12, 13]**. The goal is to describe the target by Lie algebraic data so that the theory we write down behaves formally like a gauge theory. In later sections we will see that such a description lends itself to a rigorous analysis in the BV-BRST formalism, and hence in a description of its behavior under local RG flow.

Let T_X be the sheaf of smooth vector fields and Ω^*_X be the sheaf of de Rham forms equipped with the de Rham differential. If we want to refer to the sheaf as a graded vector space, that is with the differential turned off, we use the notation $\Omega^\#_X$. If $\mathcal{E}$ is the sheaf of smooth sections of a vector bundle E with flat connection we let $\Omega^*_X(\mathcal{E})$ denote its associated de Rham complex.

The main input we need is a description of X in terms of a certain curved L_∞ algebra defined over the de Rham complex Ω^*_X. For a definition of curved L_∞ algebras and a proof of the following result we refer the reader to **[11]**.

PROPOSITION 2.1 (Lemma 4.12 in **[11]**). *Let X be a smooth manifold. Then, there is a curved L_∞ algebra $\mathfrak{g}_X$ over Ω^*_X such that*

$$\mathrm{C}^*_{\mathrm{Lie}}(\mathfrak{g}_X) \cong \Omega^*_X(\mathcal{J}_X),$$

where $\mathcal{J}_X$ is the sheaf of ∞-jets of the trivial bundle on X.

REMARK 2.2. As a graded $\Omega^\#_X$-module there is an identification $\mathfrak{g}_X \cong \Omega^\#_X \otimes_{C^\infty_X} T_X[-1]$. Moreover, the curved L_∞ algebra $\mathfrak{g}_X$ is a natural object to consider for any smooth manifold X: up to isomorphism it is unique up to a contractible choice.

We will denote by $\{\ell_k\}$ the family of L_∞ brackets of the curved L_∞ algebra $\mathfrak{g}_X$. For example, $\ell_0 \in \Omega^*_X$ is of cohomological degree two and represents the "curving" [1]. Similarly, the unary bracket $\ell_1 : \mathfrak{g}_X \to \mathfrak{g}_X$ gives $\mathfrak{g}_X$ the structure of a dg module over the dg ring Ω^*_X.

A choice of a Riemannian metric h on X determines an isomorphism $h : T_X \cong T^*_X$. This isomorphism extends to the L_∞ algebra $\mathfrak{g}_X$ in the following way.

LEMMA 2.3. *Suppose h is a Riemannian metric on X. Then, there is an isomorphism of Ω^*_X-modules*

$$h^\flat : \mathfrak{g}_X \cong \mathfrak{g}_X^\vee,$$

where $\mathfrak{g}_X^\vee$ denotes the Ω^*_X-linear dual. The inverse isomorphism is denoted $h^\#$.

PROOF. Since the isomorphism $h : T_X \cong T_X^\vee$ is of C^∞_X-modules we have an induced isomorphism of graded $\Omega^\#_X$-modules

$$h : \Omega^\#_X \otimes_{C^\infty_X} T_X[-1] \cong \Omega^\#_X \otimes_{C^\infty_X} T_X^\vee[-1].$$

□

Given any vector bundle E on X we can consider the D_X-module of smooth ∞ jets $\mathcal{J}_E$, and hence its de Rham complex Ω^*_X. If $\mathcal{E}$ denotes the sheaf of sections of E, there is a natural quasi-isomorphism

$$j_\infty : \mathcal{E} \xrightarrow{\simeq} \Omega^*_X(\mathcal{J}_E)$$

sending a smooth section to its power series expansion. For example, in the case of the trivial bundle on X this defines a quasi-isomorphism $C^\infty_X \simeq \mathrm{C}^*_{\mathrm{Lie}}(\mathfrak{g}_X)$. In general, we make the following definition.

DEFINITION 2.4. Let E be a vector bundle on X. We define the $\mathfrak{g}_X$ module, which we still denote by E, whose Chevalley-Eilenberg complex is the $\mathrm{C}^*_{\mathrm{Lie}}(\mathfrak{g}_X)$ dg module

$$\mathrm{C}^*_{\mathrm{Lie}}(\mathfrak{g}_X; E) := \Omega^*_X(\mathcal{J}_E).$$

In the case of a Riemannian manifold (X, h) we have a metric $h \in \mathrm{Sym}^2(T_X^\vee)$. In particular, we can consider its ∞-jet

$$h_\infty := j_\infty(h) \in \Omega^*_X\left(\mathcal{J}\left(\mathrm{Sym}^2(T_X^\vee)\right)\right) = \mathrm{C}^*_{\mathrm{Lie}}\left(\mathfrak{g}_X; \mathrm{Sym}^2(T_X^\vee)\right). \tag{2.1}$$

By the equality on the right-hand side, we can think of h_∞ as being a functional on $\mathfrak{g}_X$ with values in the sheaf of symmetric $(0,2)$-tensors.

2.2. The BV formalism. To write down a theory on Σ in the BV formalism it suffices to prescribe a sheaf of elliptic complexes $\mathcal{E}$ on Σ equipped with a (-1)-shifted symplectic form – the space of fields – together with a local functional on $\mathcal{E}$ – the interaction functional. For a more precise definition we refer the reader to Definition 5.4.0.3 in [**4**].

The sheaf of fields of the Riemannian σ-model is the sheaf on Σ of $\Omega^\#_X$-modules

$$\mathcal{E} = C^\infty_X \otimes_{\mathbb{C}} \mathfrak{g}_X[1] \oplus \Omega^2_X \otimes_{\mathbb{C}} \mathfrak{g}_X^\vee.$$

We write the fields as $\varphi \in C^\infty_\Sigma \otimes \mathfrak{g}_X$ and $\psi \in \Omega^2_\Sigma \otimes \mathfrak{g}_X^\vee$. The shifted symplectic pairing of degree (-1) is defined by

$$\langle \varphi, \psi \rangle = \int_\Sigma \langle \varphi, \psi \rangle_{\mathfrak{g}} \in \Omega^\#_X.$$

[1]To fix an L_∞ structure on $\mathfrak{g}_X$ it suffices to choose a connection on T_X. In this case, ℓ_0 is identified with the curvature of this connection.

We will write the classical action functional $S : \mathcal{E} \to \Omega^*_X$ in the form

$$S = S_{\text{free}} + I$$

where S_{free} is the free part that is quadratic as a functional on $\mathcal{E}$. To define this functional consider the operator

$$\partial\overline{\partial} \otimes h^\flat : C^\infty_\Sigma \otimes \mathfrak{g}_X \to \Omega^2_\Sigma \otimes \mathfrak{g}^\vee_X.$$

where $h^\flat$ is interpreted as in Lemma 2.3. There are internal differentials on $\mathfrak{g}_X$ (and $\mathfrak{g}^\vee_X$) given by ℓ_1 (and its linear dual). We let $Q = \partial\overline{\partial} \otimes h + \ell_1$. Define

$$S_{\text{free}}(\varphi, \psi) = \int_\Sigma \langle \varphi, (\partial\overline{\partial} \otimes h)(\varphi)\rangle .$$

Note that $\langle \varphi, (\partial\overline{\partial} \otimes h)\varphi\rangle = h(\varphi, \partial\overline{\partial}\varphi)$ so that this reduces to the familiar kinetic term of the σ-model. Further, we remark that the operator Q satisfies $Q^2 = \ell_1^2$ which is, in general, not zero, which reflects the fact that the L_∞ algebra is curved.

To define the interaction we first recall from (2.1) that we can view the metric as determining an element h_∞ which we view as a functional on $\mathfrak{g}_X$. We can extend this to a functional on $C^\infty_\Sigma \otimes \mathfrak{g}_X$ simply by multiplying the function component. Similarly, we can extend the brackets ℓ_k defining the L_∞ structure on $\mathfrak{g}_X$. The interaction is written as a sum of two terms $I = I^h + I^X$ where

$$I^h(\varphi) = \sum_{k\geq 1} \frac{1}{k!} \int_\Sigma h^{(k)}_\infty(\partial\varphi, \overline{\partial}\varphi; \varphi^{\otimes k})$$

and

$$I^X(\varphi, \psi) = \sum_{k\geq 2} \frac{1}{(k+1)!} \int_\Sigma \langle \psi, \ell_k(\varphi^{\otimes k}))\rangle_{\mathfrak{g}_X} .$$

For each k, $h^{(k)}_\infty$ denotes the kth term in the jet expansion and the ';' separates the tensorial indices from the jet indices.

Thus, the full action functional can be written succinctly as

$$S = \int_\Sigma h_\infty(\partial\varphi, \overline{\partial}\varphi; e^\varphi) + \int_\Sigma \langle \psi, \ell(e^\varphi)\rangle_{\mathfrak{g}_X} .$$

2.2.1. *Functionals.* To describe functionals on $\mathcal{E}$ we need to introduce the dual to the space of fields. This is defined by

$$\mathcal{E}^\vee = \text{Hom}_{\Omega^\#_X}(\mathcal{E}, \Omega^\#_X).$$

The space of fields is built from the infinite dimensional space of functions on the Riemann surface Σ, which has the structure of a topological vector space. The exact category of topological vector spaces we work with will not be important, but is discussed in depth in the Appendix of [**5**]. In the notation above, the space of homomorphisms is taken to be those that are continuous. With this convention, there is a natural embedding $\mathcal{E} \hookrightarrow \mathcal{E}^\vee[-1]$ defined by the shifted symplectic form.

The space of functionals $\mathcal{O}(\mathcal{E})$ is defined by

$$\mathcal{O}(\mathcal{E}) = \prod_{n\geq 0} \text{Hom}_{\Omega^\#_X}(\text{Sym}^n(\mathcal{E}), \Omega^\#_X).$$

There is a subspace of *local functionals* $\mathcal{O}_{\text{loc}}(\mathcal{E}) \subset \mathcal{O}(\mathcal{E})$ consisting of functionals of the form $F(\varphi, \psi) = \int \mathcal{L}$ where $\mathcal{L}$ is a Lagrangian density. For a precise definition see [**4**]. The symplectic pairing on the space of fields induces a bracket of degree $+1$ on local functionals that we denote by $\{-, -\}$.

PROPOSITION 2.5. *The function $S = S_{\text{free}} + I$ is local. Moreover, it satisfies the classical master equation*

$$QI + \frac{1}{2}\{I, I\} + F_{\ell_1} = 0,$$

where $F_{\ell_1} \in \mathcal{O}_{\text{loc}}(\mathcal{E})$ satisfies $\{F_{\ell_1}, -\} = \ell_1^2$.

PROOF. Due to the nature of the symplectic pairing it is immediate that $\{S_h, S_h\} = 0$. To see that $\{S_X, S_h\} = 0$ we observe that S_h is defined using the jet expansion of a globally defined tensor, namely the metric. Next, we note that the operator $\{S_X, -\}$ is precisely the Chevalley-Eilenberg differential for the L_∞ algebra $C^\infty(\Sigma) \otimes (\mathfrak{g}_X \ltimes \text{vol}_\Sigma \cdot \mathfrak{g}_X^\vee[-1])$ with L_∞ structure given by tensoring the algebra of functions with the L_∞ structure on the square-zero extension $\mathfrak{g}_X \ltimes \mathfrak{g}_X^\vee[-1]$. Thus $\{S_X, S_X\}(\varphi, \psi) = \langle \psi, \ell_1^2(\varphi) \rangle$ is equivalent to the closure of this curved L_∞ structure. Since ℓ_1^2 represents the curvature of the connection, we are done. □

An immediate corollary is that the operator $\{S, -\} = Q + \{I, -\}$ defines a differential on the space of local functionals (considered as an $\Omega_X^\sharp$-module).

DEFINITION 2.6. The obstruction deformation complex is defined by

$$\text{Def} = (\mathcal{O}_{\text{loc}}(\mathcal{E}), \{S, -\}) = (\mathcal{O}_{\text{loc}}(\mathcal{E}), Q + \{I, -\}).$$

2.3. Calculation of the obstruction deformation complex. The definition of the obstruction deformation complex is a very natural one from the point of view of the classical field theory: natural deformations of the theory are given by deforming the local action functional. A local action functional is specified by two pieces of data: a functional on the space of fields depending only on the infinitesimal jet data of the fields (modulo constants), and a density on the manifold. A model for this space of local functionals is given by the following tensor product:

$$\mathcal{O}_{\text{loc}}(\mathcal{E}) = \text{Dens}_M \otimes_{D_M} \mathcal{O}_{red}(\mathcal{J}\mathcal{E})$$

where D_M denotes the sheaf of differential operators on M. To make sense of this tensor product we recall that Dens_M has a natural right D-module structure and the infinite jet bundle has a canonical flat connection (and hence a left D-module structure).

A priori, the space of all deformations is very large and not very manageable. In this section, we restrict ourselves to studying deformations that respect certain symmetries of the σ-model and compute such deformations in terms of geometric quantities on the target Riemannian manifold.

We consider the theory with source manifold $\Sigma = \mathbb{C}$. In this case, we see that the theory is acted upon by the group of affine linear transformations $\text{Aff}(\mathbb{C}) = \mathbb{C}^\times \ltimes \mathbb{C}$ given by rotations and translations. This is the symmetry that we wish to impose on the deformations of the model. The subcomplex of deformations that are invariant for this symmetry is denoted $(\text{Def})^{\text{Aff}(\mathbb{C})}$.

REMARK 2.7. By $(\text{Def})^{\text{Aff}(\mathbb{C})}$ we mean the strict (underived) fixed points of the affine group. The deformation complex Def is the global sections of a $\text{Aff}(\mathbb{C})$-equivariant vector bundle on $\mathbb{C}$. This $\text{Aff}(\mathbb{C})$ representation is induced from a finite dimensional $\mathbb{C}^\times$-representation. Since $\mathbb{C}^\times$ is reductive, its derived and underived fixed points agree, so this operation is reasonable in our context.

Already, for the translation invariant subcomplex there is a vast simplification of the space of local functionals. Indeed, Lemma 6.7.1 of [**3**] implies that

$$\mathcal{O}_{\text{loc}}(\mathcal{E})^{\mathbb{C}} \simeq \mathbb{C} \cdot \mathrm{d}^2 z \otimes^{\mathbb{L}}_{\mathbb{C}[\partial_z, \partial_{\bar{z}}]} \mathcal{O}_{red}(\mathcal{J}_0\mathcal{E})$$

where $\mathcal{J}_0\mathcal{E}$ denotes the fiber of the infinite jet bundle at $0 \in \mathbb{C}$. This statement is a consequence of the fact that a translation invariant local functional is completely determined by its behavior in a neighborhood of the origin.

Before stating the main result of this section, we will set up some notation. Consider the following sheaf of dg vector spaces on the Riemannian manifold (X, h)

$$T_X \xrightarrow{\mathrm{d}} \mathrm{Sym}^2(T_X^\vee)[-1]$$

where the differential is defined by $\mathrm{d}(X) = L_X h$ where L is the Lie derivative. In fact, the Lie bracket on vector fields together with the natural action of vector fields on symmetric tensors endows this sheaf with the structure of a sheaf of dg Lie algebras. This sheaf of dg Lie algebras describes the formal neighborhood of (X, h) in the moduli space of all Riemannian structures on X. We will denote this sheaf by $\mathrm{Riem}(X, h)$.

PROPOSITION 2.8. *Consider the classical BV theory $\mathcal{E}(\mathbb{C})$ placed on the Riemann surface $\Sigma = \mathbb{C}$. There is an equivalence of Ω^*_X-modules*

$$(\mathrm{Def})^{\mathrm{Aff}(\mathbb{C})} \simeq \Omega^*_X(\mathcal{J}_F)$$

where F is the sheaf of dg vector spaces on X given by

$$F = \mathrm{Riem}(X, h)[1] \oplus \Omega^3_{cl,X}$$

where $\Omega^3_{cl,X}$ is the sheaf of closed three-forms.

As with the de Rham complex of any infinite jet bundle we observe immediately that there is an equivalence of C^∞_X-modules $(\mathrm{Def})^{\mathrm{Aff}(\mathbb{C})} \simeq F$. In fact, we can describe explicitly a quasi-isomorphism

$$\mathfrak{R} : F \to (\mathrm{Def})^{\mathrm{Aff}(\mathbb{C})}$$

as follows. Let $j_\infty : F \to \Omega^*_X(\mathcal{J}_F)$ denote the quasi-isomorphism given by taking the ∞-jet of a section. For instance, if α is a section in $\mathrm{Sym}^2(T_X)$ we have $j_\infty(\alpha) \in \Omega^*(X, \mathcal{J}\,\mathrm{Sym}^2(T_X))$. Moreover, we know that there is an isomorphism of Ω^*_X-modules $\mathrm{C}^*_{\mathrm{Lie}}(\mathfrak{g}_X; \mathrm{Sym}^2(T_X^\vee))$. We obtain a local functional via the formula

$$\mathfrak{R}(\alpha)(\varphi, \psi) = \int j_\infty(\alpha)(\partial\varphi, \bar{\partial}\varphi; e^\varphi)\, \mathrm{d}^2 z.$$

Geometrically speaking, these deformations and symmetries are apparent from the point of view of the σ-model. Indeed, we see all deformations of the Riemannian structure on the target manifold. The formal neighborhood of (X, h) in the moduli of Riemannian structures is described precisely by the dg Lie algebra $\mathrm{Riem}(X, h)$.

For the other component of the deformation complex, note that the degree zero cohomology of the sheaf $\Omega^3_{cl,X}$ can be thought of as the collection of H-fluxes. The deformation of the action is realized as follows: every closed 3-form H is locally exact, so choose a 2-form B with $dB = H$. We then have the following functional on maps $\varphi : \Sigma \to X$:

$$S_H(\varphi) = \int_\Sigma \varphi^* B.$$

2.3.1. It is convenient to introduce the notation $\mathcal{L} = \mathcal{E}[-1]$ for the L_∞ algebra obtained by shifting the space of fields up by one. In this case, the piece of the differential $\{S_X, -\}$ determines an identification

$$(\mathcal{O}_{red}(\mathcal{J}_0\mathcal{E}), \{S_X, -\}) = \mathrm{C}^*_{\mathrm{Lie,red}}(\mathcal{J}_0\mathcal{L}).$$

To handle the translation invariant piece of the deformation complex, Lemma 6.7.1 of [**3**], referenced above, says we must compute the derived tensor product $\mathbb{C} \cdot \mathrm{d}^2 z \otimes^{\mathbb{L}}_{\mathbb{C}[\partial_z, \partial_{\overline{z}}]} \mathrm{C}^*_{\mathrm{Lie,red}}(\mathcal{J}_0\mathcal{L})$. To compute this we choose a free resolution of the trivial $\mathbb{C}[\partial_z, \partial_{\overline{z}}]$ module $\mathbb{C} \cdot \mathrm{d}^2 z$ given by the complex

$$M = \left(\mathbb{C}[\epsilon, \overline{\epsilon}, \partial_z, \partial_{\overline{z}}], \mathrm{d} = \frac{\partial}{\partial \epsilon}\partial_z + \frac{\partial}{\partial \overline{\epsilon}}\partial_{\overline{z}}\right)$$

where $\epsilon, \overline{\epsilon}$ have degree -1. To remind us that we are resolving the trivial module $\mathbb{C} \cdot \mathrm{d}^2 z$ we will write $M \cdot \mathrm{d}^2 z$.

We see that the derived tensor product computing the invariant deformation complex reduces to

(2.2)

$$\begin{array}{ccc} \underline{-2} & \underline{-1} & \underline{0} \\ (\epsilon\overline{\epsilon} \cdot \mathrm{d}^2 z) \cdot \mathrm{C}^*_{\mathrm{Lie,red}}(\mathcal{J}_0\mathcal{L}) & & \mathrm{d}^2 z \cdot \mathrm{C}^*_{\mathrm{Lie,red}}(\mathcal{J}_0\mathcal{L}) \\ & \searrow^{\mathrm{d}_1} \qquad \nearrow^{\mathrm{d}_2} & \\ & (\epsilon \cdot \mathrm{d}^2 z) \cdot \mathrm{C}^*_{\mathrm{Lie,red}}(\mathcal{J}_0\mathcal{L}) \oplus (\overline{\epsilon} \cdot \mathrm{d}^2 z) \cdot \mathrm{C}^*_{\mathrm{Lie,red}}(\mathcal{J}_0\mathcal{L}) & \end{array}$$

where the differentials $\mathrm{d}_1, \mathrm{d}_2$ are determined by the action of $\mathbb{C}[\partial_z, \partial_{\overline{z}}]$ on $\mathcal{J}_0\mathcal{E}$. We have omitted the contribution of the part of the differential given by $\{S_h, -\}$.

Now, we see that the ∞-jets at $0 \in \mathbb{C}$ of the L_∞ algebra $\mathcal{L}$ are of the form

$$J_0\mathcal{L} = \mathfrak{g}_X \ltimes \mathrm{d}^2 y \cdot \mathfrak{g}_X^\vee[-1][y, \overline{y}],$$

where we have denoted the formal jet variables by y and $\overline{y}$.

We have yet to account for invariance for the other piece of the affine group $\mathbb{C}^\times$ given by scaling. Let $\lambda \in \mathbb{C}^\times$. The volume element $\mathrm{d}^2 z$ scales as $\lambda \cdot \mathrm{d}^2 z = \lambda\overline{\lambda}\,\mathrm{d}^2 z$. On the variables $\epsilon, \overline{\epsilon}$ scaling acts as $\lambda \cdot \epsilon = \lambda\epsilon$ and $\lambda\overline{\epsilon} = \overline{\lambda}\overline{\epsilon}$. Similarly, $\mathbb{C}^\times$ acts on the jet complex via $\lambda \cdot y = \lambda y$ and $\lambda \cdot \overline{y} = \overline{\lambda}\overline{y}$. Thus, in the complex (2.2) above we see that the $\mathbb{C}^\times$-invariant subcomplex will be given by summands labelled by the following $\mathbb{C}^\times$-invariant elements:

$$\epsilon\overline{\epsilon}\,\mathrm{d}^2 z \ , \ \overline{\epsilon}\,\mathrm{d}^2 z(y^\vee) \ , \ \epsilon\,\mathrm{d}^2 z(\overline{y}^\vee) \ , \ \mathrm{d}^2 z(y^\vee\overline{y}^\vee) \ , \ \mathrm{d}^2 z(\mathrm{d}^2 y)^\vee.$$

We read these terms off as follows:

(1) the term labeled by $\epsilon\overline{\epsilon}\,\mathrm{d}^2 z$ contributes the summand $\mathrm{C}^*_{\mathrm{Lie,red}}(\mathfrak{g}_X)$ in degree -2 (remembering the overall shift by -2);
(2) the term labeled by $\overline{\epsilon}\,\mathrm{d}^2 z(y^\vee)$ contributes the summand $\mathrm{C}^*_{\mathrm{Lie}}(\mathfrak{g}_X; \mathfrak{g}_X^\vee[-1])$ in degree -1;
(3) the term labeled by $\epsilon\,\mathrm{d}^2 z(\overline{y}^\vee)$ contributes the summand $\mathrm{C}^*_{\mathrm{Lie}}(\mathfrak{g}_X; \mathfrak{g}_X^\vee[-1])$ in degree -1;
(4) the term labeled by $\mathrm{d}^2 z(y^\vee\overline{y}^\vee)$ contributes the summand $\mathrm{C}^*_{\mathrm{Lie}}(\mathfrak{g}_X; \mathfrak{g}_X^\vee[-1] \otimes \mathfrak{g}_X^\vee[-1] \oplus \mathfrak{g}_X^\vee[-1])$ in degree 0;
(5) the term labeled by $\mathrm{d}^2 z(\mathrm{d}^2 y)^\vee$ contributes the summand $\mathrm{C}^*_{\mathrm{Lie}}(\mathfrak{g}_X; \mathfrak{g}_X[1])$ in degree -1.

In all, we see that the $\mathbb{C}^\times$ invariant subcomplex of (2.3) is of the form

(2.3)

$$\begin{array}{ccc} \underline{-2} & \underline{-1} & \underline{0} \\ C^*_{\text{Lie,red}}(\mathfrak{g}_X) \overset{d_1}{\to} C^*_{\text{Lie}}(\mathfrak{g}_X;\mathfrak{g}_X^\vee[-1]) \overset{d_2}{\to} & C^*_{\text{Lie}}(\mathfrak{g}_X;\mathfrak{g}_X^\vee[-1]\otimes\mathfrak{g}_X^\vee[-1]) \oplus C^*_{\text{Lie}}(\mathfrak{g}_X;\mathfrak{g}_X^\vee[-1]) \\ \searrow^{d_1} \quad \nearrow^{d_2} & \nwarrow \cdots \nearrow \\ C^*_{\text{Lie}}(\mathfrak{g}_X;\mathfrak{g}_X^\vee[-1]) & C^*_{\text{Lie}}(\mathfrak{g}_X;\mathfrak{g}_X[1]) \end{array}$$

We have seen that the differential of the deformation complex is of the form $\{S_h + S_X, -\} = \{S_h, -\} + \{S_X, -\}$. Moreover, in the proof of Proposition 2.5 we showed that $\{S_h, -\}$ and $\{S_X, -\}$ commute and are each of square zero. The differential $\{S_X, -\}$ has the effect of turning on the Chevalley-Eilenberg differential for $\mathfrak{g}_X$ (and its modules). We now study what $\{S_h, -\}$ does to the complex above.

Note that S_h is a functional purely of the φ variables. So, its Poisson bracket $\{S_h, -\}$ only acts non-trivially on the deformation complex containing functionals of the ψ field. The only term above involving the ψ field is term 5 which is given by $C^*_{\text{Lie}}(\mathfrak{g}_X;\mathfrak{g}_X[1]) \cong \Omega^*_X(\mathfrak{J}_{T_X})$. The differential $\{S_h, -\}$ maps this term to the the factor $C^*_{\text{Lie}}(\mathfrak{g}_X;\mathfrak{g}_X^\vee[-1]) \cong \Omega^*_X(\mathfrak{J}_{T^\vee X})$ in term 4. In fact, this is nothing but the isomorphism $\Omega^*_X(\mathfrak{J}_{T_X}) \cong \Omega^*_X(\mathfrak{J}_{T^\vee X})$ determined by the metric h. Thus, the part of the differential $\{S_h, -\}$ has the effect of killing term 5 and the second factor of term 4 above. We have included $\{S_h, -\}$ using the dotted arrow in the complex (2.3).

Before stating the following lemma, we recall the following basic construction. The metric h on X induces the Levi-Civita connection on the tangent bundle TX and also a covariant derivative on the cotangent bundle T^*X. At the level of sheaves, the covariant derivative is of the form

$$\nabla_h : \Omega^1_X \to \Omega^1_X \otimes \Omega^1_X.$$

LEMMA 2.9. *There is a quasi-isomorphism of (2.3) (including the dotted arrows) and the complex*

(2.4)

$$\begin{array}{ccc} \underline{-2} & \underline{-1} & \underline{0} \\ \mathfrak{J}\mathcal{O}_{red}(X) \xrightarrow{d_1} & \mathfrak{J}\Omega^1_X & \xrightarrow{d_2'} \mathfrak{J}(\Omega^1_X\otimes\Omega^1_X) \\ \searrow^{d_1} & & \nearrow^{d_2'} \\ & \mathfrak{J}\Omega^1_X & \end{array}$$

where d_2' *is equal to the infinite jet expansion of the map*

$$(\nabla_h, -\sigma\nabla_h) : \Omega^1_X \oplus \Omega^1_X \to \Omega^1_X \otimes \Omega^1_X$$

sending (ω_1, ω_2) *to* $\nabla_h\omega_1 - \sigma\nabla_h\omega_2$, *with* σ *the permutation operator sending* $\omega_1 \otimes \omega_2 \mapsto \omega_2 \otimes \omega_1$.

Note that we can make the following change of coordinates in the degree -1 part of the complex (2.4) sending $(\omega_1, \omega_2) \mapsto (\omega_1 + \omega_2, \omega_1 - \omega_2)$ so that d_2' decouples into a symmetric and antisymmetric summand $\mathrm{d}_2' = \mathrm{Sym}^2(\mathrm{d}_2') - \wedge^2 \mathrm{d}_2'$, where

$$\mathrm{Sym}^2(\mathrm{d}_2') : \mathfrak{J}\Omega^1_X \to \mathfrak{J}\,\mathrm{Sym}^2\,\Omega^1_X$$

and

$$\wedge^2(\mathrm{d}_2') : \mathfrak{J}\Omega^1_X \to \mathfrak{J}\Omega^2_X.$$

We can check locally that former differential is nothing but the Lie derivative and the former the de Rham differential. (This is classical, see for instance Proposition 2.54 of [**8**].)

We are left with the following complex

$$\mathfrak{J}\mathcal{O}^{red}_X \xrightarrow{\mathrm{d}_{dR}} \mathfrak{J}\Omega^1_X \xrightarrow{\mathrm{d}_{dR}} \mathfrak{J}\Omega^2_X$$

$$\mathfrak{J}T_X \xrightarrow{L_{(-)}h} \mathfrak{J}\,\mathrm{Sym}^2\,\Omega^1_X.$$

3. One-loop quantization

In this section we construct the one-loop quantization of the classical BV theory introduced above. In the BV formalism this means that we construct a solution to the quantum master equation modulo $\hbar^2$. We proceed in the formalism developed in [**3**–**5**] which combines the effective functional approach to the path integral with the BV-BRST formalism for analyzing gauge symmetries. There are two steps to quantization. First, we must write down an effective family of functionals satisfying the (homotopical) renormalization group equations. We utilize a regularization technique based on heat kernels to construct this effective family. Next, we must consider the *quantum master equation* (QME) for this effective family. This can be viewed as studying quantizations that respect the natural gauge symmetries of the theory.

We will only produce a quantization for the σ-model with source given by a two-dimensional disk.

3.1. Gauge fixing. Before writing down the heat kernel and propagator we must choose a gauge fixing operator. For us, this is the degree (-1) operator Q^{GF} on the complex of fields $\mathcal{E}$ given by

$$\underline{-1} \qquad\qquad \underline{0}$$

$$C^\infty_\Sigma \otimes \mathfrak{g}_X \xleftarrow{\star\otimes h^\sharp} \Omega^2_\Sigma \otimes \mathfrak{g}_X^\vee$$

where $\star$ is the Hodge star operator on Σ and $h^\sharp$ is the inverse to $h^\flat$.

The generalized Laplacian for this gauge fixing operator is

$$\begin{aligned}[Q, Q^{GF}] &= [\partial\bar{\partial} \otimes h^\flat + 1 \otimes \ell_1, \star \otimes h^\sharp] \\ &= \mathrm{D} \otimes 1 + 1 \otimes [\ell_1, h^\sharp].\end{aligned}$$

The operator D is the Laplace-Beltrami operator acting on functions and two-forms on Σ. Now, we have the freedom to choose the L_∞ structure on X to be compatible with the metric h. Indeed, we can choose the L_∞ structure coming from the Levi-Civita connection ∇ associated with h. Once we do this, the operator ℓ_1 is identified with ∇. Moreover, $[\ell_1, h^\sharp] = 0$ is equivalent to the condition $\nabla h = 0$. Thus, the generalized Laplacian is simply $\mathrm{D} \otimes 1$.

3.2. Heat kernel and the propagator. For $L > 0$ the heat kernel $K_L \in \mathcal{E} \otimes \mathcal{E}$ splits into three pieces. First, there is the scalar heat kernel for the Laplacian acting on functions on $\Sigma = \mathbb{C}$ which has the explicit form

$$k_L(x, y) = \frac{1}{4\pi L} e^{-|x-y|^2/4L}.$$

Second, there is the term $\mathrm{d}^2 x \otimes 1 - 1 \otimes \mathrm{d}^2 y$ which accounts for the natural pairing between functions and two-forms. Finally, there is the element $\mathrm{id}_{\mathfrak{g}_X} + \mathrm{id}_{\mathfrak{g}_X^\vee} \in \mathfrak{g}_X \otimes \mathfrak{g}_X^\vee \oplus \mathfrak{g}_X^\vee \otimes \mathfrak{g}_X$ accounting for the natural pairing between $\mathfrak{g}_X$ and its dual. In all, we write the heat kernel as

$$K_L(x, y) = k_L(x, y)(\mathrm{d}^2 x \otimes 1 - 1 \otimes \mathrm{d}^2 y)(\mathrm{id}_{\mathfrak{g}_X} + \mathrm{id}_{\mathfrak{g}_X^\vee}).$$

The defining property of the heat kernel is that it satisfies the relation

$$\langle K_L(x, y), \Phi(y) \rangle = \left(e^{-L[Q, Q^{GF}]} \Phi \right)(x)$$

for any $\Phi \in \mathcal{E}$.

The propagator is defined by

$$\begin{aligned} P_{\epsilon \to L}(x, y) &= \int_{t=\epsilon}^{L} (Q^{GF} \otimes 1) K_t(x, y) \, \mathrm{d}\, t \\ &= \int_{t=\epsilon}^{L} \frac{1}{4\pi t} e^{-|x-y|^2/4t} \, \mathrm{d}\, t \otimes h. \end{aligned}$$

As we discuss in Appendix A, the propagator gives rise to a homotopy RG flow operator $W(P_{\epsilon \to L}, -)$ and the corresponding homotopy Renormalization Group Equation (hRGE). As we will see in Section 4, the limit as $\epsilon \to 0$ of the hRG flow $W(P_{\epsilon \to L}, I)$ is not well-defined in general, and we must introduce counterterms as in [**3**]. In *loc. cit.*, it is shown that counterterms exist (given a choice of renormalization scheme), they are local functionals, and they only depend on the UV parameter ϵ. Identification of the one-loop counterterms will play a critical role in the computation of the β-function for our theory; this is explained in the subsequent sections.

We can now define our naïve one-loop quantization.

Definition 3.1. The *naïve* one-loop quantization of the two-dimensional σ-model of maps $\mathbb{C} \to X$ is

$$I[L] := \lim_{\epsilon \to 0} \sum_{\Gamma} W(P_{\epsilon \to L}, I + I^{CT}(\epsilon))$$

where the sum is over all genus zero and one graphs Γ, and $I^{CT}(\epsilon)$ denotes a specific choice of counterterms dependent on fixing a renormalization scheme.

This one-loop quantization indeed defines a *pre-theory* (modulo $\hbar^2$) in the parlance of [**3**], i.e., the family of functionals $\{I[L]\}$ satisfies the hRGE (modulo $\hbar^2$) and is asymptotically local in the $L \to 0$ limit. Further, $\{I[L]\}$ is a *pre-quantization* (to one loop) of the σ-model, as

$$\lim_{L \to 0} I[L] \equiv I^h + I^X \quad \text{modulo } \hbar.$$

3.3. The quantum master equation. To define a consistent QFT, the collection of interactions must satisfy the *quantum master equation*. This quantum master equation should be thought of as a quantization of the classical master equation we have already encountered.

The starting point is to introduce the operator $Q + \hbar\Delta_L$ where $\Delta_L : \mathcal{O}(\mathcal{E}) \to \mathcal{O}(\mathcal{E})$ is the scale L BV Laplacian defined by contraction with the scale L heat kernel K_L. The issue here is that this operator is not square zero, in fact for every $L > 0$ one has

$$(Q + \hbar\Delta_L)^2 = \ell_1^2.$$

To rectify this, we introduce the modified Q-differential

$$Q_L = Q + \ell_1^2 \int_{t=0}^{L} Q^{GF} e^{-t\mathrm{D}}\,\mathrm{d}t.$$

LEMMA 3.2. *For any $L > 0$, the operator Q_L satisfies*

$$\left(Q_L + \hbar\Delta_L + \frac{1}{\hbar}F_{\ell_1}\right)^2 = C$$

for some $C \in \Omega^\sharp_X$.

PROOF. This follows from the observations that $Q_L^2 = -\{F_{\ell_1}, -\} = -\ell_1^2$ and $[\Delta_L, F_{\ell_1}] = \ell_1^2$. □

DEFINITION 3.3. The family of functionals $\{I[L]\}$ is said to satisfy the scale L quantum master equation (QME) if the operator

$$Q_L + \hbar\Delta_L + \{I[L], -\}_L$$

is square zero acting on $\mathcal{O}(\mathcal{E})$.

LEMMA 3.4. *The family of functionals $I[L]$ satisfies the scale L QME if and only if*

$$Q_L I[L] + \hbar\Delta_L I[L] + \frac{1}{2}\{I[L], I[L]\}_L + F_{\ell_1}$$

is zero modulo constant terms, i.e. elements in the ring $\Omega^\sharp_X$.

3.4. The obstruction. Not every naïve quantization of a classical field theory defines a quantization. The obstruction is precisely the failure to satisfy the quantum master equation we have just defined. In this section, we show that to order $\hbar$ the obstruction, though not identically zero, determines a trivial class in cohomology. This is enough to determine the existence of a one-loop quantization of the classical σ-model.

PROPOSITION 3.5. *The obstruction to satisfying the quantum master equation modulo $\hbar^2$ is cohomologically trivial in the deformation complex* Def.

By Lemma C.5 in [**13**], we see that the obstruction satisfies

$$\Theta[L] = \lim_{\epsilon \to 0} \sum_{\Gamma, e} W_{\Gamma,e}(P_\epsilon^L, K_\epsilon - K_0, I)^{sm}$$

where "*sm*" denotes the smooth component of the functional. The sum is over one loop connected graphs Γ and edges e of the graph. The notation $W_{\Gamma,e}(A, B, I)$ where $A, B \in \mathrm{Sym}^2(\mathcal{E})$ means that we put A on each edge besides the distinguished edge, where we put B.

We find that the tadpole diagram is purely singular, hence does not contribute to $\Theta[L]$. Moreover, the diagrams involving three or more vertices vanish by type reasons. That leaves the wheel with two vertices with internal edges labeled by $P_{\epsilon\to L}$ and $K_\epsilon - K_0$, as shown here:

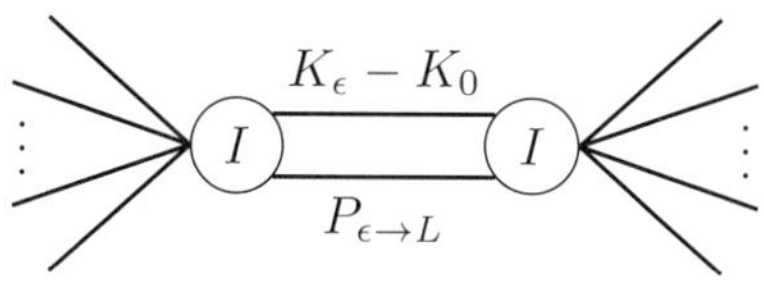

The non-vanishing part of the weight of the diagram is of the form

$$\int_{z,w}\int_{t=\epsilon}^{L} \frac{1}{4\pi\epsilon}\frac{1}{4\pi t} e^{-|z-w|^2/4t}\varphi(z,w)\,\mathrm{d}^2 z\,\mathrm{d}^2 w\,\mathrm{d}\, t$$

where φ is some compactly supported function on $\mathbb{C}^2$ depending on the inputs of the diagram. Applying Wick's formula, we see that as $L \to 0$ the smooth part of the diagram can be written as the following local functional

$$-\frac{\log 2}{4\pi}\Theta(\varphi,\psi)$$

where Θ is represented by the diagram:

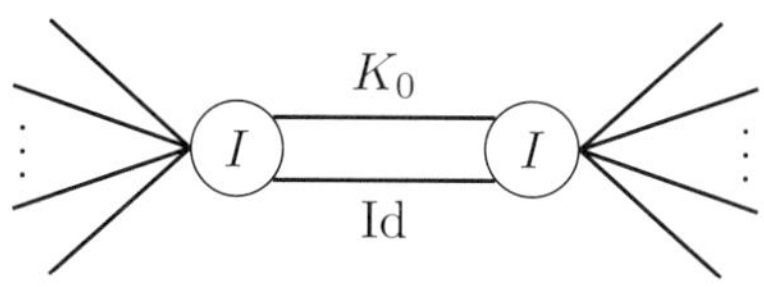

It suffices to show that we can make this functional exact in the deformation complex. For this, consider the following local functional

$J(\varphi,\psi) \quad =$ 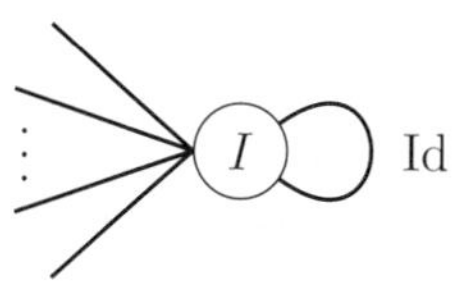

The differential of the deformation complex is of the form $\{S,-\} = Q + \{I,-\}$. Applying this operator to the functional J we find

(6) $\qquad QJ + \{I,J\} = Q$ 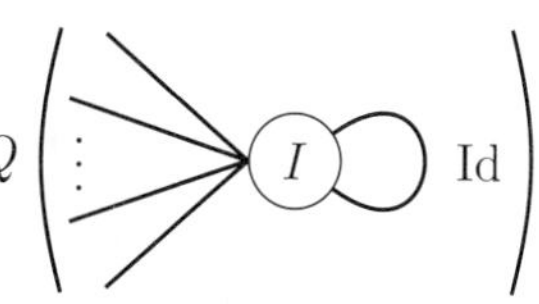

$+$ 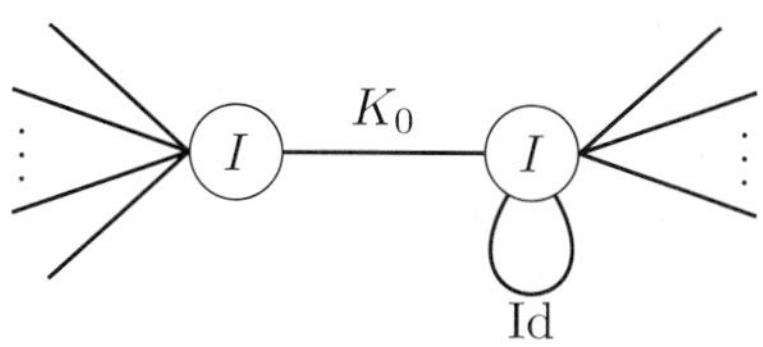

Note that the functional in the parentheses is equal to $\widetilde{\Delta}(I)$ where $\widetilde{\Delta}$ is the BV Laplacian. Moreover, one has

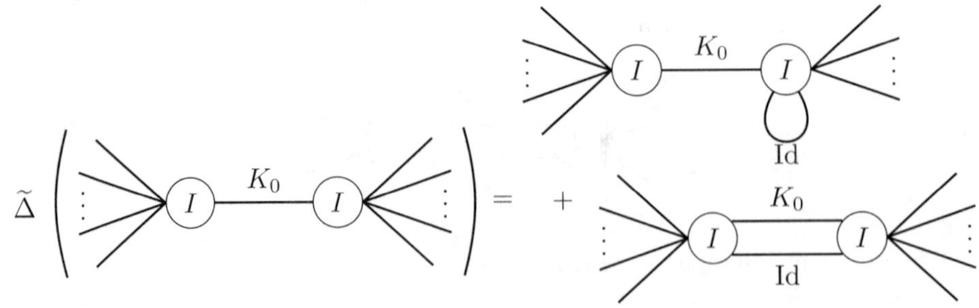

Since Q and $\widetilde{\Delta}$ commute, we find that the right-hand side of Equation (6) becomes

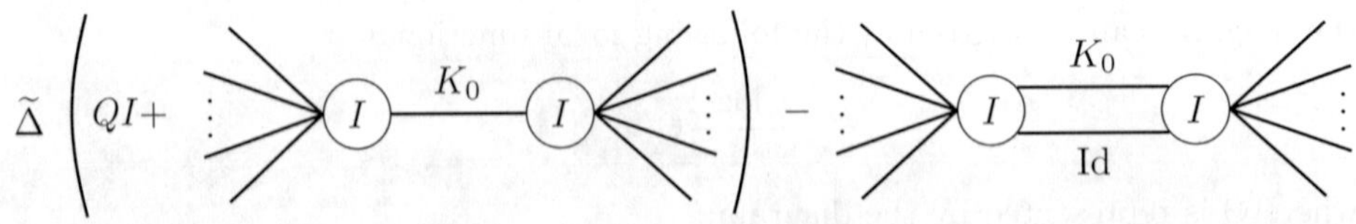

Finally, we note that the expression inside the parentheses is zero by the classical master equation. We conclude that

$$(Q + \{I, -\})\left(\frac{\log 2}{2\pi} J\right) = \Theta,$$

as desired. We note that J has scaling weight zero.

To obtain the quantization from the naïve quantization we consider the family of functionals

$$\{I[L] + \hbar J[L]\}.$$

Here, $I[L]$ is, as above, equal to a sum over graphs of genus ≤ 1. Since we are working modulo $\hbar^2$, the quantity $\hbar J[L]$ is equal to the sum of the weights of trees where at a single vertex we put the functional $\hbar J$ and at the remaining vertices we put the functional I. It is an immediate consequence of the compatibility of homotopy RG-flow and the quantum differential that this family of functionals satisfies the quantum master equation modulo $\hbar^2$.

4. β-function generalities

Having defined and quantized (to one-loop) the σ-model, we now recall some generalities and computational lemmas regarding β-functions. We compute the β-function for our quantization in Section 5 below.

Consider a translation invariant BV theory (E, Q, I) on $\mathbb{R}^n$. We denote by E_0 the fiber of the bundle E over $0 \in \mathbb{R}^n$. The group $\mathbb{R}_{>0}$ acts on the fields $\mathcal{E} = C^\infty(\mathbb{R}^n) \otimes E_0$ via the action induced by rescaling $\mathbb{R}^n$. Further, $\mathbb{R}_{>0}$ acts on the space of functionals, $\mathcal{O}(\mathcal{E})$, and preserves the subspace of local functionals $\mathcal{O}_{loc}(\mathcal{E})$. We will denote this action of $\mathbb{R}_{>0}$ on $\mathcal{O}(\mathcal{E})$ by ρ_λ. Making this action explicit, one can easily check the following :

$$\rho_\lambda(I_k) = \lambda^{n+\frac{k(2-n)}{2}} I_k, \quad \text{where} \quad I_k(\phi) = \int_{\mathbb{R}^n} \phi(x)^k. \tag{4.1}$$

DEFINITION 4.1. A classical field theory (E, Q, I) is *scale-invariant* if $\rho_\lambda(I) = I$ for all $\lambda \in \mathbb{R}_{>0}$.

From equation (4.1) above, one can see that ϕ^4 theory is scale-invariant on $\mathbb{R}^4$ and ϕ^3 theory is scale-invariant on $\mathbb{R}^6$. Scale-invariance is a stronger condition than simply being *(classically) renormalizable*, which is the condition that $\rho_\lambda(I)$ flows to a fixed point as $\lambda \to 0$.

We now recall how to extend the action of $\mathbb{R}_{>0}$ to QFTs in the BV formalism. This action is called the *(local) renormalization group flow*, or simply RG flow. We start with a classical BV theory given by the data (E, Q, I). Following Costello [**3**], a quantum field theory is a collection of functionals $\{I[L]\}$ where $L > 0$ that satisfy homotopy RG flow (the renormalization group equation) and satisfy the quantum master equation. It is a quantization if modulo $\hbar$ the family of functionals converges to the classical interaction as $L \to 0$. It is also necessary to equip our quantum BV theory with a gauge-fixing operator Q^{GF}. Further, suppose that the action of $\mathbb{R}_{>0}$ on Q^{GF} is as follows:

$$\rho_\lambda \cdot Q^{GF} := \rho_\lambda Q^{GF} \rho_{\lambda^{-1}} = \lambda^k Q^{GF}.$$

Here the *scaling dimension* of Q^{GF} is allowed to be a rational number, i.e., $k \in \mathbb{Q}$. In the scalar theories below, $k = n/2 - 1$.

DEFINITION 4.2. Let $\{I[L]\}$ be a family of translation-invariant effective interactions. Define a rescaled effective family $\{I_\lambda[L]\}$ by

$$I_\lambda[L] := \rho_\lambda \cdot I[\lambda^{-k} L].$$

It can be shown explicitly, Lemma 3.4 of [**6**], that $\{I[L]\}$ satisfies homotopy RG flow and the QME if and only if the rescaled family $\{I_\lambda[L]\}$ does.

We are (finally) in a position to define the β-functional. We will distinguish between the β-functional and β-function as the latter is the cohomology class of the former in the BV algebra of (quantum) observables.

DEFINITION 4.3. Let $\{I[L]\}$ be an effective quantization of the BV theory (E, Q, I) on $\mathbb{R}^n$. For $L \in \mathbb{R}_{>0}$, define the *scale L β-functional* to be the functional

$$\mathcal{O}_\beta[L] := \lim_{\lambda \to 1} \lambda \frac{d}{d\lambda} I_\lambda[L].$$

We can expand the β-functional in powers of $\hbar$:

$$\mathcal{O}_\beta[L] = \mathcal{O}_\beta^{(0)}[L] + \hbar \mathcal{O}_\beta^{(1)}[L] + O(\hbar^2).$$

The superscript indicates the loop depth, e.g., $\mathcal{O}_\beta^{(1)}[L]$ is the *one-loop β-functional.* For a theory where the classical theory is scale invariant, the zero-loop β-functional, $\mathcal{O}_\beta^{(0)}[L]$, is identically zero.

PROPOSITION 4.4 (Corollary 3.12 of [**6**]). *Let $\{I[L]\}$ be an effective quantization of the translation and scale invariant BV theory (E, Q, I) on $\mathbb{R}^n$. Then*

$$\mathcal{O}_\beta^{(1)} := \lim_{L \to 0} \mathcal{O}_\beta^{(1)}[L]$$

exists and determines a closed element in $\mathcal{O}_{loc}(\mathcal{E})$.

REMARK 4.5. The higher loop β-functionals, $\mathcal{O}_\beta^{(k)} = \lim_{L\to 0} \mathcal{O}_\beta^{(k)}[L]$, are not necessarily well-defined. Even if one naïvely defines $\mathcal{O}_\beta^{(k)}$ to be the $\log \epsilon$ divergence at k-loops (compare Proposition 4.9), these functionals are not closed with respect to the BRST differential $Q + \{I, -\}$. One can prove that if $\mathcal{O}_\beta^{(i)} \equiv 0$ for $i < k$, then $\mathcal{O}_\beta^{(k)}[L]$ satisfies homotopy RG flow and is BRST closed; in particular, the k-loop β-functional, $\mathcal{O}_\beta^{(k)}$, exists.

DEFINITION 4.6. Let $\{I[L]\}$ be an effective quantization of a translation invariant BV theory (E, Q, I) on $\mathbb{R}^n$. The *scale L β-function*, $\beta[L]$, is the cohomology class

$$\beta[L] := [\mathcal{O}_\beta[L]] \in H^0(\mathcal{O}(\mathcal{E}) \otimes C^\infty((\epsilon, L))).$$

REMARK 4.7. Typically, the β-function cannot be decomposed by $\hbar$ degrees; the complex of functionals is only filtered, not graded.

In light of the previous remark, we do have the following (a corollary of Proposition 4.4).

PROPOSITION 4.8. *Let $\{I[L]\}$ be an effective quantization of the translation and scale invariant BV theory (E, Q, I) on $\mathbb{R}^n$. Then*

$$\beta^{(1)} := \left[\mathcal{O}_\beta^{(1)}\right] \in H^0(\mathcal{O}_{loc}(\mathcal{E}))$$

is well-defined.

4.1. As a local functional. Consider a BV theory (E, Q, I) and the associated BRST complex $(\mathcal{O}_{loc}(\mathcal{E}), Q + \{I, -\})$. Cohomology classes in this complex define first order deformations of the theory. Hence, we could think of $H^0(\mathcal{O}_{loc}(\mathcal{E}))$ as the space of *coupling constants* for the given theory. Now if (E, Q, I) is translation and scale invariant, then the one-loop β-functional is BRST closed. Consequently, we can think of the one-loop β-function as a function on $H^0(\mathcal{O}_{loc}(\mathcal{E}))$.

In order to derive an explicit formula, we should fix *bare values* (a basis) for the space $H^0(\mathcal{O}_{loc}(\mathcal{E}))$, and we should also check how $\beta^{(1)}$ transforms under a change of basis (Proposition 3.18 of [**6**]). Moreover, we typically would like to distinguish a finite dimensional subspace of $H^0(\mathcal{O}_{loc}(\mathcal{E}))$ and realize $\beta^{(1)}$ as a function of a $\mathbb{R}^n$ valued parameter $\mathbf{c}$. In some good cases, e.g., ϕ^3 theory below, we can find a one-dimensional subspace and $\beta^{(1)}$ is a function of a real number c.

Ideally, we would like to choose the distinguished subspace to be the one spanned by the classical interaction I under homotopy *RG* flow. However, as we discuss in Section 4.3.4, homotopy RG flow may not actually preserve this subspace and can introduce *dynamic coupling constants.* We can however restrict to the subspace spanned by functionals which are $\log \epsilon$ divergent in the $\epsilon \to 0$ limit to obtain a well-defined function of just a few coupling constants.

4.2. Computational lemmas. The gloss above illustrates the development and interpretation of the β-function/functional. However, if one is interested in computations at one-loop, they can utilize the following result.

PROPOSITION 4.9 (Proposition 3.23 of [**6**]). *Let (E, Q, I) be a translation and scale-invariant BV theory and $\{I[L]\}$ an effective quantization at one-loop. Let k denote the scaling dimension of Q^{GF}. Then the one-loop β-functional satisfies*

$$\mathcal{O}_\beta^{(1)} = k I_{\log}^{CT}.$$

Finally, we note a convenient property of the β-function: homotopy invariance.

PROPOSITION 4.10 (Proposition 3.20 of [**6**]). *The β-function is locally constant on the space of quantum field theories with fixed classical BV complex and gauge fixing operator.*

As an example, the first and second order formulations of Yang-Mills theory are homotopy equivalent, so the β-function can be computed using the first order formalism.

4.3. First examples: Scalar field theory. Building on [**3**], we describe the β-functional/function for certain scalar field theories on $\mathbb{R}^n$. That is, our theory has field content

$$\mathcal{E} : C_c^\infty(\mathbb{R}^n) \xrightarrow{D} C_c^\infty(\mathbb{R}^n)[-1],$$

where D is the Laplacian, and action

$$S(\phi) = \int_{\mathbb{R}^n} \phi D \phi + I(\phi), \text{ for } \phi \in \mathcal{E}.$$

In this setting the propagator is determined by the heat kernel $K_\ell(x_1, x_2)$ on $\mathbb{R}^n$.

We will choose a renormalization scheme where ϵ^{-1} and $\log \epsilon$ are purely singular. Moreover, we will consider the space $\mathcal{S} \subset H^0(\mathcal{O}_{loc}(\mathcal{E}))$ spanned by those functionals which are $\log \epsilon$ divergent in the $\epsilon \to 0$ limit. In the theories below, one can verify that $\mathcal{S}$ is one-dimensional and we will choose the classical interaction functional I as a basis vector, i.e., $\mathcal{S} = \{cI : c \in \mathbb{R}\}$.

4.3.1. *ϕ^4 theory on $\mathbb{R}^4$.* Consider the scalar theory on $\mathbb{R}^4$ with interaction

$$I(\phi) = I_{0,4} = \frac{1}{4!} \int_{x \in \mathbb{R}^4} \phi(x)^4.$$

The only term that contributes to one-loop $\log \epsilon$ divergence, and hence the one-loop β-functional via Proposition 4.9, comes from the following graph.

γ_4 :

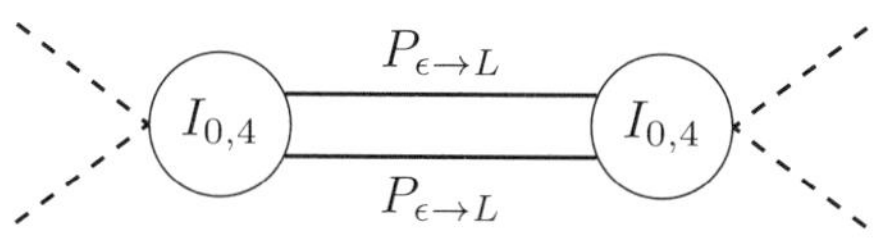

The corresponding counterterm is computed by Costello in Section 4, Chapter 4 of [**3**]. Therefore we have,

$$\mathcal{O}_\beta^{(1)} = I_{1,4}^{CT}(\epsilon) = -\pi^{-2} 2^{-8} c^2 \log \epsilon \int_{x \in \mathbb{R}^4} \phi(x)^4.$$

4.3.2. *ϕ^3 theory on $\mathbb{R}^6$.* Again, in the case

$$I(\phi) = I_{0,3} = \frac{1}{3!} \int_{x \in \mathbb{R}^6} \phi(x)^3,$$

there is only one term that contributes to the one-loop $\log \epsilon$ divergence: the wheel with three vertices.

γ_3 :

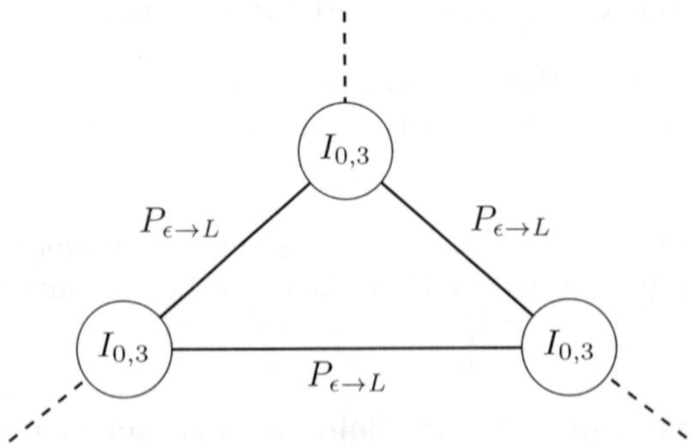

Let us compute the weight of this graph with respect to the propagator $P_{\epsilon\to L}$.

$$
\begin{aligned}
&\omega_{\gamma_3}\\
=&\int_{\ell_1,\ell_2,\ell_3\in[\epsilon,L]}\int_{x_1,x_2,x_3\in\mathbb{R}^6} K_{\ell_1}(x_1,x_2)K_{\ell_2}(x_2,x_3)K_{\ell_3}(x_3,x_1)d\mathbf{x}d\ell\\
=&\frac{1}{(4\pi)^9}\int_{\ell_1,\ell_2,\ell_3\in[\epsilon,L]}\int_{x_1,x_2,x_3\in\mathbb{R}^6}(\ell_1\ell_2\ell_3)^{-3}\\
&\qquad\times e^{-\|x_1-x_2\|^2/4\ell_1-\|x_2-x_3\|^2/4\ell_2-\|x_3-x_1\|^2/4\ell_3}d\mathbf{x}d\ell\\
\overset{(1)}{=}&\frac{3}{8(4\pi)^9}\int_{\ell_1,\ell_2,\ell_3\in[\epsilon,L]}\int_{y,z_1,z_2\in\mathbb{R}^6}(\ell_1\ell_2\ell_3)^{-3}e^{-\|z_1\|^2/\ell_1-\|z_2\|^2/\ell_2-\|z_1+z_2\|^2/\ell_3}dyd\mathbf{z}d\ell\\
\overset{(2)}{=}&\frac{3}{2^{21}\pi^3}\int_{\ell_1,\ell_2,\ell_3\in[\epsilon,L]}(\ell_1\ell_2\ell_3)^{-3}\left(\ell_1^{-1}\ell_2^{-1}+\ell_1^{-1}\ell_3^{-1}+\ell_2^{-1}\ell_3^{-1}\right)^{-3}d\ell\\
=&\frac{3}{2^{21}\pi^3}\int_{\ell_1,\ell_2,\ell_3\in[\epsilon,L]}\frac{1}{(\ell_1+\ell_2+\ell_3)^3}d\ell.
\end{aligned}
$$

In the equality (1) we have used the coordinate substitution

$$z_1=\frac{1}{2}(x_2-x_3),\quad z_2=\frac{1}{2}(x_3-x_1),\quad y=\frac{1}{2}(x_1+x_2+x_3).$$

Further, in equality (2) we have used the equality

$$\int_{y,z_1,z_2\in\mathbb{R}^6}(\ell_1\ell_2\ell_3)^{-3}e^{-\|z_1\|^2/\ell_1-\|z_2\|^2/\ell_2-\|z_1+z_2\|^2/\ell_3}dyd\mathbf{z}=\pi^6(\det A)^{-3},$$

where A is the matrix corresponding to the quadratic form

$$\|z_1\|^2/\ell_1+\|z_2\|^2/\ell_2+\|z_1+z_2\|^2/\ell_3.$$

Actually, all we care about is the $\log\epsilon$ divergence of the graph weight ω_{γ_3}:

$$\mathrm{sing}_{\log\epsilon}\omega_{\gamma_3}=\frac{3}{2^{22}\pi^3}\log\epsilon.$$

Therefore, after accounting for graph automorphisms, we find that the $\log\epsilon$ counterterm and hence the one loop β functional is given by

$$\mathcal{O}_\beta^{(1)}=2I_{1,3}^{CT}(\epsilon)=2^{-21}\pi^{-3}c^3\log\epsilon\int_{x\in\mathbb{R}^6}\phi(x)^3.$$

4.3.3. *Turning on a mass.* When one introduces a mass term, $m^2 \int_{\mathbb{R}^n} \phi(x)^2$, the theory is often no longer scale invariant. One could choose to ignore this subtlety and formally compute the one loop β-functional as the $\log \epsilon$ divergence at one loop.

In a massive theory, the heat kernel (and hence propagator) changes as follows

$$K_\ell^m(x_1, x_2) = K_\ell^{m=0}(x_1, x_2)e^{-tm^2},$$

where K_ℓ^m is the kernel for the massive theory and $K_\ell^{m=0}$ is the kernel for the massless theory.

One can then show directly that for any admissible graph γ and interaction I, we have

$$\mathrm{sing}_{\{\epsilon^{-1}, \log \epsilon\}} \omega(P^m_{\epsilon \to L}, I)(\phi) = \mathrm{sing}_{\log \epsilon} \omega(P^{m=0}_{\epsilon \to L}, I)(\phi) + O(\epsilon^{-1}).$$

Consequently, a mass introduces no further $\log \epsilon$ corrections and hence doesn't change the one loop β-functional/function.

4.3.4. *Dynamic coupling constants.* We note that renormalization/homotopy renormalization group flow can introduce *dynamic coupling constants.* That is, homotopy RG flow does not preserve the $\mathbb{R}$-linear subspace of $\mathcal{O}_{loc}(\mathcal{E})$ spanned by the original (classical) interaction $I^{CL} = \sum_n I_{0,n}$. This phenomenom already appears in ϕ^3 theory on $\mathbb{R}^6$ (and ϕ^4 theory on $\mathbb{R}^4$).

Indeed, in ϕ^3 theory on $\mathbb{R}^6$, we find that

$$I^{CT}_{1,2}[\epsilon] \sim \epsilon^{-1}\phi(x)^2.$$

This term could be interpreted as a *dynamic mass.* As noted above, these mass terms are not scale invariant; in ϕ^3 theory this term has scaling weight 2.

There are (at least) two ways to account for this defect/feature of renormalization in terms of the β-functional. First, one could impose additional symmetry so that these dynamic terms are not invariant. Secondly, one could note that flow generated by the β-functional (suitably interpreted as a vector field on $\mathcal{O}_{loc}(\mathcal{E})$) preserves the subspace of local functionals which are $\log \epsilon$ divergent in the $\epsilon \to 0$ limit. We will take the second approach in what follows; the former way is illustrated in the case of Yang-Mills in [**6**].

5. The one-loop β-function for the σ-model

We now use the framework from the preceding section to the compute the one-loop β function for our theory. Recall that the classical action consists of two pieces, $S = S_h + S_X$. Therefore, there are two flavors of classical interaction terms (vertex types): $I^h_{0,k}$ and $I^X_{0,\ell}$. As noted in Section 3.2, the propagator only pairs the ϕ fields, and one can check that the (singular) BV operator Δ_L pairs ϕ and ψ fields. Diagramatically, we can represent these two functionals/operators as follows.

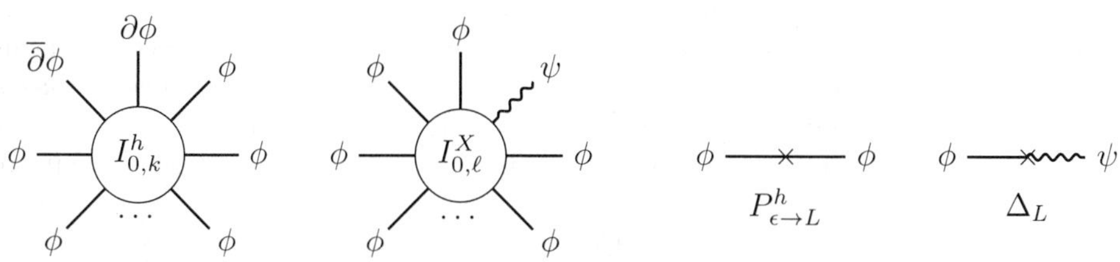

5.1. Reduction to β_h: A cohomological calculation. We decompose the one-loop β function into two pieces and show that only the component depending on the metric h survives at the level of cohomology.

Note that in the proof of Proposition 2.5, we showed that the two pieces of the action functional bracket to zero, i.e., $\{S_X, S_h\} = 0$. In particular, the differentials $\{S_X, -\}$ and $\{S_h, -\}$ commute. The following is an immediate corollary.

COROLLARY 5.1. *We can decompose $\beta^{(1)}$ as*

$$\beta^{(1)} = \beta_h^{(1)} + \beta_X^{(1)} \in H^0(\mathcal{O}_{loc}(\mathcal{E})),$$

where $\beta_h^{(1)}$ (resp. $\beta_X^{(1)}$) only involves diagrams with vertex type $I_{0,k}^h$ (resp. $I_{0,k}^X$).

PROPOSITION 5.2. *The term $\beta_X^{(1)}$ vanishes. Consequently,*

$$\beta^{(1)} = \beta_h^{(1)} \in H^0(\mathcal{O}_{loc}(\mathcal{E})).$$

To prove the proposition, we will proceed via a spectral sequence argument, where we consider

$$(\mathcal{O}_{loc}(\mathcal{E}), \{S, -\}) = (\mathcal{O}_{loc}(\mathcal{E}), \{S_x, -\} + \{S_h, -\})$$

as a double complex.

LEMMA 5.3. *For the decomposition $\beta^{(1)} = \beta_h^{(1)} + \beta_X^{(1)}$, we have*

(a) $\left\{S_X, \beta_h^{(1)}\right\} = 0$;

(b) $\left\{S_h, \beta_h^{(1)}\right\} = \pm\left\{S_X, \beta_X^{(1)}\right\}$.

PROOF OF PROPOSITION 5.2. Consider the spectral sequence of the relevant double complex with E_1 page given by $H^*(\mathcal{O}_{loc}(\mathcal{E}), \{S_X, -\})$. Via the preceding lemma (part (a)), we see that $\beta_h^{(1)}$ is closed and hence survives to the E_1 page. Part (b) of the lemma further implies that $[\beta_h^{(1)}]_{E_1}$ is closed with respect to the differential and hence survives to the E_2 page. The E_2 page is concentrated on a single row, so the spectral sequence collapses. Hence, we have

$$\beta^{(1)} = \left[\beta_h^{(1)} + \beta_X^{(1)}\right]_{E_\infty} = \left[\beta_h^{(1)}\right]_{E_2} = \beta_h^{(1)}.$$

□

5.2. Computation of β_h. We now compute β_h by reducing to a single Feynman weight computation and identifying the result in terms of geometric data on the target manifold X.

Recall that the one-loop quantization consisted of a family of functionals of the form $\{I[L] + \hbar J[L]\}$ where $I[L]$ was the naive quantization defined in terms of the weight expansion of the classical interaction via the propagators, and $J[L]$ was a quantum correction term we introduced to ensure the theory satisfied the QME. Since we are working modulo $\hbar^2$, the quantity $\hbar J[L]$ is defined in terms of a Feynman expansion over genus zero graphs, that is, trees. Hence, it has a well-defined $\epsilon \to 0$ limit and does not contribute to the divergences.

Thus, the only part of the theory that contributes to the β-function is the naïve quantization

$$I[L] = \lim_{\epsilon \to 0} W(P_{\epsilon \to L}, I + I^{CT}(\epsilon)).$$

PROPOSITION 5.4. *If Γ is a wheel with at least two vertices then the weight $W_\Gamma(P_{\epsilon\to L}, I)$ has well-defined $\epsilon \to 0$ limit. That is, these graphs are UV finite, and hence do not contribute any counterterms.*

PROOF. The result follows from a simple power counting argument. The singular behavior of the diagrams is the same as scalar field theory in dimension 2. It is then a computation–completely analogous to those of Section 4.3–that such a wheel with k vertices has $\epsilon \to 0$ asymptotic behavior $\epsilon^{k-1}\log\epsilon + O(\epsilon^{k-1})$ and thus converges absolutely for $k \geq 2$. □

Hence, we are left to consider the weight of following type of one-loop diagrams. The index k indicates the number of external edges at the vertex. We denote the ϵ-dependent weight by $\mathcal{T}_k(\epsilon)$.

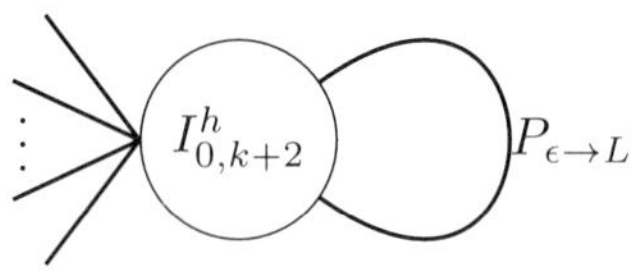

$\mathcal{T}_k(\epsilon)$

LEMMA 5.5. *As elements of $(\mathcal{O}(\mathcal{E}) \otimes C^\infty((\epsilon, L)), \{S, -\})$, the following are cohomologous*

$$\sum_k \mathcal{T}_k(\epsilon) \sim \mathcal{T}_2(\epsilon).$$

PROOF. Recall the infinite jet map

$$j_\infty : \mathrm{Sym}^2(T_X^\vee) \hookrightarrow \Omega^*_X\left(\mathcal{J}\left(\mathrm{Sym}^2(T_X^\vee)\right)\right) = \mathrm{C}^*_{\mathrm{Lie}}\left(\mathfrak{g}_X; \mathrm{Sym}^2(T_X^\vee)\right),$$

which is actually a quasi-isomorphism. By construction (see Section 2),

$$\sum_k I^h_{0,k} = \int_\Sigma j_\infty(h)(\partial\phi, \bar{\partial}\phi; e^\phi).$$

Further, bracketing with the Lie theoretic part of the action, S_X, recovers the Chevalley-Eilenberg differential, i.e., $\{S_X, -\} = d_{\mathrm{Lie}}$. Finally, as $\{S_X, -\}$ and $\{S_h, -\}$ commute, the proposition follows from the fact that j_∞ is a quasi-isomorphism. □

Next, it is a straightforward computation with scalar heat kernels to see that $\mathcal{T}_2(\epsilon)$ is actually purely $\log\epsilon$ divergent.

LEMMA 5.6. *The Feynman weight $\mathcal{T}_2(\epsilon)$ is purely $\log\epsilon$ divergent, so*

$$\mathcal{T}_2(\epsilon) = \log(\epsilon)\mathcal{T}_2$$

for some local functional $\mathcal{T}_2$.

PROOF. The analytic part of the propagator is of the form $\int_{t=\epsilon}^L \frac{1}{4\pi t} e^{-|x-y|^2/4t}\, \mathrm{d}t$. Thus, the analytic contribution to the weight of the tadpole diagram (where we set $x = y$) is purely $\log\epsilon$ divergent. □

We can summarize the preceding arguments as follows:

$$\beta^{(1)} = \left[\mathcal{O}_\beta^{(1)}\right] = [\mathrm{sing}_{\log\epsilon}\mathcal{T}_2(\epsilon)] = [\mathcal{T}_2] \in H^0(\mathcal{O}_{loc}(\mathcal{E})).$$

We would now like to realize the β-function in terms of the geometry of the target Riemannian manifold (X, h). To begin, following Section 4.1, we could describe a basis for $H^0(\mathcal{O}_{loc}(\mathcal{E}))$. Alternatively, we recall the map

$$\mathfrak{R} : \mathrm{Sym}^2(T_X^\vee) \to \mathcal{O}_{loc}(\mathcal{E})$$

we constructed in Section 2.3.

THEOREM 5.7. *As elements of $H^0(\mathcal{O}_{loc}(\mathcal{E}))$, we have an equality*

$$\beta^{(1)} = -\frac{1}{12\pi}\left[\mathfrak{R}(\mathrm{Ric})\right].$$

PROOF. First note that in the $\epsilon \to 0$ limit, $P_{\epsilon\to L}$ becomes a delta function on the diagonal: that is, we take a trace over pairs of jet indices. Next, up to a constant, the covariant curvature tensor $R_{ikj\ell}$ corresponding to the metric h determines the 4-valent vertex $I^h_{0,4}$; for $k > 4$, the $I^h_{0,k}$ are just the jet expansions of this tensor. Combining the previous two observations, we have an equivalence of local functionals

$$\begin{aligned}\mathcal{T}_2 = -\frac{1}{4\pi}\int_\Sigma j_2(h)(\partial\phi, \overline{\partial}\phi; \phi_k, \phi_l)\delta_{kl} &= -\frac{1}{4\pi}\int_\Sigma \left(\frac{1}{3}\right) R_{ikj\ell}(\partial\phi, \phi, \overline{\partial}\phi, \phi)\delta_{k\ell} \\ &= -\frac{1}{12\pi}\int_\Sigma R_{ikjk}(\partial\phi, \phi, \overline{\partial}\phi, \phi).\end{aligned}$$

Thus we have

$$\mathcal{T}_2 = -\frac{1}{12\pi}\int_\Sigma \mathrm{Ric}(\partial\phi, \overline{\partial}\phi).$$

The same argument as in Lemma 5.5 proves that

$$-\frac{1}{12\pi}\mathfrak{R}(\mathrm{Ric}) \quad \text{and} \quad -\frac{1}{12\pi}\int_\Sigma \mathrm{Ric}(\partial\phi, \overline{\partial}\phi)$$

are cohomologous as elements of $\mathcal{O}_{loc}(\mathcal{E})$. □

Using the image of the map $\mathfrak{R} : \mathrm{Sym}^2(T_X^\vee) \to \mathcal{O}_{loc}(\mathcal{E})$, we can rephrase the theorem as follows.

COROLLARY 5.8. *With respect to the renormalization scheme where ϵ^{-1} and $\log\epsilon$ are purely singular, we have*

$$\beta^{(1)}(h) = -\frac{1}{12\pi}\mathrm{Ric}(h) \in \mathrm{Sym}^2(T_X^\vee) \subset H^0(\mathcal{O}_{loc}(\mathcal{E})).$$

REMARK 5.9. One could give an alternative proof of Theorem 5.7 as follows. In Section 2.3, we have identified $H^0(\mathcal{O}_{loc}(\mathcal{E}))$ as a direct sum. Next, argue locally on the target: locally the L_∞ algebra $\mathfrak{g}_X$ is curved, but abelian, so there are no interesting vertices of type I_X. Consequently, we know that $\beta^{(1)}$ is a tensor on the target. Finally, we choose geodesic normal coordinates and compute this tensor at a point in X.

Appendix A. Homotopy renormalization group flow

The homotopy renormalization group flow equation can be described in terms of Feynman graphs. Note that our description is for an arbitrary functional on a space of fields $\mathcal{E}$. Further, we will work relative to an arbitrary dg algebra $\mathcal{A}$ equipped with a nilpotent ideal $\mathcal{I}$.

DEFINITION A.1. A graph $\mathcal{G}$ consists of the following data:

(1) A finite set of vertices $V(\mathcal{G})$;
(2) A finite set of half-edges $H(\mathcal{G})$;
(3) An involution $\sigma : H(\mathcal{G}) \to H(\mathcal{G})$. The set of fixed points of this map is denoted by $T(\mathcal{G})$ and is called the set of tails of $\mathcal{G}$. The set of two-element orbits is denoted by $E(\mathcal{G})$ and is called the set of internal edges of $\mathcal{G}$;
(4) A map $\pi : H(\mathcal{G}) \to V(\mathcal{G})$ sending a half-edge to the vertex to which it is attached;
(5) A map $g : V(\mathcal{G}) \to \mathbb{Z}_{\geqslant 0}$ assigning a genus to each vertex.

It is clear how to construct a topological space $|\mathcal{G}|$ from the above abstract data. A graph $\mathcal{G}$ is called *connected* if $|\mathcal{G}|$ is connected. The genus of the graph $\mathcal{G}$ is defined to be

$$g(\mathcal{G}) := b_1(|\mathcal{G}|) + \sum_{v \in V(\mathcal{G})} g(v),$$

where $b_1(|\mathcal{G}|)$ denotes the first Betti number of $|\mathcal{G}|$. Let

$$\mathcal{O}^+(\mathcal{E}) \subset \mathcal{O}(\mathcal{E})[[\hbar]]$$

be the subspace consisting of those functionals which are at least cubic modulo $\hbar$ and the nilpotent ideal $\mathcal{I}$ in the base ring $\mathcal{A}$. Let $F \in \mathcal{O}^+(\mathcal{E})$ be a functional, which can be expanded as

$$F = \sum_{g,k \geq 0} \hbar^g F_g^{(k)}, \quad F_g^{(k)} \in \mathcal{O}^{(k)}(\mathcal{E}).$$

We view each $F_g^{(k)}$ as an S_k-invariant linear map

$$F_g^{(k)} : \mathcal{E}^{\otimes k} \to \mathcal{A}.$$

With the propagator $P_{\epsilon \to L}$, we will describe the *(Feynman) graph weights*

$$W_{\mathcal{G}}(P_{\epsilon \to L}, F) \in \mathcal{O}^+(\mathcal{E})$$

for any connected graph $\mathcal{G}$. We label each vertex v in $\mathcal{G}$ of genus $g(v)$ and valency k by $F_{g(v)}^{(k)}$. This defines an assignment

$$F(v) : \mathcal{E}^{\otimes H(v)} \to \mathcal{A},$$

where $H(v)$ is the set of half-edges of $\mathcal{G}$ which are incident to v. Next, we label each internal edge e by the propagator

$$P_e = P_{\epsilon \to L} \in \mathcal{E}^{\otimes H(e)},$$

where $H(e) \subset H(\mathcal{G})$ is the two-element set consisting of the half-edges forming e. We can then contract

$$\otimes_{v \in V(\mathcal{G})} F(v) : \mathcal{E}^{H(\mathcal{G})} \to \mathcal{A}$$

with

$$\otimes_{e \in E(\mathcal{G})} P_e \in \mathcal{E}^{H(\mathcal{G}) \setminus T(\mathcal{G})}$$

to yield a linear map

$$W_{\mathcal{G}}(P_{\epsilon\to L}, F) : \mathcal{E}^{\otimes T(\mathcal{G})} \to \mathcal{A}.$$

DEFINITION A.2. We define the (homotopy) RG flow operator with respect to the propagator $P_{\epsilon\to L}$

$$W(P_{\epsilon\to L}, -) : \mathcal{O}^+(\mathcal{E}) \to \mathcal{O}^+(\mathcal{E}),$$

by

$$W(P_{\epsilon\to L}, F) := \sum_{\mathcal{G}} \frac{\hbar^{g(\mathcal{G})}}{|\mathrm{Aut}(\mathcal{G})|} W_{\mathcal{G}}(P_{\epsilon\to L}, F) \tag{A.1}$$

where the sum is over all connected graphs.

Equivalently, it is useful to describe the (homotopy) RG flow operator formally via the simple equation

$$e^{W(P_{\epsilon\to L},F)/\hbar} = e^{\hbar\partial_{P_{\epsilon\to L}}} e^{F/\hbar}.$$

DEFINITION A.3. A family of functionals $F[L] \in \mathcal{O}^+(\mathcal{E})$ parametrized by $L > 0$ is said to satisfy the homotopy renormalization group flow equation (hRGE) if for each $0 < \epsilon < L$

$$F[L] = W(P_{\epsilon\to L}, F[\epsilon]).$$

Acknowledgments

It's a pleasure to thank Kevin Costello, Chris Elliott, Owen Gwilliam, Steve Rosenberg, and Philsang Yoo for helpful feedback and discussion. We owe an especially large debt to Si Li and thank him for his insight and suggestions over the course of this project. We appreciate Eric Berry's careful reading of a preliminary version of this work that helped correct numerous typos. In addition, we thank the referee for providing useful feedback that greatly improved the organization and exposition of the paper.

References

[1] Mauro Carfora, *Renormalization group and the Ricci flow*, Milan J. Math. **78** (2010), no. 1, 319–353, DOI 10.1007/s00032-010-0110-y. MR2684783

[2] Kevin Costello, *A geometric construction of the Witten genus, II.* available at arXiv:1112.0816.

[3] Kevin Costello, *Renormalization and effective field theory*, Mathematical Surveys and Monographs, vol. 170, American Mathematical Society, Providence, RI, 2011. MR2778558

[4] Kevin Costello and Owen Gwilliam, *Factorization algebras in quantum field theory. Vol. 2.* available at `http://people.mpim-bonn.mpg.de/gwilliam/vol2may8.pdf`.

[5] Kevin Costello and Owen Gwilliam, *Factorization algebras in quantum field theory. Vol. 1*, New Mathematical Monographs, vol. 31, Cambridge University Press, Cambridge, 2017. MR3586504

[6] Chris Elliott, Brian Williams, and Philsang Yoo, *Asymptotic freedom in the BV formalism*, J. Geom. Phys. **123** (2018), 246–283, DOI 10.1016/j.geomphys.2017.08.009. MR3724786

[7] Daniel Harry Friedan, *Nonlinear models in $2+\varepsilon$ dimensions*, Ann. Physics **163** (1985), no. 2, 318–419, DOI 10.1016/0003-4916(85)90384-7. MR811072

[8] Sylvestre Gallot, Dominique Hulin, and Jacques Lafontaine, *Riemannian geometry*, 3rd ed., Universitext, Springer-Verlag, Berlin, 2004. MR2088027

[9] Vasily Gourbonov, Owen Gwilliam, and Brian Williams, *Chiral differential operators via Batalin-Vilkovisky quantization.* available at arXiv:1610.09657.

[10] Ryan Grady and Owen Gwilliam, *One-dimensional Chern–Simons theory and the $\hat{A}$ genus*, Algebr. Geom. Topol. **14** (2014), no. 4, 419–497. MR3331615

[11] Ryan Grady and Owen Gwilliam, *L_∞ spaces and derived loop spaces*, New York J. Math. **21** (2015), 231–272. MR3358542

[12] Ryan E. Grady, Qin Li, and Si Li, *Batalin-Vilkovisky quantization and the algebraic index*, Adv. Math. **317** (2017), 575–639, DOI 10.1016/j.aim.2017.07.007. MR3682678

[13] Qin Li and Si Li, *On the B-twisted topological sigma model and Calabi–Yau geometry*, J. Differential Geom. **102** (2016), no. 3, 409–484. MR3466804

[14] Timothy Nguyen, *Quantization of the nonlinear sigma model revisited*, J. Math. Phys. **57** (2016), no. 8, 082301, 40, DOI 10.1063/1.4961153. MR3539697

[15] Pavol Ševera and Fridrich Valach, *Ricci flow, Courant algebroids, and renormalization of Poisson-Lie T-duality*, Lett. Math. Phys. **107** (2017), no. 10, 1823–1835, DOI 10.1007/s11005-017-0968-5. MR3690034

Department of Mathematical Sciences, Montana State University, Bozeman, Montana 59717
Email address: `ryan.grady1@montana.edu`

Department of Mathematics, Northwestern University, Evanston, Illinois 60208
Email address: `bwill@math.northwestern.edu`

Contemporary Mathematics
Volume **718**, 2018
https://doi.org/10.1090/conm/718/14481

The holomorphic bosonic string

Owen Gwilliam and Brian Williams

ABSTRACT. We describe and analyze a holomorphic version of the bosonic string in the formalism of quantum field theory developed by Costello and collaborators, which provides a powerful combination of renormalization theory and the Batalin-Vilkovisky formalism. Our focus here is on the case in which the target space-time is a vector space. We identify the critical dimension as an obstruction to satisfying the quantum master equation, and when the obstruction vanishes, we construct a one-loop exact quantization. Moreover, we show how the factorization algebra of observables recovers the BRST cohomology of the string and use this perspective to give a new construction of its Gerstenhaber structure. Finally, we show how the factorization homology along closed manifolds encodes the determinant line bundle over the moduli space of Riemann surfaces. An auxiliary goal of this paper is to give an exposition of this QFT formalism with the holomorphic bosonic string theory as the running example.

CONTENTS

1. Introduction

Two intertwined goals govern our exposition. First, we want to describe a two-dimensional field theory, which we view as a holomorphic version of bosonic string theory, and its perturbative quantization. We'll see that this theory encodes the chiral sector of a bosonic string with linear target space, justifying our interpretation. Second, we want to use this theory as the running example for key ideas and techniques in the formalism for quantum field theory developed by Costello

---

2010 *Mathematics Subject Classification.* Primary 81T30, 81T70, Secondary 58D27, 17B69.

and collaborators [**Cos11, CG17, CG, LL16, GG14, GLL17, Li**]. We hope to give readers a feel for how to use this formalism by exhibiting it with a beautiful theory.

Our focus is thus on narrative rather than detailed argumentation. That is, we work systematically according the natural flow of the formalism. Along the way we emphasize the motivations behind each step rather than the nitty-gritty computations. Precedence is given to communicating the essence of an argument, over spelling everything out. We do give detailed citations where such arguments can be found in the literature, but we defer some not-yet-extant details to a forthcoming work on this theory with curved target space [**GW**].

None of the results here about string theory is new, as the bosonic string has been under intensive study for several decades, but this formalism recovers them in a single, systematic process, often giving a novel argument or perspective. It is compelling to have a direct path from the action functional to such sophisticated constructions as the semi-infinite cohomology of a vertex algebra. In fact, since so many of these results are familiar, the reader may see more clearly what's distinctive and illuminating about this approach to field theory.

There are many references on the bosonic string that have influenced us. In the physics literature there are the classic sources [**GSW12a, GSW12b, Pol98**] that explain perturbative string theory. In addition, there is an extensive mathematically-oriented treatment of perturbative string theory in [**DP88**], as well as D'Hoker's notes in Volume II of [**DEF$^+$99**]. Our approach, while intimately related, starts with a "first-order" description of the bosonic string.

Given the vastness of the string theory literature, it should not be a surprise that there is already work along these lines, notably by Losev, Marshakov, and Zeitlin [**LMZ06**]. One could view this paper as attempting to communicate many of their insights to those with an intuition growing out of homotopical algebra and the functorial approach to geometry. Again, we note that the formalism of Costello provides a mathematical articulation and verification of many ideas long known to physicists, such as the Wilsonian view of renormalization and the Batalin-Vilkovisky (BV) approach to gauge and gravity theories.[1] This machinery allows us to revisit such prior work in a manner particularly amenable to mathematicians.

1.1. Overview. The central figure of this paper is a holomorphic analogue of the bosonic string. We proceed, as usual in physics, from the classical to the quantum.

Hence, we begin by introducing the classical theory, expressed both in the BV formalism and also in terms of an action functional. We take some time to identify this theory as the chiral sector of a limit of the bosonic string, where the Kähler metric of the target is made very large. We also interpret the theory in the language of derived geometry.

We then turn to analyzing the deformations of this classical theory, which by Costello's work admits a nice description in terms of a type of Gelfand-Fuks cohomology. This perspective naturally leads to a discussion of string backgrounds.

With a firm grip on the classical theory, we turn to constructing the perturbative quantization. We first work with a disk or $\mathbb{C}$ as the source manifold, and we review relevant features of Costello's approach to renormalization. The usual

[1] We also note that given the literature's size, and our relative and unfortunate ignorance of much of it, we have chosen to mention a reference when we feel its description is particularly useful for us, even if it is not the original or standard reference for the result.

dimensional Weyl anomaly appears as an obstruction to satisfying the quantum master equation, a key condition in the BV formalism. At this stage, the anomaly appears as a computation with Feynman diagrams.

The next section describes the vertex algebra of the quantized theory, using the machinery of factorization algebras of [**CG17**, **CG**]. We find this piece of the formalism particularly illuminating, as it lets a mathematician understand how to read off the OPE from path integral manipulations.

We then turn to the case of a compact Riemann surface as the source manifold. Here we discuss how the formalism relates to the global approach to computing anomalies using, for instance, the Grothendieck-Riemann-Roch formula. We also discuss conformal blocks in this formalism.

Finally, we sketch how to modify the approach here to allow a complex manifold as the target. This paper can be viewed as an expository precursor to future work, which pushes into new territory (particularly in describing the vertex algebra).

1.2. Lessons to bear in mind. Before turning to our example, we want to expound some key ideas of the Costello formalism so that the reader is alert to them when proceeding through the text. That is, we wish to articulate here the structural features of this BV/renormalization package that make the arguments below conceptual.

For instance, in a gauge theory we know that connections provide the "naive" fields and that one must identify connections that are gauge-equivalent. A mathematician would say the true fields are a *stacky* quotient of the naive fields. Similarly, the critical locus of the action functional S is the zero locus of its differential $\mathrm{d}\,S$ (ignoring some subtleties of the variational set-up), which is the intersection of $\mathrm{d}\,S$ with the zero section of the cotangent space of the fields. But in mathematics it is better to take *derived* intersections.

LESSON 1.1 (Part 1, [**CG**]). The classical BV formalism is a method for computing the derived critical locus of the action functional on the derived stack of fields. Ghosts appear to describe the direction along which one quotients—the stacky direction—while the antifields appear to describe the direction along which one intersects—the derived direction.

We will describe our theory in the usual way, involving fields and ghosts, but we will also sketch its meaning in terms of global derived geometry, which we find illuminates the deep connections between string theory and algebraic geometry.

Path integral quantization amounts to trying to put a kind of measure or volume form on the derived stack of fields. When the fields form a linear space, there is a natural quantization that is translation-invariant along the fields, which is the analogue of the Lebesgue measure on an ordinary vector space.

LESSON 1.2 ([**GH18**]). Linear BV quantization is functorial, and it behaves much like a determinant functor. Hence, when one takes the fiberwise quantization of a family of linear theories, one typically obtains a determinant line bundle over the base.

This situation is relevant to us because the theory we study arises from a simple free theory, the free $\beta\gamma$ system, which lives on any Riemann surface. Hence the quantization of the free $\beta\gamma$ system makes sense over the moduli of Riemann surfaces and naturally produces a line bundle.

To be more specific, our classical theory of interest arises by gauging the natural action of holomorphic vector fields on the free $\beta\gamma$ system. As holomorphic vector fields are infinitesimal biholomorphisms, one can say that we couple the $\beta\gamma$ system to holomorphic gravity. But then we recognize a natural consequence of our prior lessons.

LESSON 1.3 (§5.11, [**Cos11**]). Gauging a classical theory corresponds to taking a stacky quotient of the original fields. To quantize the gauged theory corresponds to descending the quantization to the quotient. Hence, an anomaly that prevents quantization should be understood as an obstruction to descent.

The formalism of Costello makes this relationship manifest, as the anomaly that appears in trying to produce a BV quantization—which is a Feynman diagram construction—is a cocycle in a dg Lie algebra determined by the classical field theory. Thus, the anomaly determines an element of a natural Lie algebra cohomology group (in this case, Gelfand-Fuks cohomology), whose descent-theoretic meaning is typically easy to recognize. Here we will discover the famed Weyl, or conformal, anomaly, which requires the target space to be real 26-dimensional.

Anomalies are often characteristic classes, and this BV/renormalization package offers a structural explanation. Most classical field theories—at least most of broad interest—make sense on a class of manifolds, and so the anomaly ought to be determined by the local geometry of this class. In more mathematical language we have the following.

LESSON 1.4 ([**GGW**]). If a classical theory determines a sheaf on some site of manifolds (such as the site of Riemann surfaces and local biholomorphisms), then to quantize the theory over the whole site, it suffices to check on a generating cover (typically given by disks with geometric structure) but compatibly with all automorphisms.

In particular, the BV anomaly is a cocycle for the Lie algebra of automorphisms of the *formal* disk equipped with such geometric structures. In other words, it lives in some kind of Gelfand-Fuks cohomology, which gives deep and informative connections with foliation theory and topology.

So far, everything we have mentioned is well-known in field theory, albeit often expressed in a different dialect of mathematics. We now turn to the main new notion of this framework: factorization algebras, which provide an efficient and powerful way to organize the local-to-global structure of the observables of a field theory.

LESSON 1.5 ([**CG17**, **CG**]). Every BV theory produces a factorization algebra. The local structure encodes the OPE algebra, so that for a chiral CFT, one recovers a vertex algebra. On compact manifolds, the global structure often has finite-dimensional cohomology because solutions to the equations of motion are typically finite-dimensional. For a chiral CFT, one recovers the conformal blocks in this way.

A technical result of [**CG17**] gives a precise articulation of this lesson, and we will apply it to identify the vertex algebra arising from our holomorphic version of the bosonic string.

2. The classical holomorphic bosonic string

There is a basic format for a string theory, at least in the perturbative approach. One starts with a nonlinear σ-model, whose fields are smooth maps from a Riemann surface to a target manifold X; in this setting we want the theory to make sense for an arbitrary Riemann surface as the source manifold. In the usual bosonic string theory, this nonlinear σ-model picks out the harmonic maps from a Riemannian 2-manifold to a Riemannian manifold. In our holomorphic setting, the nonlinear σ-model picks out holomorphic maps from a Riemann surface to a complex manifold. One then quotients the space of fields (and solutions to the equations of motion) with respect to reparametrization. In the usual bosonic string, one quotients by diffeomorphisms and Weyl rescalings (i.e., multiplying the metric by a positive real function), which can thus change the metric on the source. In our setting, we quotient by biholomorphisms, which act on the complex structure on the source.

In this section we begin by describing our theory in the BV formalism. We do not expect the reader to find the action functional immediately clear, so we devote some time to analyzing what it means and how it arises from concrete questions. We then turn to interpreting this classical BV theory using dg Lie algebras and derived geometry (i.e., we identify the moduli space it encodes). Finally, we conclude by sketching how our theory appears as the chiral sector of a degeneration of the usual bosonic string when the target is a complex manifold with a Hermitian metric. Our theory thus does provide insights into the usual bosonic string; moreover, it clarifies why so many aspects of the bosonic string, like the anomalies or B-fields, have holomorphic analogues.

2.1. The theory we study. Let V denote a complex vector space (the target), and let $\langle -,-\rangle_V$ denote the evaluation pairing between V and its linear dual $V^\vee$. Let Σ denote a Riemann surface (the source). Let $T^{1,0}_\Sigma$ denote the holomorphic tangent bundle on Σ, let $\langle -,-\rangle_T$ denote the evaluation pairing between $T^{1,0}_\Sigma$ and its vector bundle dual $T^{1,0*}_\Sigma$. These are the key geometric inputs.

In a BV theory, the fields are $\mathbb{Z}$-graded; we call this the *cohomological grading*. We have four kinds of fields:

field	-1	0	1	2
γ		$\Omega^{0,0}(\Sigma)\otimes V$	$\Omega^{0,1}(\Sigma)\otimes V$	
β		$\Omega^{1,0}(\Sigma)\otimes V^\vee$	$\Omega^{1,1}(\Sigma)\otimes V^\vee$	
c	$\Omega^{0,0}(\Sigma, T^{1,0}_\Sigma)$	$\Omega^{0,1}(\Sigma, T^{1,0}_\Sigma)$		
b			$\Omega^{1,0}(\Sigma, T^{1,0*}_\Sigma)$	$\Omega^{1,1}(\Sigma, T^{1,0*}_\Sigma)$

More accurately, we have eight different kinds of fields, but we view each row as constituting a single type since each given row consists of the Dolbeault forms of a holomorphic vector bundle. For instance, the field γ is a $(0,*)$-form with values in the trivial bundle with fiber V, and the field b is a $(0,*)$-form with values in the bundle $T^{1,0*}\otimes T^{1,0*}$.

To orient oneself it is helpful to start by examining the fields of cohomological degree zero, since these typically have a manifest physical meaning. For instance, the degree zero γ field is a smooth V-valued function and hence the natural field for the nonlinear σ-model into V. The degree zero c field is a smooth $(0,1)$-form with values in vector field "in the holomorphic direction," and hence encodes an infinitesimal change of complex structure of Σ. The degree -1 part of c contains the gauge fields of the theories, vector fields. The equations of motion dictate that these vector fields be holomorphic, so we are seeing the infinitesimal version of the symmetry by biholomorphisms we mentioned above. These constitute the obvious fields to introduce for a holomorphic version of the bosonic string. The fields β and b are less obvious but appear as "partners" (or antifields) whose role is clearest once we have the action functional and hence equations of motion.

The action functional is

$$S(\gamma,\beta,c,b)=\int_\Sigma\langle\beta,\bar{\partial}\gamma\rangle_V+\int_\Sigma\langle b,\bar{\partial}c\rangle_T+\int_\Sigma\langle\beta,[c,\gamma]\rangle_V+\int_\Sigma\langle b,[c,c]\rangle_T. \tag{1}$$

(We discuss below how to think about fields with nonzero cohomological degrees as inputs.) The equations of motion are thus

$$\begin{aligned}0&=\bar{\partial}\gamma+[c,\gamma] &\qquad 0&=\bar{\partial}\beta+[c,\beta]\\0&=\bar{\partial}c+\tfrac{1}{2}[c,c] &\qquad 0&=\bar{\partial}b+[c,b]\end{aligned}$$

Note that these equations are familiar in complex geometry. For instance, the equation purely for c encodes a deformation of complex structure on Σ; concretely, it modifies the $\bar{\partial}$ operator to $\bar{\partial}+c$. The other equations then amount to solving for holomorphic sections (of the relevant bundle) withe respect to this deformed complex structure. For instance, the equation in γ picks out holomorphic maps from Σ, with the c-deformed complex structure, to V.

The field b can be understood as an "antifield" to the ghost field c; in other words, it is an *antighost.* In that sense, b does not have any intrinsic, physical meaning by itself.

REMARK 2.1. Just looking at this action functional, one might notice that if one drops the last two terms, which are cubic in the fields, then one obtains a free theory

$$S_{free}(\gamma,\beta,c,b)=\int_\Sigma\langle\beta,\bar{\partial}\gamma\rangle_V+\int_\Sigma\langle b,\bar{\partial}c\rangle_T, \tag{2}$$

which is known as the *free $bc\beta\gamma$ system.* Thus, one may view the holomorphic bosonic string as a deformation of this free theory by "turning on" those interaction terms. We will repeatedly try a construction first with this free theory before tackling the string itself, as it often captures important information with minimal work. For instance, we will examine the vertex algebra for the free theory before seeing how the interaction affects the operator products. Similarly, one can identify the anomaly already at the level of the free theory.

This viewpoint of arriving at the bosonic string as a deformation of a free CFT is central to the analysis of the string in the physics literature [**GSW12a**] and Chapter 2 of [**Pol98**]. See also the work in [**Sch75**].

REMARK 2.2. It is easy to modify this action functional to allow a curved target, i.e., one can replace the complex vector space V with an arbitrary complex manifold X. The fields b,c remain the same. The degree 0 field γ still encodes smooth maps into X, but now the degree 1 field is a section of $\Omega^{0,1}(\Sigma,\gamma^*T^{1,0}_X)$. Similarly, β is now a section of $\Omega^{1,*}(\Sigma,\gamma^*T^{1,0*}_X)$. The action is then

$$S(\gamma,\beta,c,b)=\int_\Sigma\langle\beta,\bar{\partial}\gamma\rangle_{T_X}+\int_\Sigma\langle b,\bar{\partial}c\rangle_{T_\Sigma}+\int_\Sigma\langle\beta,[c,\gamma]\rangle_{T_X}+\int_\Sigma\langle b,[c,c]\rangle_{T_\Sigma}. \tag{3}$$

In Section 7 we will indicate how the results with linear target generalize to this situation.

2.2. From the perspective of derived geometry. We would like to explain what this theory is about in more conceptual terms, rather than simply by formulas and equations. Thankfully this theory is amenable to such a description. We will be informal in this section and not specify a particular geometric context (e.g., derived analytic stacks), except when we specialize to the deformation-theoretic situation (i.e., perturbative setting) that is our main arena.

Let $\mathcal{M}$ denote the moduli space of Riemann surfaces, so that a surface Σ determines a point in $\mathcal{M}$. Let $\mathrm{Maps}_{\bar{d}}(\Sigma, V)$ denote the space of holomorphic maps from Σ to V, and hence a bundle $\mathrm{Maps}_{\bar{d}}(-, V)$ over $\mathcal{M}$ by varying Σ. For our equations of motion, the γ and c fields of a solution determine a point in this bundle $\mathrm{Maps}_{\bar{d}}(-, V)$. The commutative algebra $\mathcal{O}(\mathrm{Maps}_{\bar{d}}(\Sigma, V))$ of functions on the space encodes the observables of the classical theory.

This construction makes sense on noncompact Riemann surfaces as well. Let $\mathcal{RS}$ denote the category whose objects are Riemann surfaces and whose morphisms are holomorphic embeddings. There is a natural site structure: a cover is a collection of maps $\{S_i \to \Sigma\}_i$ such that the union of the images is all of Σ. Then $\mathrm{Maps}_{\bar{d}}(-, V)$ defines a sheaf of spaces over $\mathcal{RS}$. The observables for the classical theory is, in essence, the *co*sheaf of commutative algebras $\mathcal{O}(\mathrm{Maps}_{\bar{d}}(-, V))$, and hence provides a factorization algebra.

In fact, it is better to work with the derived version of these spaces. One important feature of derived geometry is that the appropriate version of a tangent space at a point is, in fact, a cochain complex. In our setting, a point (c, γ) in $\mathrm{Maps}_{\bar{d}}(-, V)$ determines a complex structure $\bar{d} + c$ on Σ—we denote this Riemann surface by Σ_c—and γ a V-valued holomorphic function on Σ_c. The tangent complex of $\mathrm{Maps}_{\bar{d}}(-, V)$ at (c, γ) is precisely

$$\Omega^{0,*}(\Sigma_c, T^{1,0})[1] \oplus \Omega^{0,*}(\Sigma_c, V).$$

The first summand is the usual answer from the theory of the moduli of surfaces (recall, for example, that the ordinary tangent space is the sheaf cohomology $H^1(\Sigma, T_\Sigma)$ of the holomorphic tangent sheaf), and the second is usual elliptic complex encoding holomorphic maps.

REMARK 2.3. It is useful to bear in mind that the degree zero cohomology of the tangent complex will recover the "naive" tangent space. In our case, we have

$$H^1(\Sigma_c, T_{\Sigma_c}) \oplus H^0(\Sigma_c, V),$$

which encodes deformations of complex structure and holomorphic maps. Negative degree cohomology of the tangent complex detects infinitesimal automorphisms (and automorphisms of automorphisms, etc) of the space. For instance, here we see $H^0(\Sigma_c, T_{\Sigma_c})$ appear in degree -1, since a holomorphic vector field is an infinitesimal automorphism of a complex curve. These negative directions are called "ghosts" (or ghosts for ghosts, and so on) in physics. The positive degree cohomology detects infinitesimal relations (and relations of relations, and so on).

Note that the underlying graded spaces of this tangent complex are the c and γ fields from the BV theory described above. We emphasize that the tangent complex is only specified up to quasi-isomorphism, but it is compelling that a natural representative is the BV theory produced by the usual physical arguments. This behavior, however, is typical of the relationship between derived geometry and

BV theories: when physicists write down a classical BV theory, the underlying free theory is essentially always the tangent complex of a nice derived stack.

The reader has probably noticed that, yet again, we have postponed discussing the β and b fields. From a derived perspective, the full BV theory describes the shifted cotangent bundle $\mathbb{T}^*[-1]\,\mathrm{Maps}_{\bar{d}}(-,V)$. At the level of a tangent complex, the shifted cotangent direction contributes

$$\Omega^{1,*}(\Sigma_c, T^{1,0*})[-1] \oplus \Omega^{1,*}(\Sigma_c, V^\vee),$$

whose underlying graded spaces are the β and b fields. These "antifields" are added so that the overall space of fields has a 1-shifted symplectic structure when Σ is closed, and a shifted Poisson structure when Σ is open.

2.3. Relationship to the Polyakov action functional. This holomorphic bosonic string has a natural relationship with the usual bosonic string. We sketch it briefly, only considering a linear target.

We begin with a bosonic string theory where the source is a 2-dimensional smooth oriented Riemannian manifold Σ and the target is a Hermitian vector space (V,h). The "naive" action functional is

$$S_{Poly}^{\mathrm{naive}}(\varphi, g) = \int_\Sigma h(\varphi, \Delta_g \varphi)\, \mathrm{dvol}_g$$

where the field g is a Riemannian metric on Σ and the field φ is a smooth map from Σ to V. The notation Δ_g denotes the Laplace-Beltrami operator on Σ.

Note that S_{Poly}^{naive} is invariant under the diffeomorphism group $\mathrm{Diff}(\Sigma)$ and under rescalings of the metric (i.e., the theory is classically conformal). Typically we express these Weyl rescalings as $g \mapsto e^f g$ with $f \in C^\infty(\Sigma)$. As we are interested in a string theory, we want to gauge these symmetries. In geometric language, we want to think about the quotient stack obtained by taking solutions to the equations of motion and quotienting by these symmetry groups.

Our focus is perturbative, so that we want to study the behavior of this action near a fixed solution to the equations of motion. In other words, we want to work with the Taylor expansion of the true action near some solution. Hence, we work around a fixed metric g_0 on Σ, and we substitute for the field g, the term $g_0 + \alpha$ where $\alpha \in \Gamma(\Sigma, \mathrm{Sym}^2(T_\Sigma))$. That is, we will consider deformations of g_0. As φ is linear, we just consider expanding around the zero map. Thus our new fields are $\varphi \in C^\infty(\Sigma, V)$ and $\alpha \in \Gamma(\Sigma, \mathrm{Sym}^2(T_\Sigma))$.

There are also ghost fields associated to the symmetries we gauge. First, there are infinitesimal diffeomorphisms, which are described by vector fields on Σ. We denote this ghost field by $X \in \Gamma(\Sigma, T_\Sigma)$. It acts on the fields by the transformation

$$(\varphi, \alpha) \mapsto (\varphi + X \cdot \varphi, \alpha + L_X \alpha),$$

where L_X denotes the Lie derivative on tensors. Second, there are infinitesimal rescalings of the metric, such as $\alpha \mapsto \alpha + f\alpha$, with ghost field $f \in C^\infty(\Sigma)$. The rescaling does not affect φ. The two symmetries are compatible: given f and X, then $L_X(f\alpha) = X(f)\alpha + f L_X \alpha$ for any $\alpha \in \mathrm{Sym}^2(T_\Sigma)$.

To summarize, we have the following graded vector space of fields:

field/antifield	-1	0	1	2
$\varphi, \varphi^\vee$		$\Omega^0(\Sigma)\otimes V$	$\Omega^2(\Sigma)\otimes V$	
$\alpha, \alpha^\vee$		$\Omega^0(\Sigma, \mathrm{Sym}^2(T_\Sigma))$	$\Omega^2(\Sigma; \mathrm{Sym}^2(T^*_\Sigma))$	
$X, X^\vee$	$\mathrm{Vect}(\Sigma)$			$\Omega^2(\Sigma; T^*_\Sigma)$
$f, f^\vee$	$C^\infty(\Sigma)$			$\Omega^2(\Sigma)$.

The BV action functional is of the form:

$$S^{BV}_{Poly}(\varphi,\alpha,X,f) = \int_\Sigma h(\varphi, \Delta_{g_0}\varphi)\,\mathrm{dvol}_{g_0} + \sum_{n\geq 1}\frac{1}{n!}\int_\Sigma h(\varphi, D_n(\alpha)\varphi)\,\mathrm{dvol}_{g_0} \tag{4}$$

$$+\int_\Sigma h(\varphi, X\cdot\varphi)\,\mathrm{dvol}_{g_0} \tag{5}$$

$$+\,S'(X,f,\alpha) \tag{6}$$

The right hand side of the first line amounts to expanding out the Laplace-Beltrami operator $\Delta_{g_0+\alpha}$ as a function of α. Hence, the D_n are differential operators of the form $D_n : \left(\mathrm{Sym}^2(T_\Sigma)\right)^{\otimes n} \to \mathrm{Diff}^{\leq 2}(\Sigma)$ where $\mathrm{Diff}^{\leq 2}(\Sigma)$ are order ≤ 2 differential operators on Σ. Thus, for each section α of $\mathrm{Sym}^2(T_\Sigma)$, we get a second-order differential operator $D_n(\alpha)$ acting on functions on Σ. (This term is the nth term in the Taylor expansion, so its homogeneous of order n: $D_n(t\alpha) = t^n D_n(\alpha)$ for a scalar t.) The second line encodes how vector fields act on the maps of the σ-model. The third line $S'(X,f,\alpha)$ is independent of φ and only depends on the fields f, X, α and their antifields (denoted with checks $(-)^\vee$). It is of the form

$$S'(f,X,\alpha) = \int_\Sigma \langle \alpha^\vee, L_X(g_0+\alpha) + f(g_0+\alpha)\rangle + \int_\Sigma \langle X^\vee, [X,X]\rangle + \int_\Sigma \langle f^\vee, X\cdot f\rangle\,.$$

The first term encodes how vector fields and Weyl transformations act on the perturbed metric $g_0+\alpha$ and the remaining terms are required to ensure the gauge symmetry is consistent (satisfies the classical master equation).

An explicit formula for $D_n(\alpha,\ldots,\alpha)$ is a rather involved exercise (and not needed here). For instance, if we are working locally on $\Sigma = \mathbb{R}^2$ with the g_0 the flat metric, then the operator $D_1(\alpha)$ is sum of a first-order and a second-order differential operator

$$D_1(\alpha) = \frac{1}{2}\frac{\partial}{\partial x^i}(\mathrm{tr}(\alpha))\frac{\partial}{\partial x^i} + \frac{1}{2}\mathrm{tr}(\alpha)\frac{\partial}{\partial x^i}\frac{\partial}{\partial x^i},$$

or in a more coordinate-free notation,

$$D_1(\alpha) = \frac{1}{2}\star\mathrm{d}\,(\mathrm{tr}(\alpha)\star\mathrm{d})\,.$$

Here, we use the natural trace map $\mathrm{tr} : \mathrm{Sym}^2 T\Sigma \to C^\infty(\Sigma)$ of symmetric 2×2 matrices.

There is an important parameter in this action functional: the Hermitian inner product h. We can consider scaling it $h \to th$, with $t \in (0,\infty)$. The "infinite volume limit" as $t\to\infty$ admits a nice description, provided one rewrites the action functional in a first-order formalism (i.e., adjoins fields so that only first-order differential operators appear in the action, which is a sort of action functional analogue of working with phase space).

LEMMA 2.4. *In this infinite volume limit the bosonic string becomes equivalent to a BV theory whose action functional has the form*

$$S(\beta,\gamma,b,c)+\overline{S}(\overline{\beta},\overline{\gamma},\overline{b},\overline{c}),$$

where $S(\beta,\gamma,b,c)$ is the action functional for the holomorphic bosonic string in Equation (1) and $\overline{S}$ is its anti-holomorphic conjugate.

REMARK 2.5. The action functional $\overline{S}$ is similar to S where the fields γ,β,b,c are replaced by sections in the the relevant conjugate bundles. For example, $\beta\in\Omega^{1,*}(\Sigma)$ becomes $\overline{\beta}\in\Omega^{*,1}(\Sigma)$. Moreover, the operator $\bar{\partial}$ is replaced by the holomorphic Dolbeault operator ∂. Another way of saying this is that $\overline{S}$ is the holomorphic string on $\overline{\Sigma}$, which is the conjugate complex structure to Σ.

REMARK 2.6. For physics references that study the holomorphic splitting of the Polyakov action (and more general conformal field theories), we refer to the original sources [**BK86**, **KLS91**].

OUTLINE OF PROOF. There are two things that may cause alarm in the statement of the claim. First, the space of fields of the Polyakov string (in the BV language) and those of the holomorphic bosonic string do not match up. Second, the infinite volume limit $t\to\infty$ is naively ill-defined using the action functional (4). It turns out that these two issues are solved by the same maneuver.

We begin with the first term in the first line of (4). Notice that it is simply the action functional for the σ-model of maps from (Σ,g_0) to (V,h). It is shown in Appendix 21 of [**GGW**] how to make sense of the infinite volume limit of this usual σ-model. The idea is to rewrite this theory in the *first order formalism*. This amounts to introducing a new field $B\in\Omega^1(\Sigma)\otimes V^\vee$ and action functional

$$\int_\Sigma\langle B,\mathrm{d}\varphi\rangle_V-\frac{1}{2}\int_\Sigma h^\vee(B,\star B)$$

where $\langle-,-\rangle_V$ represents the evaluation pairing between V and its dual, $\star$ is the Hodge star operator for the metric g_0, and $h^\vee$ denotes the dual metric on V. This action functional is equivalent to the original σ-model; one can compare the equations of motion. Moreover, since $(th)^\vee=(1/t)h^\vee$, the dual $(th)^\vee$ goes to 0 in the infinite volume limit $t\to\infty$, and hence kills the second term in the first order action. The remaining theory splits as the direct sum of the free $\beta\gamma$ system with target V and its anti-holomorphic conjugate. At the level of fields, the original field φ corresponds to $\gamma+\overline{\gamma}$ in the first order description, and B corresponds to $\beta+\overline{\beta}$.

We now consider the remaining terms in the first line of the action (4). Note that this action only depends on the conformal class of the metric, i.e., on the metric up to Weyl rescaling. Hence this feature remains true in the infinite volume limit, which simplifies the situation considerably, as we now explain.

It is convenient to work in holomorphic coordinates for the complex structure determined by the background metric g_0. With respect to this complex structure, the tensor square of the cotangent bundle splits as

$$T^*\otimes T^*=T^{*0,1}\otimes T^{1,0}\oplus T^{*1,0}\otimes T^{0,1}\oplus T^{*1,0}\otimes T^{*0,1}\oplus T^{*0,1}\otimes T^{*1,0}.$$

Sitting inside of this bundle is the symmetric square, where the field α lives. With respect to this splitting, we write sections as $\alpha=c+\overline{c}+fg_0$, where $f\in C^\infty(\Sigma)$. But since the action only depends on the conformal class of the metric, only the fields c and $\overline{c}$ are relevant. In the first-order formalism of the preceding paragraph, we thus find that the remaining terms in the first line of (4) read

$$\int_\Sigma \langle \beta, [c,\gamma]\rangle_V + \int_\Sigma \langle \overline{\beta}, [\overline{c},\overline{\gamma}]\rangle_V .$$

This first term is precisely the third term in the holomorphic string action functional (1), which describes how deformations of complex structure couple to the fields of the σ-model.

In the infinite volume limit, the term $S'(f, X, \alpha)$ recovers the terms

$$\int_\Sigma \langle b, \overline{\partial} c\rangle_T + \int_\Sigma \langle b, [c,c]\rangle_T$$

in the action of the holomorphic string, plus their conjugates. The arguments are similar to those we have just sketched. □

REMARK 2.7. Another approach to arrive at the holomorphic theory we consider comes from considering supersymmetry. Without gravity, the pure holomorphic σ-model can be viewed as the *holomorphic twist* of the $N = (2,0)$ supersymmetric σ model (in this case the target is required to be Kähler). Moreover, the $\beta\gamma bc$ system is the holomorphic twist of the $N = (2,2)$ model. Conjecturally, we expect the holomorphic theory of gravity we consider to be the holomorphic twist of two-dimensional $N = 2$ supergravity.

REMARK 2.8. In this infinite volume limit, one can put the dependence of the target metric back into the theory by choosing a certain background to work in. In the BV formalism this amounts to choosing a certain deformation parameter, which in this instance corresponds to infinitesimal deformations of the target metric. Note that to deform the metric on the target we leave the world of "holomorphic field theory" as the deformation involves both z and $\overline{z}$ dependent terms. It would be interesting to study how to formulate the theory with finite target metric in the BV formalism.

3. Deformations of the theory and string backgrounds

Whenever one is studying a theory, it is helpful to understand how it can be modified and how features of the theory change as one adjusts natural parameters of the theory, such as coupling constants of the action functional. In other words, one wants to understand the theory in the moduli space of classical theories.

In the BV formalism, because we are working homologically, this moduli space is derived, and there is a tangent complex to our theory in the moduli of classical BV theories. We call it the *deformation complex* of the theory. A systematic discussion can be found in Chapter 5 of [**Cos11**].

As a gloss, the underlying graded vector space of this deformation complex consists of the local functionals on the jets of fields, i.e., Lagrangian densities. (Note that we allow local functionals of arbitrary cohomological degree.) There is also a shifted Lie bracket $\{-,-\}$, which arises from the pairing $\int_\Sigma \langle -,-\rangle$ on the fields. It is, in essence, the shifted Poisson bracket corresponding to that shifted symplectic pairing on the fields. The differential on the local functionals is then $\{S,-\}$, where S is the classical action. All together, the deformation complex forms a shifted dg Lie algebra. Observe that if we find a degree zero element I such that

$$0 = \{S+I, S+I\} = 2\{S,I\} + \{I,I\},$$

then I is a shifted Maurer-Cartan element and hence determines a new classical BV theory whose action functional is $S+I$. In particular, degree 0 cocycles determine first-order deformations of the classical BV theory. Cocycles in degree -1 encode local symmetries of the classical theory; and obstructions to satisfying the quantum master equation end up being degree 1 cocycles.

In this section, we will explain why the deformation complex $\mathrm{Def}_{\mathrm{string}}$ of the holomorphic string can be expressed in terms of Gelfand-Fuks cohomology [**Fuk86**]. Along the way we will see how the usual backgrounds for the bosonic string (a target metric, dilaton term, and so on) appear as elements in this complex of local functionals and hence as deformations of the classical action.

Right now, we will focus on the case $\Sigma = \mathbb{C}$, and in Section 6 we will consider arbitrary Riemann surfaces. We restrict ourselves to examining *translation-invariant* local functionals (which will allow us to descend to a theory defined on an elliptic curve). Unpacking what this means will lead swiftly to Gelfand-Fuks cohomology.

3.1. Deformations for the classical theory. As a local functional is given by integration of a Lagrangian density, translation invariance requires the density to be the Lebesgue measure $\mathrm{d}^2 z$, up to rescaling, and requires the Lagrangian to be specified by its behavior at one point. Hence, a translation-invariant local functional on $\mathbb{C}$ is determined by a function of the jet (i.e., Taylor expansion) of the fields at the origin in $\mathbb{C}$.

It is particularly easy to understand what we mean in the case of the free $bc\beta\gamma$ system. For instance, the γ fields live in the Dolbeault complex $\Omega^{0,*}(\mathbb{C}; V)$, and their jets at the origin are $(V[[z,\overline{z}]][\mathrm{d}\,\overline{z}], \bar{\partial})$, where $\bar{\partial}$ is the formal Dolbeault differential. An example of an element is thus $\widehat{\gamma} = \sum_{m,n} \frac{1}{m!n!} g_{mn} z^m \overline{z}^n$, which is just a formal power series with values in V. An example of a functional is

$$F(\widehat{\gamma}) = g_{10} + g_{21} = (\partial_z \widehat{\gamma})\,|_0 + \left(\partial_z^2 \partial_{\overline{z}} \widehat{\gamma}\right)|_0,$$

which corresponds to the local functional

$$F(\gamma) = \int_{\mathbb{C}} \partial_z \gamma + \partial_z^2 \partial_{\overline{z}} \gamma \; \mathrm{d}^2 z.$$

We call the first kind of term a *chiral* interaction, as it only depends on holomorphic derivatives.

By the $\bar{\partial}$-Poincaré lemma, this complex $(V[[z,\overline{z}]][\mathrm{d}\,\overline{z}], \bar{\partial})$ is quasi-isomorphic to $V[[z]]$, concentrated in degree zero. This observation is actually quite concrete: it simply says that for a solution γ to the equation of motion $\bar{\partial}\gamma = 0$, its Taylor expansion is just a power series in z and it is independent of $\overline{z}$. In consequence, if we consider translation-invariant Lagrangians depending only on the γ field, then up to quasi-isomorphism these are $\mathrm{Sym}(V^\vee[z^\vee])$. In other words, only chiral interactions yield distinct modifications of the action, when one takes into account the equation of motion.

Note that we have chosen to work with functionals of the fields that are polynomials built out of continuous linear functionals $V^\vee[z^\vee]$ of the jets. This choice is the standard and natural one for variational problems. We note as well that constant functionals are irrelevant, so we want to use $\mathrm{Sym}^{>0}(V^\vee[z^\vee])$ to describe translation-invariant local functionals.

An analogous argument applies to the c field. It shows there is a quasi-isomorphism of dg Lie algebras between the jet at the origin of the Dolbeault complex $\Omega^{0,*}(\mathbb{C}; T^{1,0}_{\mathbb{C}})$ of holomorphic vector fields and the Lie algebra of formal vector fields $W_1 = \mathbb{C}[[z]]\partial_z$. The translation-invariant Lagrangians depending only on the c field are thus quasi-isomorphic to $C^*_{\mathrm{Lie,red}}(W_1)$, by which we mean the (reduced) *continuous* Lie algebra cohomology, often known as the Gelfand-Fuks cohomology Similar arguments work for the β and b fields.

If we take all the fields into account together and consider the full equations of motion for the holomorphic string, which couple the c field to the others, then these arguments yield the following.

LEMMA 3.1. *There is a quasi-isomorphism*

$$\mathrm{Def}_{\mathrm{string}}(\mathbb{C}, V)^{\mathbb{C}} \simeq C^*_{\mathrm{Lie,red}}(W_1, \mathrm{Sym}(V^\vee[z^\vee] \oplus V[z^\vee]\,\mathrm{d}\,z^\vee \oplus W_1^{\mathrm{ad}}[2]))[2]$$

between the deformation complex of translation-invariant local functionals for the holomorphic string and a certain Gelfand-Fuks cochain complex.

This lemma already substantially simplifies our lives, as one can invoke the literature on Gelfand-Fuks cohomology. But before we do, we will take advantage of another symmetry condition to simplify the situation.

3.2. Dilating cotangent fibers. We have already seen how to think of the holomorphic bosonic string theory as corresponding to the shifted cotangent bundle $\mathbb{T}^*[-1]\mathrm{Maps}_{\bar{\partial}}(-, V)$, as a bundle over the moduli of Riemann surfaces. There is a natural action of the group $\mathbb{C}^\times$ on this space by scaling the shifted cotangent fibers, and we will use the notation $\mathbb{C}^\times_{\mathrm{cot}}$ to indicate this appearance of the multiplicative group.

This group action can be seen on the level of the field theory as follows: we give the γ and c fields—the base of the cotangent bundle—weight 0 and give the β and b fields—the cotangent fiber—weight 1. Note that, in consequence, the pairing $\langle -, - \rangle$ on fields thus has weight -1. In these terms, the classical action functional is weight 1. Thus, we focus on weight 1 deformations of the action for the holomorphic bosonic string, as we are interested in local functionals of the same kind. That means we consider the subcomplex of weight 1 local functionals inside the deformation complex.

REMARK 3.2. Although this action S has weight 1, its role in the cochain complex of classical observables is to define the differential $\{S, -\}$. Observe that the shifted Poisson bracket $\{-, -\}$ has weight -1, because it is determined by the pairing, and so the differential has weight 0.

This subcomplex admits a nice description in terms of the geometry of the target.

LEMMA 3.3. *There is a* $\mathrm{GL}(V)$*-equivariant quasi-isomorphism*

$$\mathrm{Def}_{\mathrm{string}}(\mathbb{C})^{\mathbb{C},\mathrm{wt}(1)} \simeq \mathrm{Sym}(V^*) \otimes V[1]$$

between the weight 1, translation-invariant deformation complex and the polynomial vector fields on V*, placed in degree -1.*

Concretely, this result says that there are no weight zero interactions that are not not trivialized by an automorphism of the theory. This claim is a consequence of the fact that the zeroth cohomology group vanishes. On the other hand, this lemma says the theory admits a large group of symmetries, namely diffeomorphisms of the target, which appears as the degree -1 cohomology.

The $\mathrm{GL}(V)$ equivariance takes into account the natural symmetries of the target. It also is the first step in the approach to studying the deformation complex with general curved target. We will discuss this further in the section on string backgrounds.

3.3. Interaction terms that appear at one loop. As we will see in Section 4, the quantization of the holomorphic string only involves local functional of weight zero for this $\mathbb{C}^\times_{\text{cot}}$-action. (Concretely, this restriction appears because the one-loop Feynman diagrams only have external legs for c and γ fields.) Hence, it behooves us to compute the weight zero subcomplex of the deformation complex as well.

LEMMA 3.4. *There is a* $\mathrm{GL}(V)$*-equivariant quasi-isomorphism*

$$\mathrm{Def}_{\text{string}}(\mathbb{C})^{\mathbb{C},\mathrm{wt}(0)} \simeq \mathbb{C}[-1] \oplus \Omega^2_{cl}(V)[1] \oplus \Omega^1(V) \oplus \Omega^1_{cl}(V)[-1]$$

between the weight 0, translation-invariant deformation complex and natural complexes related to the geometry of the target.

Before explaining the key steps of the proof, we remark that there is another, more structural way to see that only weight zero local functionals should be relevant. A quick physical argument would say that we want the path integral measure $\exp(-S/\hbar)$ to be weight zero, which forces $\hbar$ to have weight one to cancel out with the weight of the classical action. But the one-loop term I_1 in the quantized action $S^{\mathrm{q}} = S + \hbar I_1 + \cdots$ must then have weight zero.

There is a BV analogue of this argument. It notes that the differential of the quantum observables has the form $\{S^{\mathrm{q}}, -\} + \hbar\Delta$, where Δ denotes the BV Laplacian. (See Section 4.2 for a discussion of these objects.) As the BV Laplacian has weight -1 because it is determined by the bracket, we must give $\hbar$ weight 1 to ensure the total differential has weight zero. Again the one-loop interaction is forced to have weight zero.

3.3.1. *Sketch of proof.* We have already mentioned that we can identify the full translation-invariant deformation complex with a certain Gelfand-Fuks cohomology. In terms of this Gelfand-Fuks cohomology we find that the cotangent weight zero piece is identified with

$$\mathrm{Def}_{\text{string}}(\mathbb{C})^{\mathbb{C},\mathrm{wt}(0)} = \mathrm{C}^*_{\text{Lie,red}}\left(\mathrm{W}_1; \mathrm{Sym}(V^\vee[z^\vee])\right)[2].$$

We will drop the overall shift by 2 until the end of the proof.

Any symmetric algebra has a natural maximal ideal: for any vector space W,

$$\mathrm{Sym}(W) = \mathbb{C} \oplus \mathrm{Sym}^{\geq 1}(W).$$

Thus, we can decompose our complexes as

$$\mathrm{C}^*_{\text{Lie,red}}\left(\mathrm{W}_1; \mathrm{Sym}(V^\vee[z^\vee])\right) = \mathrm{C}^*_{\text{Lie,red}}(\mathrm{W}_1) \oplus \mathrm{C}^*_{\text{Lie}}\left(\mathrm{W}_1; \mathrm{Sym}^{\geq 1}(V^\vee[z^\vee])\right).$$

The first summand is the reduced Gelfand-Fuks cohomology of formal vector fields with values in the trivial module. It is well-known that $H^3_{\text{red}}(\mathrm{W}_1) \cong \mathbb{C}[-3]$, i.e., this cohomology is one-dimensional and concentrated in degree 3.

We now proceed to computing the second summand. Denote by $\{L_n = z^{n+1}\partial_z\}$ the standard basis for the Lie algebra of formal vector fields W_1. Notice that the Euler vector field $L_0 = z\partial_z$ induces a grading on W_1, that we will call *conformal dimension*. Note that L_n has conformal dimension n. This grading extends naturally to the Chevalley-Eilenberg complex of W_1 with coefficients in any module.

Let $\lambda_n \in \mathrm{W}_1^\vee$ be the dual vector to L_n. (We work with the continuous dual vector space, as in the setting of Gelfand-Fuks cohomology.) An arbitrary element of $V[[z]]$ is linear combination of vectors of the form $v \otimes z^k$. Write ζ_k for the dual element $(z^k)^\vee$. Thus an element of $(V[[z]])^\vee$ is a linear combination of the vectors of the form $v^\vee \otimes \zeta_k$.

LEMMA 3.5. *Let M be any W_1-module. The inclusion of the subcomplex of conformal dimension zero elements*

$$\mathrm{C}^*_{\mathrm{Lie}}(\mathrm{W}_1; M)^{\mathrm{wt}(0)} \xrightarrow{\simeq} \mathrm{C}^*_{\mathrm{Lie}}(\mathrm{W}_1; M)$$

is a quasi-isomorphism.

PROOF. For each $p-1 \geq 0$, define the operator $\iota_{L_0} : \mathrm{C}^p_{\mathrm{Lie}}(\mathrm{W}_1; M) \to \mathrm{C}^{p-1}_{\mathrm{Lie}}(\mathrm{W}_1; M)$ by sending a cochain φ to the cochain

$$(\iota_{L_0}\varphi)(X_1, \ldots, X_p) = \varphi(L_0, X_1, \ldots, X_p).$$

Let d be the differential for the complex $\mathrm{C}^*_{\mathrm{Lie}}(\mathrm{W}_1; M)$. It is easy to check that the difference $\mathrm{d}\,\iota_{L_0} - \iota_{L_0}\,\mathrm{d}$ is equal to the projection onto the dimension zero subspace. $\square$

The underlying graded vector space of this conformal dimension zero subcomplex splits as follows:

(7)

$$\mathrm{C}^{\#}_{\mathrm{Lie}}(\mathrm{W}_1)^{\mathrm{wt}(0)} \otimes \left(\mathrm{Sym}^{\geq 1}\left(V[[z]]\right)^\vee\right)^{\mathrm{wt}(0)} \oplus \mathrm{C}^{\#}_{\mathrm{Lie}}(\mathrm{W}_1)^{\mathrm{wt}(1)} \otimes \left(\mathrm{Sym}^{\geq 1}\left(V[[z]]\right)^\vee\right)^{\mathrm{wt}(-1)}$$

In the first component, the purely dimension zero part of the reduced symmetric algebra is simply $\mathrm{Sym}^{\geq 1}(V^\vee)$, i.e., power series on V with no constant term. We denote this algebra concisely as $\mathcal{O}_{red}(V)$, for reduced functions on V. Similarly, in the second component, the dimension one part of $\mathrm{Sym}^{\geq 1}\left(V[[z]]\right)^\vee$ is of the form $\mathrm{Sym}(V^\vee) \otimes z^\vee V^\vee$, which is naturally identified with $\Omega^1(V)$.

The differential in this Gelfand-Fuks complex has the form

$$\begin{array}{cccc}
\underline{0} & \underline{1} & \underline{2} & \underline{3} \\
1 \otimes \mathcal{O}_{red}(V) & \lambda^0 \otimes \mathcal{O}_{red}(V) \longrightarrow & \lambda^{-1} \wedge \lambda^1 \otimes \mathcal{O}_{red}(V) & \lambda^{-1} \wedge \lambda^1 \wedge \lambda^0 \otimes \mathcal{O}_{red}(V) \\
\quad\searrow^{\mathrm{d}_{dR}} & \quad\searrow^{\mathrm{d}_{dR}} & & \\
& \lambda^{-1} \otimes \Omega^1(V) \longrightarrow & \lambda^{-1} \wedge \lambda^0 \otimes \Omega^1(V) &
\end{array}$$

The top line comes from the first summand in (7) and the bottom line corresponds to the second summand. The top horizontal map sends λ^0 to $2 \cdot \lambda^{-1} \wedge \lambda^1$, and the bottom horizontal map sends λ^{-1} to $\lambda^{-1} \wedge \lambda^0$ (both are the identity on V). The diagonal maps are given by the de Rham differential $\mathrm{d}_{dR} : \mathcal{O}_{red}(V) \to \Omega^1(V)$. This complex is quasi-isomorphic to

$$\begin{array}{cccc}
1 \otimes \mathcal{O}_{red}(V) & & & \lambda^{-1} \wedge \lambda^1 \wedge \lambda^0 \otimes \mathcal{O}_{red}(V) \\
\quad\searrow^{\mathrm{d}_{dR}} & & & \\
& \lambda^{-1} \otimes \Omega^1(V) & \lambda^{-1} \wedge \lambda^0 \otimes \Omega^1(V) &
\end{array}$$

which, in turn, is identified with $\Omega^2_{cl}(V)[-1] \oplus \Omega^1(V)[-2] \oplus \Omega^1_{cl}(V)[-3]$. After accounting for the overall shift by 2, we arrive at the identification of the $\mathbb{C}^\times_{\mathrm{cot}}$-weight zero component of the translation-invariant deformation complex.

3.4. Interpretation as string backgrounds. We now discuss, in light of the calculations above, how to interpret string backgrounds in our approach. Since V is flat, we will see that the following deformations will be trivializable. Note that this trivializations will *not* be equivariant for the obvious $\mathrm{GL}(V)$ action (or for non-flat targets, general diffeomorphisms of the target). Thus, these deformations are relevant for the case of a curved target, and we can give an interpretation of them in terms of the usual perspective of *string backgrounds*.

We have already mentioned that we should think of the $\mathbb{C}^\times_{\mathrm{cot}}$ weight 1 local functionals as deformations of the classical theory as a cotangent theory. The cohomological degree zero deformations of the weight one deformations is $H^1(V;T_V)$. Given any such element $\mu \in H^1(V;T_V)$ we can consider the following local functional

$$\int_\Sigma \langle \beta, \mu(\gamma)\rangle_V .$$

The element μ determines a deformation of the complex structure of V, and we have prescribed an action functional encoding this deformation. We propose that this an appearance of the ordinary curved background in bosonic string theory from the perspective of the holomorphic model we work with.

There are interesting deformations that go outside of the world of cotangent theories. Consider the cohomological degree zero part of the weight 0 complex. There is a term of the form $H^1(V;\Omega^2_{cl}(V))$. It is shown in Part 2 Section 8.5 of [**GGW**] how closed holomorphic two-forms determine local functionals of the $\beta\gamma$ system with curved target. A sketch of this construction goes as follows. Locally we can write a closed holomorphic 2-form as $\mathrm{d}\,\theta$ for some holomorphic one-form $\theta \in \Omega^1(V)$. If $\gamma : \Sigma \to V$ is a map of the σ-model there is an induced map (when γ satisfies the equations of motion) $\gamma^* : \Omega^1(V) \to \Omega^1(\Sigma)$. We can then integrate $\gamma^*\theta$ along any closed cycle C in Σ and one should think of this as a residue along C. In [**GGW**] we write down a local functional that realizes this residue, and one can show that it only depends on the corresponding class in $H^1(V;\Omega^2_{cl}(V))$. We posit that this is the appearance of the B-field deformation of the ordinary bosonic string.

In future work we aim to study how our description of holomorphic string backgrounds compares to the approaches of string backgrounds in the physics literature. See for instance [**CFMP85**] for an overview.

4. Quantizing the holomorphic bosonic string on a disk

For us, quantization will mean that we use perturbative constructions in the setting of the BV formalism. Concretely, this means that we enforces the gauge symmetries using the homological algebra of the BV formalism and that we use Feynman diagrams and renormalization to obtain an approximation for the desired, putative path integral. There are toy models for this approach where one can see very clearly how it gives asymptotic expansions for finite-dimensional integrals [**GJF**]. In particular, these toy models show that this approach need not recover the true integral but does know important information about it; a similar relationship should hold between this quantization method and the putative path integral, but in this case there is no *a priori* definition of the true integral in most cases.

This notion of quantization applies to any field theory arising from an action functional, and the algorithm one applies to obtain a quantization is the following:

FIGURE 1. The $\beta\gamma$ and bc propagators

(1) Write down the integrals labeled by Feynman diagrams arising from action functional.
(2) Identify the divergences that appear in these integrals and add "counterterms" to the original action that are designed to cancel divergences.
(3) Repeat these steps until no more divergences appear in Feynman diagrams. We call this the "renormalized action."
(4) Check if the renormalized action satisfies the quantum master equation. If it does, you have a well-posed BV quantum theory, and we call the result a *quantized action*. If not, guess a way to adjust the renormalized action and begin the whole process again.

It should be clear that along the way, one makes many choices; hence if a quantization exists, it may not be unique. It is also possible that a BV quantization may not exist.

In this section we will apply the algorithm in the case of $\Sigma = \mathbb{C}$. For this theory we are lucky, however: at one-loop the integrals that appear in our quantization from the Feynman diagrams do not have divergences, so that renormalized action is easy to compute. This aspect is the subject of the first part of this section. (In Section 6 we will provide an argument based on deformation theory as to why quantizations exist on arbitrary Riemann surfaces.) Moreover, it is easy to check whether the quantum master equation is satisfied, and the answer is simple. This aspect is the subject of the second part. The results can be summarized as follows.

PROPOSITION 4.1. *The holomorphic bosonic string with source $\mathbb{C}$ and target $\mathbb{C}^d$ admits a BV quantization if $d = 13$. This quantized action only has terms of order $\hbar^0$ and $\hbar$ (i.e., it quantizes at one loop).*

4.1. The Feynman diagrams. Let us describe the combinatorics of the Feynman diagrams that appear here before we describe the associated integrals.

4.1.1. The procedure constructs graphs out of a prescribed type of vertices and edges; we must consider all graphs with such local structure. The classical action functional determines the allowed kinds of vertices and edges. The quadratic terms of the action tell us the edges; each quadratic term yields an edge whose boundary is labeled by the two fields appearing in the term. For us there are thus two types of edges: an edge that flows from β to γ, and an edge that flows from b to c displayed in Figure 1.

The nonquadratic terms tell us the vertices: each n-ary term yields a vertex with n legs, and the legs are labeled by the n types of fields appearing in the term. For us there are thus two types of trivalent vertices: a vertex with two c legs and a b leg, and a vertex with a c leg, a γ leg, and a β leg. It helpful to picture these legs as directed, so that c and γ legs flow into a vertex and b and β legs flow out. These vertices are displayed in Figure 2.

The kinds of graphs one can build with such vertices and edges are limited. We focus on connected graphs, since an arbitrary graph is just a union of connected components.

FIGURE 2. The trivalent vertices for $\int\langle\beta,[c,\gamma]\rangle$ and $\int\langle b,[c,c]\rangle$

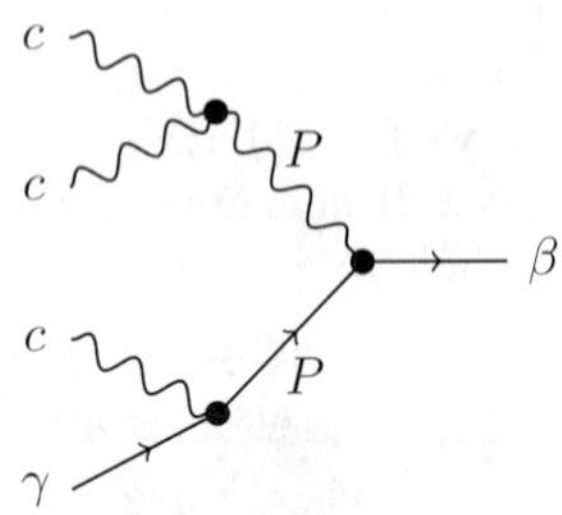

FIGURE 3. An example of a tree with four inputs and one output

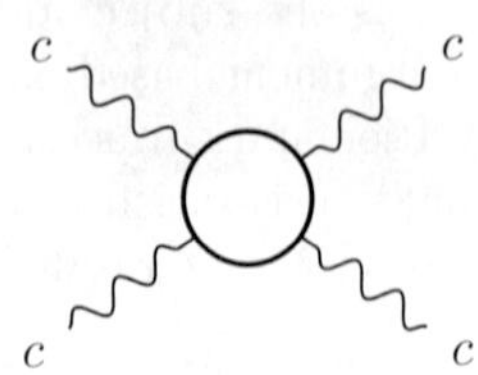

FIGURE 4. An example of a wheel with four inputs

A tree (i.e., a connected graph with no loops) must have at most one outgoing leg, which must be either a b or a β; the other legs are incoming, so each must be labeled by a c or a γ. An example of such a tree is given in Figure 3.

Note that there are two types of trees. If there is a γ leg, then there is a β leg, and there is a chain of $\gamma\beta$ edges connecting them; all other external legs are of c type. If there is a b leg, then the only other legs are c type.

A one-loop graph will consist of a wheel (i.e., a sequence of edges that form an overall loop) with trees attached. The outer legs are all of c type. Every edge along a wheel will have the same type. It is not possible to build a connected graph with more than one loop. This combinatorics is the essential reason that we can quantize at one loop. For an example of such a wheel see Figure 4.

We write $\mathbf{Graph}_{\text{string}}$ for the collection of connected graphs just described, namely the directed trees and 1-loop graphs allowed by the string action functional. Let $\mathbf{Graph}^{(0)}_{\text{string}}$ denote the 0-loop graphs (i.e., trees) and let $\mathbf{Graph}^{(1)}_{\text{string}}$ denote the 1-loop graphs (i.e., wheels with trees attached).

4.1.2. These graphs describe linear maps associated to the field. More precisely, a graph with k legs describes a linear functional on the k-fold tensor product of the space of fields. One builds this linear functional out of the data of the action functional.

As an example, a k-valent vertex corresponds to a k-ary term in the action, which manifestly takes in k copies of the fields and outputs a number. Thus, the vertex labels an element of a (continuous) linear dual of the k-fold tensor product of fields. In fact, one restricts to *compactly-supported* fields, since the action functional is rarely well-defined on all fields when the source manifold is non-compact. (Note this domain of compactly-supported fields is all one needs for making variational arguments or for constructing a BV quantization.)

An edge corresponds an element P of the 2-fold tensor product of the space of fields, often called a *propagator*. More precisely, the edge should correspond to the Green's function for the linear differential operator appearing in the associated quadratic term of the action; hence the propagator is an element of the *distributional completion* of the 2-fold tensor product. For us the $\beta\gamma$ leg should be labeled by $\bar{\partial}^{-1} \otimes \mathrm{id}_V$, where $\bar{\partial}^{-1}$ denotes an inverse to the Dolbeault operator on functions. The bc leg should be labeled by $\bar{\partial}_T^{-1}$, the inverse of the Dolbeault operator on the bundle $T^{1,0}$.

Given a graph Γ, one should contract the tensors associated to the vertices and edges. We denote the linear functional for this graph by $w_\Gamma(P, I)$, where w stands for "weight," the term P indicates we label edges by the propagator P, and the term I indicates we label vertices by the "interaction" term of the action S (i.e., the terms that are cubic and higher).

This contraction is not always well-posed, unfortunately. Each vertex labels a distributional section of some vector bundle on Σ, and each edge labels a distributional section of a vector bundle on Σ^2. Thus the desired contraction can be written *formally* as an integral over the product manifold Σ^v, where v denotes the number of vertices. In most situations this contraction is ill-defined, since one cannot (usually) pair distributions. Concretely, one sees that the integral expression is divergent.

Thus, to avoid these divergences, one labels the edges by a smooth replacement of the Green's functions. (Imagine replacing a delta function δ_0 by a bump function.) Since one can pair smooth functions and distributions, each graph yields a linear functional on fields using these mollified edges. Thus we have *regularized* the divergent expression.

But now this linear functional depends on the choice of mollifications. Hence the challenge is to show that if one picks a sequence of smooth replacements that approaches the Green's function, there is a well-defined limit of the linear functionals.

4.1.3. We will now sketch one method well-suited to complex geometry that allows us to see that no divergences appear for the holomorphic bosonic string. Our approach is an example of the renormalization method developed by Costello in [**Cos11**], which applies to many more situations.

Our primary setting in this section is $\Sigma = \mathbb{C}$. For this Riemann surface, a standard choice of Green's function for the $\bar{\partial}$ that acts on functions is

$$P(z, w) = \frac{1}{2\pi i} \frac{\mathrm{d}\,\bar{z} + \mathrm{d}\,\bar{w}}{z - w}.$$

It is a distributional one-form on $\mathbb{C}^2$ that satisfies $\bar{\partial} \otimes 1(P) = \delta_\Delta$, where δ_Δ is the delta-current supported along the diagonal $\Delta : \mathbb{C} \hookrightarrow \mathbb{C}^2$ and providing the integral

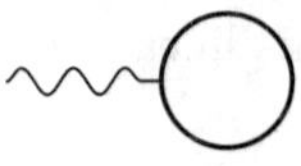

FIGURE 5. The tadpole diagram Γ_{tad}

kernel for the identity. In terms of our discussion above, we view this one-form as a distributional section of the fields γ and β: for example, for fixed w, the one-form $\mathrm{d}\,z/(z-w)$ is a β field in the z-variable as it is a $(1,0)$-form. (This propagator is for the $\beta\gamma$ fields—and one must tensor with a kernel for the identity on V—but a similar formula provides a propagator for the bc fields.)

4.1.4. We will now describe the integral associated to a simple diagram. For simplicity, we assume $V = \mathbb{C}$ so that the γ and β fields are simply functions and 1-forms on $\mathbb{C}$, respectively. Consider a "tadpole" diagram, Figure 5, Γ_{tad} whose outer legs are c fields (i.e., vector fields on $\mathbb{C}$).

There is only one vertex here, corresponding to the cubic function on fields

$$w_{\Gamma_{\text{tad}}}(P, I_{\text{string}}) = \int_{z\in\mathbb{C}} \beta \wedge c\gamma.$$

If the field c is of the form $f(z)\,\mathrm{d}\,\overline{z}\partial_z$, with f compactly supported, then our integral is

$$\int_{z\in\mathbb{C}} \beta \wedge f(z)(\partial_z\gamma)\,\mathrm{d}\,\overline{z}.$$

(Note that a general cubic function could be described as an integral over $\mathbb{C}^3$, but our function is supported on the small diagonal $\mathbb{C} \hookrightarrow \mathbb{C}^3$.) The linear functional for this tapole diagram should be given by inserting the propagator P in place of the β and γ fields. Hence it ought to be given by the following integral over $\mathbb{C}$:

$$\int_{z\in\mathbb{C}} c(z)P(z,w)|_{z=w} = \int_{z\in\mathbb{C}} f(z)\partial_z\left(\frac{1}{2\pi i}\frac{\mathrm{d}\,z + \mathrm{d}\,w}{z-w}\right)|_{z=w}\,\mathrm{d}\,\overline{z}.$$

This putative integral is manifestly ill-defined, since the distribution is singular along the diagonal.

4.1.5. We smooth out the propagator P using familiar tools from differential geometry. Fix a Hermitian metric on Σ, which then associates provides an adjoint $\bar{\partial}^*$ to the Dolbeault operator $\bar{\partial}$. For the usual metric on $\mathbb{C}$, we have

$$\bar{\partial}^* = -2\frac{\partial}{\partial(\mathrm{d}\,\overline{z})}\frac{\partial}{\partial z}.$$

In physics one calls a choice of the operator $\bar{\partial}^*$ a *gauge-fix*. The commutator $[\bar{\partial}, \bar{\partial}^*]$, which we will denote D, is equal to $\frac{1}{2}\Delta$, where Δ is the Laplace-Beltrami operator for this metric. In the physics literature, explicit gauge fixes for the bosonic string can be found in [**Boc87**].

We introduce a smoothed version of the propagator using the heat kernel e^{-tD}, which is a notation that denotes a solution to the heat equation $\partial_t f(t,z) + Df(t,z) = 0$. For $\mathbb{C}$ with the Euclidean metric, the standard heat kernel is

$$e^{-tD}(z,w) = \frac{1}{4\pi t}e^{-|z-w|^2/4t}(\mathrm{d}\,z - \mathrm{d}\,w) \wedge (\mathrm{d}\,\overline{z} - \mathrm{d}\,\overline{w}).$$

For $0 < \ell < L < \infty$, we define

$$P_\ell^L = \bar{\partial}^* \int_\ell^L e^{-tD}\, \mathrm{d}\, t.$$

We compute

$$\bar{\partial} P_\ell^L = D \int_\ell^L e^{-tD}\, \mathrm{d}\, t = \int_\ell^L \frac{d}{dt} e^{-tD}\, \mathrm{d}\, t = e^{-LD} - e^{-\ell D}.$$

In the limit as $\ell \to 0$ and $L \to \infty$, the operator P_ℓ^L goes to a propagator (or Green's function) P for $\bar{\partial}$. To see this, consider an eigenfunction f of D where $Df = \lambda f$ where λ is a non-negative real number. Then

$$(\bar{\partial} P_\ell^L) f = (e^{-L\lambda} - e^{-\ell\lambda}) f,$$

which goes to f as $L \to \infty$ and $\ell \to 0$. Thus, if one works with the correct space of functions, P_ℓ^L is almost an inverse to $\bar{\partial}$; moreover, it is a smooth function on $\Sigma \times \Sigma$.

4.1.6. We now return to the tadpole diagram and put P_ℓ^L on the edge instead of P. (We again assume $V = \mathbb{C}$ for simplicity.) The propagator is

$$P_\ell^L(z,w) = \int_\ell^L \mathrm{d}\, t \frac{\partial}{\partial(\mathrm{d}\, \bar{z})} \frac{\partial}{\partial z} \left(\frac{1}{4\pi t} e^{-|z-w|^2/4t} (\mathrm{d}\, z - \mathrm{d}\, w) \wedge (\mathrm{d}\, \bar{z} - \mathrm{d}\, \bar{w}) \right) \tag{8}$$

$$= \int_\ell^L \mathrm{d}\, t \frac{1}{4\pi t} \frac{\bar{z} - \bar{w}}{2t} e^{-|z-w|^2/4t} (\mathrm{d}\, z - \mathrm{d}\, w). \tag{9}$$

Note that it is smooth everywhere on $\mathbb{C}^2$. The integral for the tadpole diagram is

$$\begin{aligned} w_{\Gamma_{\mathrm{tad}}}(P_\ell^L, I_{\mathrm{string}}) &= \int_{z \in \mathbb{C}} c(z) P_\ell^L(z,w)|_{z=w} \\ &= \int_{z\in\mathbb{C}} \int_\ell^L \mathrm{d}\, t f(z) \partial_z \left(\frac{1}{4\pi t} \frac{\bar{z}-\bar{w}}{2t} e^{-|z-w|^2/4t} (\mathrm{d}\, z - \mathrm{d}\, w) \right) |_{z=w}\, \mathrm{d}\, \bar{z} \\ &= \int_{z\in\mathbb{C}} \int_\ell^L \mathrm{d}\, t f(z) \left(\frac{1}{4\pi t} \left(\frac{\bar{z}-\bar{w}}{2t} \right)^2 e^{-|z-w|^2/4t} (\mathrm{d}\, z - \mathrm{d}\, w) \right) |_{z=w}\, \mathrm{d}\, \bar{z} \\ &= 0, \end{aligned}$$

since the integrand vanishes along the diagonal. Note that this integral is independent of ℓ and L and hence the limit is zero.

4.1.7. By explicitly analyzing the $\ell \to 0$ limit for the integral associated to every Feynman diagram, we find the following result.

PROPOSITION 4.2. *For any graph $\Gamma \in$ **Graph**$_{\mathrm{string}}$ allowed by the combinatorics of the string action functional and for any $L > 0$, there is a well-defined limit* $\lim_{\ell \to 0} w_\Gamma(P_\ell^L, I_{\mathrm{string}})$.

We denote this limit by $w_\Gamma(P_0^L, I_{\mathrm{string}})$. The necessary manipulations and inequalities referenced below are very close to those used in [**Cos**, **GGW**].

OUTLINE OF PROOF. When Γ is a tree, there is never an issue with divergences; we could even use the Green's function $\bar{\partial}^{-1}$ on each edge. To see this, note that one can view a tree as having a distinguished root, given by the leg that is either of β or b type. One can then see the tree as describing a multilinear map from the leaves (i.e., legs that are not roots) to the root. Indeed, one can view each cubic vertex as such an operator. For instance, $\langle b, [c, c]\rangle$ corresponds to the Lie bracket of vector fields, since we view $\langle b, -\rangle$ as an element of the c fields. For a tree, one can then input arbitrary elements into the leaves, apply the operations labeled by the vertices, apply the operator labeled by the edge, and so on, until one reaches the root. The composite multilinear operator sends smooth sections to smooth sections, even if the edges are labeled by distributional sections, since the associated operator sends smooth sections to smooth sections.

When Γ is a one-loop graph, it consists of a wheel with trees attached to the outer legs. By the preceding argument, we know those trees do not introduce singularities; hence any divergences are due solely to the wheel. It thus suffices to consider pure wheels (i.e., those with no trees attached).

Let the wheel have n vertices. The kth vertex has a coordinate z_k on $\mathbb{C}$; the kth external leg has input $c_k = f_k(z_k, \overline{z}_k)\,\mathrm{d}\,\overline{z}_k\,\partial_{z_k}$, where f_k is a compactly-supported smooth function. Then the integral has the form

$$\int_{(z_1,\dots,z_n)\in\mathbb{C}^n} \mathrm{d}^n\,\overline{z}\,(f_1\partial_{z_1}P^L_\ell(z_1,z_n))(f_2\partial_{z_2}P^L_\ell(z_2,z_1))\cdots(f_n\partial_{z_n}P^L_\ell(z_n,z_{n-1})),$$

since the kth input will act on one of the propagators entering the kth vertex. One needs to show that this expression has a finite $\ell\to 0$ limit.

Let us prove this limit exists for the case $n=2$. Then we have

$$\int_{z_1,z_2\in\mathbb{C}} \mathrm{d}\,\overline{z}_1\,\mathrm{d}\,\overline{z}_2 \int_\ell^L \mathrm{d}\,t_1 \int_\ell^L \mathrm{d}\,t_2\, f_1(z_1)f_2(z_2)\partial_{z_1}\left(\frac{1}{4\pi t_1}\frac{\overline{z}_1-\overline{z}_2}{2t_1}e^{-|z_1-z_2|^2/4t_1}(\mathrm{d}\,z_1-\mathrm{d}\,z_2)\right)$$
$$\times\,\partial_{z_2}\left(\frac{1}{4\pi t_2}\frac{\overline{z}_1-\overline{z}_2}{2t_2}e^{-|z_1-z_2|^2/4t_2}(\mathrm{d}\,z_2-\mathrm{d}\,z_1)\right),$$

which is already a bit lengthy. As our focus is on showing a limit exists, we will throw out unimportant factors and simplify the expression. First, note that taking the partial derivative ∂_{z_i} will simply multiply the integrand by $(\overline{z}_1-\overline{z}_2)/2t_i$. Moreover, we change coordinates to $u = z_1 - z_2$ and $v = z_2$. Then the integral is proportional to

$$\int_\ell^L \mathrm{d}\,t_1 \int_\ell^L \mathrm{d}\,t_2 \int_{\mathbb{C}^2} \mathrm{d}^2\,u\,\mathrm{d}^2\,v\,f_1 f_2 \frac{\overline{u}^4}{t_1^3 t_2^3} e^{-|u|^2(\frac{1}{t_1}+\frac{1}{t_2})}.$$

We take the integral over v last; it will be manifestly well-behaved after we take the other integrals.

Thus consider the integral just over $u\in\mathbb{C}$, so that we are computing the expected value of $F = f_1 f_2$ against a Gaussian measure whose variance is determined by t_1 and t_2. (Namely, the variance is $t_1t_2/(t_1+t_2)$.) We might as well focus on values of t_i that are very small, as those would be the source of divergences when $\ell\to 0$. For small t_i, we only care about the behavior of F near the origin as the measure is concentrated near the origin. Thus, consider a partial Taylor expansion of F. The polynomial part can be computed quickly since the expected values of monomials against a Gaussian measure (i.e., the moments) have a simply expression in terms of the variance. The first nonzero contribution would come from the u^4 term in the Taylor expansion of F, and it contributes a factor of the form $(t_1t_2/(t_1+t_2))^5$, up to constant that we ignore. We are left with

$$\int_\ell^L \mathrm{d}\,t_1 \int_\ell^L \mathrm{d}\,t_2 \frac{(t_1t_2)^3}{(t_1+t_2)^5} \le \int_\ell^L \mathrm{d}\,t_1 \int_\ell^L \mathrm{d}\,t_2\, 2^{-5}\sqrt{t_1t_2} = 2^{-5}(L^{3/2}-\ell^{3/2})^2,$$

where we use the arithmetic-geometric mean inequality $\sqrt{t_1t_2}/(t_1+t_2) \leq 1/2$ in the middle. This expression has a finite limit as $\ell \to 0$. The higher terms in the Taylor expansion contribute bigger powers of the variance and hence have $\ell \to 0$ limits. Finally, the expected value of the error term of our partial Taylor expansion, which vanishes to some positive order at the origin, can be bounded in such as way that an $\ell \to 0$ limit exists. $\square$

We can now define the effective theory that we consider for the string.

DEFINITION 4.3. The *renormalized action functional* at scale L for the holomorphic bosonic string is

$$I[L] = \sum_{\Gamma \in \mathbf{Graph}^{(0)}_{\text{string}}} w_\Gamma(P_0^L, I_{\text{string}}) + \hbar \sum_{\Gamma \in \mathbf{Graph}^{(1)}_{\text{string}}} w_\Gamma(P_0^L, I_{\text{string}}).$$

We denote the first summand—the tree-level expansion—by $S_0[L]$ and the second summand—the one-loop expansion—by $S_1[L]$. We use the notation $S[L] = S_{free} + I[L]$ where S_{free} is the classical free part of the action functional.

REMARK 4.4. For any functional J, let $w(P_\ell^L, J)$ denote the sum over all graphs as above with the smooth propagator P_ℓ^L placed at the edges and J placed at the vertices. Then, the family $\{I[L]\}$ satisfies the *homotopy RG equation*

$$I[L] = w(P_\ell^L, I[\ell]).$$

The operator $w(P_\ell^L, -)$ defines a homotopy equivalence between the theory at scale ℓ, defined using $S[\ell]$, and the theory at scale S, defined using $S[L]$.

4.2. The quantum master equation. In the BV formalism the basic idea is to replace integration against a path integral measure $e^{-S(\phi)/\hbar}\mathcal{D}\phi$ with a cochain complex. In this cochain complex, we view a cocycle as defining an observable of the theory, and its cohomology class is viewed as its expected value against the path integral measure. For toy models of finite-dimensional integration, see [**GJF**]; these examples are always cryptomorphically equivalent to a de Rham complex, which is a familiar homological approach to integration.

Hence the content of the path integral, in this approach, is encoded in the differential. A key idea is that the differential is supposed to behave like a divergence operator for a volume form: recall that given a volume form μ on a manifold, its divergence operator maps vector fields to functions by the relationship

$$\text{div}_\mu(\mathcal{X})\mu = L_{\mathcal{X}}\mu.$$

This relationship, in conjunction with Stokes lemma, implies that if a function f is a divergence $\text{div}_\mu(\mathcal{X})$, then $\int f\mu = 0$, i.e., its expected value against the measure μ is zero. The BV formalism axiomatizes general properties of divergence operators; a putative differential must satisfy these properties to provide a BV quantization.

When following the algorithm of Section 1.1, we want the renormalized action

$$S = S^{\text{cl}} + \hbar S_1 + \hbar^2 S_2 + \cdots$$

to determine a putative differential d_S^q on the graded vector space of observables. To explain this operator, we need to describe further algebraic properties on the observables that the BV formalism uses.

First, in practice, the observables are the symmetric algebra generated by the continuous linear duals to the vector spaces of fields. There is also a pairing on fields that is part of the data of the classical BV theory, between each field and its "anti-field." (This pairing is a version of the action of constant vector fields on functions in the toy models.) In our case, there is the pairing between b and c and between β and γ, respectively. It behaves like a "shifted symplectic" pairing as it has cohomological degree -1, and hence it determines a degree 1 Poisson bracket $\{-,-\}$ on the graded algebra of observables. Finally, the pairing also determines a second-order differential operator Δ_{BV} on the algebra of observables by the condition that

$$\Delta_{BV}(FG) = (\Delta_{BV}F)G + (-1)^F F(\Delta_{BV}G) + \{F, G\}.$$

(This equation is a characteristic feature of divergence operators with respect to the product of polyvector fields.)

With these structures in hand, we can give the formula

$$\mathrm{d}^q_S = \{S, -\} + \hbar\Delta_{BV}$$

for the putative differential. As S has cohomological degree 0, the operator $\{S,-\}$ has degree 1. We remark that modulo $\hbar$, one recovers the differential $\{S^{\mathrm{cl}}, -\}$ on the classical observables; the zeroth cohomology of the classical observables is functions on the critical locus of the classical action S^{cl}.

By construction, this putative differential d^q_S satisfies the conditions of behaving like a divergence operator. The only remaining condition to check is that it is square-zero. This condition ends up being equivalent to S satisfying the *quantum master equation*

$$\hbar\Delta_{BV}S + \frac{1}{2}\{S, S\} = 0. \tag{10}$$

More accurately, d^q_S is a differential if and only if the right hand side is a constant.

4.2.1. We now turn to examining this condition in our setting. It helps to understand it is diagrammatic terms.

As the bracket is determined by a linear pairing, it admits a simple diagrammatic description as an edge. For instance, given an observable F that is a homogeneous polynomial of arity m and an observable G of arity n, then $\{F, G\}$ has arity $m + n - 2$. It can be expressed as a Feynman diagram where the edge connecting F and G is labeled by a 2-fold tensor K.

The BV Laplacian acts by attaching an edge labeled by K as a loop in all possible ways. This diagrammatic behavior corresponds to the fact that Δ_{BV} is a constant-coefficient second-order differential operator.

The tensor K determined by the pairing on fields is distributional. As one might expect from our discussion of divergences above, these diagrammatic descriptions of the BV bracket and Laplacian are thus typically ill-defined. In other words, the quantum master equation is *a priori* ill-posed for the same reason that the initial Feynman diagrams are ill-defined. We can apply, however, the same cure of mollification.

4.2.2. Costello's framework [**Cos11**] provides an approach to renormalization built to be compatible with the BV formalism. A key feature is that for each "length scale" $L > 0$, there is a BV bracket $\{-,-\}_L$ and BV Laplacian Δ_L. The scale L renormalized action $S[L]$ satisfies the scale L quantum master equation (QME)

$$\hbar\Delta_L S[L] + \frac{1}{2}\{S[L], S[L]\}_L = 0$$

if and only if $S[L']$ satisfies the scale L' quantum master equation for every other scale L', see Lemma 9.2.2 in [**Cos11**]. Hence, we say a renormalized action satisfies the quantum master equation if its solves the scale L equation for some L.

Thus it remains for us to describe the scale L bracket and BV Laplacian in our setting, so that we can examine whether the renormalized action satisfies the quantum master equation.

DEFINITION 4.5. The *scale L bracket* $\{-,-\}_L$ is given by pairing with the scale L heat kernel

$$K_L(z,w) = \frac{1}{4\pi L} e^{-|z-w|/4L} (\mathrm{d}\, z - \mathrm{d}\, w) \wedge (\mathrm{d}\, \overline{z} - \mathrm{d}\, \overline{w}).$$

The *scale L BV Laplacian* Δ_L is given by the contraction ∂_{K_L}.

These definitions mean that testing the quantum master equation leads to diagrams whose integrals are similar to those we encountered earlier. We explain the diagrammatics and sketch the relevant integrals in the proof of the following result, which characterizes when the string action admits a BV quantization.

We emphasize that up to now, we have not indicated explicitly which vector space V is the target space for our string. But the action functional explicitly depends on this choice, so here we will write S_V for the action with target V.

PROPOSITION 4.6. *The obstruction to satisfying the quantum master equation is the functional*

$$Ob_V[L] = \hbar\Delta_L S_V[L] + \frac{1}{2}\{S_V[L], S_V[L]\}_L.$$

It has the form

$$Ob_V[L] = \hbar(\dim_{\mathbb{C}}(V) - 13)F[L],$$

where $F[L]$ is a functional independent of V.

In short, the failure to satisfy the QME is a linear function of the dimension of the target space V. In particular, when $V \cong \mathbb{C}^{13}$, the obstruction vanishes and the renormalized action *does* satisfy the QME, giving us an immediate corollary. (Note that we do *not* need to know $F[L]$ to recognize that the obstruction vanishes!)

COROLLARY 4.7. *When the target vector space is 13-dimensional (i.e., has 26 real dimensions), the holomorphic bosonic string admits a BV quantization.*

PROOF. It is a general feature of Costello's formalism that the tree-level term $S_0[L]$ of the renormalized action satisfies the scale L equation

$$\{S_0[L], S_0[L]\}_L = 0,$$

known as the classical master equation. Hence the first obstruction to satisfying the QME can only appear with positive powers of $\hbar$. We can also see quickly that no terms of $\hbar^2$ appear: the one-loop term $S_1[L]$ is only a function of the c field, so

$$\{S_1[L], S_1[L]\}_L = 0 \quad \text{and} \quad \Delta_L S_1[L] = 0.$$

Hence the obstruction to satisfying the QME is precisely

$$\hbar\left(\{S_0[L], S_1[L]\} + \Delta_L S_0[L]\right).$$

Thus we see that the obstruction is a multiple of $\hbar$. For simplicity, we will divide out that factor and let Ob_V denote the term inside the parenthesis.

Consider the term $\{S_0[L], S_1[L]\}_L$. Diagrammatically, it corresponds to attaching a tree with a b "root" to a wheel using an edge labeled by K_L. Arguments similar to Lemma 16.0.3 of [**Cos**] carry over to account for the vanishing of this term in the $L \to 0$ limit.

Now consider the term $\Delta_L S_0[L]$. Diagrammatically, it corresponds to turning a tree into a wheel by using an edge—labeled by K_L—to attach the root to an incoming leaf. There are thus two kinds of wheels that appear, since there are two kinds of trees. There are the wheels where the K edge is for bc fields. Note that these wheels are the same for every choice of target V as they only depend on the bc fields, i.e., are independent of the $\beta\gamma$ fields. These will contribute a term $F[L]$ to the obstruction. On the other hand, there are the wheels where the K edge is for $\beta\gamma$ fields. These depend on V but in a very simple way: the distribution K is just the heat kernel tensored with the identity on V, and hence the contraction amounts to taking $\dim_{\mathbb{C}}(V)$ copies of the $V = \mathbb{C}$ value. In other words, the $\beta\gamma$ wheels contribute a term $\dim_{\mathbb{C}}(V)G[L]$ to the obstruction, where $G[L]$ is the value for $V = \mathbb{C}$. The last part of the proof of the theorem is a direct calculation of the functionals $F[L]$ and $G[L]$. So as to not diverge from our track of thought we include this calculation in Appendix A where we show that $F[L], G[L]$ are both independent of L and satisfy $F = -13G$, thus completing the proof. □

REMARK 4.8. One can consider coupling the $\beta\gamma$ system to any tensor bundle on the Riemann surface. For instance, suppose γ is a section of $T_\Sigma^{\otimes n}$ and hence β is a section of $T_\Sigma^{*\otimes n+1}$. In this case, one can show that the part of the obstruction with internal edges labeled by the $\beta\gamma$ propagators contributes a factor $(6n^2 + 6n + 1)G$, with G the same functional above.

5. OPE and the string vertex algebra

Vertex algebras are mathematical objects that axiomatize the behavior of local observables (i.e., point-like observables) of a chiral conformal field theory (CFT), such as the $bc\beta\gamma$ system or the holomorphic bosonic string. In particular, the operator product expansion (OPE) for these local observables—which is of paramount importance in understanding a chiral CFT—is encoded by the vertex operator of the vertex algebra of the CFT. (We will not review vertex algebras here as there are many nice expositions [**FHL93, FBZ01**].)

In this section we will explain how to extract the vertex algebra of the holomorphic bosonic string, using machinery developed in [**CG17, Li, GGW**]. The answer we recover is precisely the chiral sector of the usual bosonic string.

5.1. Some context. In the BV formalism one constructs a cochain complex of observables, for both the classical and the quantized theory, if it exists. The cochain complexes are local on the source manifold of a theory: on each open set U in that manifold Σ, one can pick out the observables with support in U by asking for the observables that vanish on fields with support outside U. Furthermore, you can combine observables that have support on disjoint open sets. It is the central result of [**CG17, CG**] that the observables also satisfy a local-to-global property, akin to the sheaf gluing axiom. Such a structure is known as a *factorization algebra* on Σ.

We will not need that general notion here. Instead, we will use vertex algebras. Theorem 5.2.3.1 of [**CG17**] explains how a factorization algebra F on $\Sigma = \mathbb{C}$ yields a vertex algebra $\mathbb{V}\mathrm{ert}(F)$, under natural hypotheses on F. It assures us that the observables of a chiral CFT determine a vertex algebra.

In particular, Section 5.3 of [**CG17**] examines the free $\beta\gamma$ system in great detail. Its main result is that the well-known $\beta\gamma$ vertex algebra $\mathcal{V}_{\beta\gamma}$ is recovered by the two-step process of BV quantization, which yields a factorization algebra, and then the extraction of a vertex algebra. The exact same arguments apply to the free bc system, recovering the vertex algebra $\mathcal{V}_{bc}$; and of course, the exact same arguments apply to the free $bc\beta\gamma$ system.

Let V denote the vector space appearing in the $\beta\gamma$ contribution of the holomorphic bosonic string theory, as introduced in Section 2. Let $\mathrm{Obs}^{\mathrm{q}}_{free}$ denote the observables of this theory on $\Sigma = \mathbb{C}$. As a quantization of a free field theory, it is a factorization algebra valued in the category of $\mathbb{C}[\hbar]$-modules. In particular, the associated vertex algebra $\mathbb{V}\mathrm{ert}(\mathrm{Obs}^{\mathrm{q}}_{free})$ is also valued in $\mathbb{C}[\hbar]$-modules. Putting the claims together, we have the following.

PROPOSITION 5.1. *For $n = \dim_{\mathbb{C}}(V)$, there is an isomorphism of vertex algebras*

$$\mathbb{V}\mathrm{ert}(\mathrm{Obs}^{\mathrm{q}}_{free})_{\hbar=2\pi i} \cong \mathcal{V}_{bc} \otimes \mathcal{V}_{\beta\gamma}^{\otimes n}$$

where on the left-hand side we have set $\hbar = 2\pi i$.

5.2. A reminder on the chiral algebra of the string. We now provide a brief review of the vertex algebra for the chiral sector of the bosonic string. For a detailed reference we refer the reader to [**LZ93, LZ96**]. The construction builds a *differential graded vertex algebra*, which is simply a vertex algebra in the category of cochain complexes. The underlying graded vertex algebra has a state space of the form

$$\mathcal{V}_{\beta\gamma}^{\otimes 13} \otimes \mathcal{V}_{bc},$$

where $\mathcal{V}_{\beta\gamma}$ and $\mathcal{V}_{bc}$ are the $\beta\gamma$ and bc vertex algebras, respectively. The β and γ generators are in grading degree zero, the c generator is in grading degree -1, and the b is in grading degree 1. In the physics literature it is referred to as the *BRST* grading or *ghost number*.

Forgetting the cohomological (or BRST) grading, this vertex algebra is a conformal vertex algebra of central charge zero (by construction). In particular, this means that the vertex algebra has a stress-energy tensor. Explicitly, it is of the form

$$T_{\text{string}}(z) = \left(\sum_{i=1}^{13} \beta_i(z)\partial_z\gamma_i(z) + \partial_z\beta_i(z)\gamma_i(z)\right) + \left(b(z)\partial_z c(z) + 2\partial_z b(z)c(z)\right).$$

Note that T_{string} is of cohomological degree zero. The first parenthesis is interpreted as the stress-energy tensor of the vertex algebra $\mathcal{V}_{\beta\gamma}^{\otimes 13}$ and the second term is the stress-energy tensor of $\mathcal{V}_{bc}$.

We have not yet described the differential on the graded vertex algebra. The BRST differential is defined to be the vertex algebra derivation obtained by taking the following residue

$$Q^{BRST} = \oint c(z)T_{\text{string}}(z). \tag{11}$$

By construction this operator satisfies $(Q^{BRST})^2 = 0$.

DEFINITION 5.2. The *string vertex algebra* is the dg vertex algebra

$$\mathcal{V}_{\text{string}} = \left(\mathcal{V}_{\beta\gamma}^{\otimes 13} \otimes \mathcal{V}_{bc}, \, Q^{BRST} \right).$$

There is another grading on $\mathcal{V}_{\text{string}}$ coming from the eigenvalues of the vertex algebra derivation c_0 called the *conformal dimension*. In particular, this determines a filtration and we can consider the associated graded Gr $\mathcal{V}_{\text{string}}$. The conformal weight grading preserves the cohomological grading so that this object still has the structure of a dg vertex algebra.

Note that the cohomology of a dg vertex algebra is an ordinary (graded) vertex algebra. The cohomology of the string vertex algebra is called the *BRST cohomology* of the bosonic string. In the remainder of this section we will show how we recover the string vertex algebra from the quantization of the holomorphic bosonic string.

5.3. The case of the string. The holomorphic bosonic string is a chiral CFT and so the machinery of [**CG17**] applies to it. One can extract a vertex algebra directly by this method, as one does with the free $bc\beta\gamma$ theory.

But there is a slicker approach, using Li's work [**Li**], which studies chiral deformations of *free* chiral BV theories such as the free $bc\beta\gamma$ system. Recall that a deformation of a classical field theory is given by a local functional. We have seen that this is essentially the data of a Lagrangian density, which is a density valued multilinear functional that depends on (arbitrarily high order) jets of the fields. In other words, for a field φ, a Lagrangian density is of the form

$$\mathcal{L}(\varphi) = \sum (D_{k_1}\varphi) \cdots (D_{k_m}\varphi) \cdot \text{vol}_\Sigma$$

for $C^\infty(\Sigma)$-valued differential operators D_{k_i}. By a *chiral* Lagrangian density we mean a Lagrangian for which the differential operators D_{k_i} are all holomorphic. For instance, on $\Sigma = \mathbb{C}$, we require D_{k_i} to be a sum of operators of the form $f(z)\partial_z^n$ where $f(z)$ is a holomorphic function. On $\Sigma = \mathbb{C}$ we will also require the chiral Lagrangian to be translation invariant. This means that all differential operators D_{k_i} are of the form ∂_z^n. Thus, a *translation-invariant chiral deformation* is a local functional of the form

$$I(\varphi) = \sum \int (\partial_z^{k_1}\varphi) \cdots (\partial^{k_m}\varphi) \, \mathrm{d}^2 z.$$

Such a deformation stays within the class of chiral CFTs.

One of Li's main results is that for a free chiral BV theory with action S_{free} and associated vertex algebra $\mathcal{V}_{\text{free}}$, one has the following:

- For any chiral interaction I, the action $S = S_{\text{free}} + I$ yields a renormalized action functional $I[L] = \lim_{\ell \to 0} W(P_\ell^L, I)$ that requires no counterterms. That is, the weights of all Feynman diagrams are finite (compare to Proposition 4.2),
- If the renormalized action $\{I[L]\}$ satisfies the quantum master equation, then it determines a vertex algebra derivation D_I of $\mathcal{V}_{\text{free}}$ of the form

$$D_I = \oint I^q \, \mathrm{d}\, z$$

that is square-zero and of cohomological degree one. Here, $I^q = \lim_{L \to 0} I[L]$, where $I[L]$ is the renormalized action functional. Modulo $\hbar$, it agrees with the chiral interaction I, but it has $\hbar$-dependent terms that provide the "quantum corrections" to the classical action.

- The dg vertex algebra $\mathcal{V}_I$ for such an action $\{I[L]\}$ has the same underlying graded vertex algebra $\mathcal{V}_{\text{free}}$ but it is equipped with the differential $\oint I^q \, \mathrm{d}\, z$.

This construction significantly reduces the work of constructing the vertex algebra for the chiral deformation, as one need not analyze the factorization algebra directly.

REMARK 5.3. The fact that I satisfies the quantum master equation implies that one has a map, for each open set $U \subset \mathbb{C}$, from the free factorization algebra evaluated on U to the factorization algebra of the deformed theory evaluated on U:

$$e^{I/\hbar} : \mathrm{Obs}^q_{free}(U) \to \mathrm{Obs}^q_I(U).$$

This map sends an observable $O \in \mathrm{Obs}^q_{free}(U)$ to $O \cdot e^{I/\hbar}$. In fact, this map is an isomorphism with inverse given by $O \mapsto O \cdot e^{-I/\hbar}$. So, open by open, the factorization algebra assigns the same vector space for the deformed theory. This isomorphism is *not* compatible with the factorization product, so we do get a different factorization algebra in the presence of a deformation.

The holomorphic bosonic string with target $V = \mathbb{C}^{13}$ provides a concrete example of this situation. The free theory is the $bc\beta\gamma$ system, the holomorphic bosonic string is a chiral deformation of it, and we have seen that the renormalized action of the string satisfies the QME. Hence we obtain the following.

PROPOSITION 5.4. *Let* $\mathrm{Obs}^{\mathrm{q}}_{\mathrm{string}}$ *be the factorization algebra on* $\Sigma = \mathbb{C}$ *of the holomorphic bosonic string with target* $V = \mathbb{C}^{13}$. *Let* $\mathbb{V}\mathrm{ert}(\mathrm{Obs}^q_{\mathrm{string}})$ *be the dg vertex algebra (defined over* $\mathbb{C}[\hbar]$*) obtained via Li's construction. There is an isomorphism*

$$\mathcal{V}_{\text{string}} \cong \mathbb{V}\mathrm{ert}(\mathrm{Obs}^q_{\text{string}})|_{\hbar = 2\pi i}$$

of dg vertex algebras. Moreover, this vertex algebra is isomorphic to the chiral sector of the bosonic string as in Section 5.2.

The factorization algebra $\mathrm{Obs}^{\mathrm{q}}_{\mathrm{string}}$ is also a quantization of the factorization algebra $\mathrm{Obs}^{\mathrm{cl}}_{\mathrm{string}}$ of classical observables. We have noted that the classical observables of any theory has the structure of a P_0 factorization algebra, and the $\hbar \to 0$ limit of $\mathrm{Obs}^{\mathrm{q}}_{\mathrm{string}}$ is isomorphic to $\mathrm{Obs}^{\mathrm{cl}}_{\mathrm{string}}$ as P_0 factorization algebras. By definition, the classical observables are simply functions on the solutions to the classical equations of motion. The P_0 structure is induced from the symplectic pairing of degree -1 on the fields. The classical factorization algebra still has enough structure to determine a vertex algebra $\mathbb{V}\mathrm{ert}(\mathrm{Obs}^{\mathrm{cl}}_{\mathrm{string}})$. Moreover, the P_0 bracket on the classical observables determines the structure of a *Poisson vertex algebra* on $\mathbb{V}\mathrm{ert}(\mathrm{Obs}^{\mathrm{cl}}_{\mathrm{string}})$.

COROLLARY 5.5. *In the classical limit, there is an isomorphism*

$$\mathbb{V}\mathrm{ert}(\mathrm{Obs}^{\mathrm{cl}}_{\mathrm{string}}) \cong \mathrm{Gr}\, \mathcal{V}_{\mathrm{string}}$$

of Poisson vertex algebras.

PROOF OF PROPOSITION 5.4. By Proposition 5.1 we know that the vertex algebra of the associated free theory is identified with the $bc\beta\gamma$ vertex algebra. The thing we need to check is that the differential induced from the quantization of the holomorphic string agrees with the differential of the string vertex algebra. In fact, we observe that the induced differential $\oint I \; \mathrm{d}\, z$ from the classical interaction of the holomorphic bosonic string agrees with the BRST charge in Equation (11). To see that this persists at the quantum level we need to check that there are no quantum corrections. Indeed, this follows from the fact that the quantum master equation holds identically (as opposed to holding up to an exact term in the deformation complex) provided $\dim_{\mathbb{C}} V = 13$. □

5.4. The E_2 algebra and descent. In this section we highlight a remarkable feature of the vertex algebra associated to the bosonic string. At first glance, the theory we have constructed is far from being topological. Indeed, the classical theory depends delicately on the complex structure of the two-dimensional source. Nevertheless, the local observables of the bosonic string behave like the observables of a *topological* field theory (TFT). In particular, as noted perhaps first by [**Get94**], the observables of a 2-dimensional TFT have the structure of a *Gerstenhaber algebra*. In this section we provide two equivalent methods for extracting this algebra. The first is intuitive from the point of view of factorization algebras, but has the disadvantage of not giving a concrete description of the algebra. The second approach gives an explicit formula for the bracket and is based on the formalism of "descent" for local operators.

5.4.1. *The E_2 algebra.* We continue to consider the theory on the Riemann surface $\Sigma = \mathbb{C}$. In this section we show how to produce, from the point of view of factorization algebras, the structure of a Gerstenhaber algebra on the BRST cohomology of the bosonic string.

Recall that a Gerstenhaber algebra is another name for an algebra over the operad given by the homology of the little 2-diska operad. Hence, our approach is to see why the factorization algebra naturally exhibits the structure of a algebra of little 2-disks. Here we use an important result of Lurie (namely Theorem 5.4.5.9 of [**Lur**]): a *locally constant* factorization algebra on $\mathbb{R}^n$ is equivalent to an algebra over the little n-disks operad, i.e., an E_n-algebra.

PROPOSITION 5.6. *The factorization algebra* $\mathrm{Obs}^{\mathrm{q}}_{\mathrm{string}}$ *is locally constant, and hence it determines an E_2 algebra.*

In particular, the cohomology $H^*(\mathrm{Obs}^{\mathrm{q}}_{\mathrm{string}})$ is an algebra over the cohomology of the E_2 operad and hence a Gerstenhaber algebra. See Section 5.4.4 for more discussion of this point.

REMARK 5.7. When a topological field theory arises from an action functional (e.g., Chern-Simons theories), the factorization algebra is locally constant. Hence such a TFT in n real dimensions produces an E_n-algebra, by Lurie's result. (This claim holds true, at least, for all the examples with which we are familiar.) In this sense, holomorphic bosonic string theory is a 2-dimensional topological field theory. Moreover, by work of Scheimbauer [**Sch**], every E_n algebra determines a fully-extended framed n-dimensional TFT in the functorial sense, albeit with values in an unusual target (∞, n)-category. In this sense, at least, the holomorphic bosonic string determines a functorial 2-dimensional TFT.

PROOF. We need to show that for any inclusion of open disks $D \hookrightarrow D'$, the natural map

$$\mathrm{Obs}^{\mathrm{q}}_{\mathrm{string}}(D) \to \mathrm{Obs}^{\mathrm{q}}_{\mathrm{string}}(D')$$

is a quasi-isomorphism.

We first show that the classical observables are locally constant. We have already mentioned that the classical observables are the commutative algebra of functions on the space of solutions to the classical equations of motion. This space of solutions forms a sheaf on Σ, since satisfying a PDE is a local condition. We find it convenient to encode the equations of motion as the Maurer-Cartan equation of a sheaf of dg Lie algebras:

$$\Omega^{0,*}(\Sigma;\mathcal{T}_\Sigma) \ltimes \left(\Omega^{0,*}(\Sigma;V)[-1] \oplus \Omega^{1,*}(\Sigma;V^*)[-1] \oplus \Omega^{1,*}(\Sigma;\mathcal{T}_\Sigma^*)[-2]\right).$$

(Note that the underlying graded space is simply the fields shifted up by one degree, which is a generic phenomenon in the BV formalism.) The dg Lie algebra $\Omega^{0,*}(\Sigma;\mathcal{T}_\Sigma)$ is simply a sheaf-theoretic resolution of holomorphic vector fields, with the usual Lie bracket. Our large dg Lie algebra is a square-zero extension of $\Omega^{0,*}(\Sigma;\mathcal{T}_\Sigma)$, by the dg module inside the parentheses. The vector fields act by the Lie derivative on the space

$$\Omega^{0,*}(\Sigma;V)[-1] \oplus \Omega^{1,*}(\Sigma;V^*)[-1] \oplus \Omega^{1,*}(\Sigma;\mathcal{T}_\Sigma^*)[-2],$$

which is simply a copy of the $\beta\gamma$ system with target vector space V, plus the b-field part of the classical theory.

For simplicity, we write $\mathcal{L} = \Omega^{0,*}(\Sigma;\mathcal{T}_\Sigma)$ and write $\mathcal{M}$ for the module inside the parentheses. In this language, the space of classical observables supported on an open set $U \subset \Sigma$ is the Chevalley-Eilenberg cochain complex

$$\mathrm{Obs}^{\mathrm{cl}}_{\mathrm{string}}(U) = \mathrm{C}^*_{\mathrm{Lie}}\left(\mathcal{L}(U) \ltimes \mathcal{M}(U)\right) = \mathrm{C}^*_{\mathrm{Lie}}\left(\mathcal{L}(U);\ \mathrm{Sym}(\mathcal{M}(U)^*[-1])\right),$$

where $\mathcal{M}(U)^*$ denotes the continuous linear dual of $\mathcal{M}(U)$.

Consider now the case that the open set is a disk $U = D$, which we can assume is centered at zero. By the $\bar{\partial}$-Poincaré lemma there is a quasi-isomorphism of dg Lie algebras $\mathcal{T}^{hol}(D) \hookrightarrow \mathcal{L}(D)$ where $\mathcal{T}^{\mathrm{hol}}(D)$ is the vector space of holomorphic vector fields on D. Thus, we have a quasi-isomorphism

$$\mathrm{C}^*_{\mathrm{Lie}}\left(\mathcal{T}_{hol}(D);\ \mathrm{Sym}(\mathcal{M}(D)^*[-1])\right) \simeq \mathrm{Obs}^{\mathrm{cl}}_{\mathrm{string}}(D).$$

This quasi-isomorphism clearly holds for any disks (and is compatible with inclusions of disks), so it suffices to check that the left-hand side is a quasi-isomorphism for an inclusion of disks.

Consider the composition of Lie algebras

$$\mathrm{W}_1^{\mathrm{poly}} \hookrightarrow \mathcal{T}_{hol}(D) \to \mathrm{W}_1$$

where $\mathrm{W}_1^{\mathrm{poly}}$ are the holomorphic vector fields with *polynomial* coefficients, and W_1 is the Lie algebra with *power series* coefficients (i.e., formal vector fields). The second map is the power series expansion, at zero, of a holomorphic vector field. We will compare Lie algebra cohomology using these different Lie algebras.

Let $\mathcal{A}(D)$ denote $\mathrm{Sym}(\mathcal{M}(D)^*[-1])$. It determines a module over $\mathrm{W}_1^{\mathrm{poly}}$ by restriction, which we will abusively denote $\mathcal{A}(D)$ as well. Likewise, if $j_0^\infty\mathcal{M}$ denotes the infinite jet of the sheaf $\mathcal{M}$ at the origin of the disk D, then it determines a natural module over W_1. Then $\mathrm{Sym}(\mathcal{M}(D)^*[-1])$ determines a W_1-module that we will also abusively denote by $\mathcal{A}(D)$.

The inclusion $D \hookrightarrow D'$ then yields a commutative diagram

$$\begin{array}{ccccc} C^*_{\mathrm{Lie}}\left(W_1^{\mathrm{poly}}; \mathcal{A}(D)\right) & \longleftarrow & C^*_{\mathrm{Lie}}\left(\mathcal{T}_{\mathrm{hol}}(D); \mathcal{A}(D)\right) & \longleftarrow & C^*_{\mathrm{Lie}}\left(W_1; \mathcal{A}(D)\right) \\ \downarrow & & \downarrow & & \downarrow \\ C^*_{\mathrm{Lie}}\left(W_1^{\mathrm{poly}}; \mathcal{A}(D')\right) & \longleftarrow & C^*_{\mathrm{Lie}}\left(\mathcal{T}_{\mathrm{hol}}(D'); \mathcal{A}(D')\right) & \longleftarrow & C^*_{\mathrm{Lie}}\left(W_1; \mathcal{A}(D')\right). \end{array}$$

By Lemma 3.5 (and an analogous result for polynomial vector fields), these complexes $C^*_{\mathrm{Lie}}(W_1; \mathcal{M})$ and $C^*_{\mathrm{Lie}}(W_1^{\mathrm{poly}}; \mathcal{M})$ are quasi-isomorphic to the subcomplex consisting of conformal dimension zero elements, i.e., to the constants. As the conformal dimension zero subcomplex does not depend on the size of the disk, we conclude that vertical arrows on the outside of the commutative diagram are quasi-isomorphisms. It follows that the middle vertical arrow is as well, thus showing that $\mathrm{Obs}^{\mathrm{cl}}_{\mathrm{string}}(D) \to \mathrm{Obs}^{\mathrm{cl}}_{\mathrm{string}}(D')$ is a quasi-isomorphism, as desired.

To finish the proof, we need to prove the quasi-isomorphism for *quantum* observables. Consider the spectral sequence induced from the filtration of the module $\mathrm{Sym}\, \mathcal{M}(D)$ by symmetric polynomial degree. The E_1 page of this spectral sequence is the classical observables above, and it converges to the cohomology of the quantum observables. As the map of factorization algebras induced by the inclusion $D \hookrightarrow D'$ preserves this filtration, we obtain a map of spectral sequences, which is quasi-isomorphism on the first page. Hence, $\mathrm{Obs}^{\mathrm{q}}_{\mathrm{string}}(D) \to \mathrm{Obs}^{\mathrm{q}}_{\mathrm{string}}(D')$ is also a quasi-isomorphism. □

5.4.2. *The stress-energy tensor.* In [**Wit88**], where the notion of a TFT was introduced, Witten characterized a topological field theory as a theory whose stress-energy tensor is (homotopy) trivial. We now verify that property of the holomorphic bosonic string. That is, we want to show that the translation symmetries of the holomorphic bosonic string act trivially on the cohomology of the observables.

As a first step, consider the action of the differential operators $\frac{\mathrm{d}}{\mathrm{d}\,z}$ and $\frac{\mathrm{d}}{\mathrm{d}\,\overline{z}}$ on the Dolbeault complex $\Omega^{0,*}(\mathbb{C})$. This action extends to an action on the fields of the holomorphic bosonic string, and hence to their classical observables as well. By Noether's theorem any symmetry of a theory determines classical observables: for these symmetries, these are simply the zz and $\overline{zz}$ components of the stress-energy tensor T_{zz}, $T_{\overline{zz}}$. In the case of the bosonic string, we will now show that the stress-energy tensor is cohomologically trivial on the quantum observables. (Similar but simpler arguments apply to the classical case.)

For each open $U \subset \mathbb{C}$, the differential operators lift to cochain maps on the quantum level

$$\frac{\mathrm{d}}{\mathrm{d}\,z}, \frac{\mathrm{d}}{\mathrm{d}\,\overline{z}} : \mathrm{Obs}^{\mathrm{q}}_{\mathrm{string}}(U) \to \mathrm{Obs}^{\mathrm{q}}_{\mathrm{string}}(U)$$

because the BV Laplacian is translation-invariant. These cochain maps intertwine with the structure maps of the factorization algebra in the sense that they define *derivations* of the factorization algebra. (See Definition 7.3.2 of [**CG17**] for a discussion of this notion.) Note that these operators preserve the cohomological degree.

Consider now the operator

$$\overline{\eta} = \frac{\partial}{\partial(\mathrm{d}\,\overline{z})}$$

acting on Dolbeault forms. This operator $\overline{\eta}$ extends to a derivation of degree -1 on the factorization algebra $\mathrm{Obs}^{\mathrm{q}}_{\mathrm{string}}$. It satisfies the relation

$$[\overline{\partial} + \hbar\Delta + \{I^{\mathrm{q}}, -\}, \overline{\eta}] = \frac{\mathrm{d}}{\mathrm{d}\,\overline{z}} \tag{12}$$

as endomorphisms of the factorization algebra, as we now explain. One observes first that $[\overline{\partial}, \overline{\eta}] = \frac{\mathrm{d}}{\mathrm{d}\,\overline{z}}$. Moreover, since I^{q} is a chiral deformation, we also have $\overline{\eta} \cdot I^{\mathrm{q}} = 0$. Finally, since the pairing defining the -1-shifted symplectic structure is holomorphic, we see that $\overline{\eta}$ also commutes with the BV Laplacian $[\overline{\eta}, \Delta] = 0$. Hence we have shown the following, by relation (12).

LEMMA 5.8. *The operator* $\frac{\mathrm{d}}{\mathrm{d}\,\overline{z}}$ *acts homotopically trivial on* $\mathrm{Obs}^{\mathrm{q}}_{\mathrm{string}}$.

This fact ensures that the stress-energy tensor vanishes in the $\overline{zz}$ direction.

We now turn to $\mathrm{d}/\mathrm{d}\,z$. View this vector field $\frac{\mathrm{d}}{\mathrm{d}\,z}$ as a constant c-field. Consider the linear local functional of cohomological degree -2:

$$O_{\frac{\mathrm{d}}{\mathrm{d}\,z}}(\beta, \gamma, b, c) = \int \left\langle b, \frac{\mathrm{d}}{\mathrm{d}\,z} \right\rangle,$$

It only depends on the b-field. Note that for this integral to be nonzero, the field b must live in $\Omega^{1,1}(\Sigma, T_\Sigma^{1,0*})$. (In fact, b must also be compactly supported for the integral to be well-defined.) Using the BV bracket, we obtain a derivation of the factorization algebra

$$\eta = \{O_{\frac{\mathrm{d}}{\mathrm{d}\,z}}, -\}$$

of cohomological degree -1. It might help to draw this bracket diagrammatically, so one can see that it is a derivation that acts linearly on the generators (i.e., linear functionals on the fields).

LEMMA 5.9. *The derivation* η *satisfies*

$$[\overline{\partial} + \hbar\Delta + \{I^{\mathrm{q}}, -\}, \eta] = \frac{\mathrm{d}}{\mathrm{d}\,z}. \tag{13}$$

PROOF. The derivation η commutes with both $\overline{\partial}$ and Δ. Thus, the left-hand side reduces to

$$[\{I, -\}, \eta] = \{\{I, O_{\frac{\mathrm{d}}{\mathrm{d}\,z}}\}, -\}.$$

The only part of the interaction that contributes is $\int \langle \beta, c \cdot \gamma \rangle + \int \langle b, [c, c] \rangle$, and one computes that

$$\{I, O_{\frac{\mathrm{d}}{\mathrm{d}\,z}}\} = \int \langle \beta, \partial_z \gamma \rangle + \int \langle b, [\partial_z, c] \rangle\,.$$

Bracketing with this local functional encodes applying $\frac{\mathrm{d}}{\mathrm{d}\,z}$ to the inputs, as desired. (In diagrammatic terms, this feature is almost immediately visible.) □

Together these two lemmas ensure that translations act trivially on the cohomological observables.

5.4.3. *Descent for local operators.* We will now sketch an important consequence of the work above. As we will see, it gives both an approach to the method of descent (expositions of this method, as related to two-dimensional gravity, can be found in [**WZ92, DVV91**]) as well as another explicit description of the E_2 algebra associated to the quantum observables of the bosonic string.

The key role here is played by observables that are local, in the sense discussed in Section 3, where they appeared in our description of the deformation complex, but we revisit now the main idea in a more useful form for our current purposes. We will focus first on the classical theory, where the constructions manifestly make sense, before discussing what needs to be modified in the quantum setting.

Let JE_{string} denote the ∞-jet bundle of the classical fields of the holomorphic bosonic string. Concretely, a fiber of JE_{string} at a point x corresponds to all the possible Taylor series at x of fields of the bosonic string. In consequence, the ∞-jet of a γ field determines a section of JE_{string}, as does the ∞-jet of any other field in the theory. This bundle JE_{string} is equipped with a canonical flat connection ∇^{jet} such that horizontal sections are precisely the ∞-jets of classical fields; and so JE_{string} is a left D-module on the Riemann surface, where D means the sheaf of smooth differential operators. (See the appendix of [**GG18**] for expository background oriented toward the approach here.)

Lagrangians can be expressed naturally in terms of JE_{string}, as sections of the bundle

$$\mathrm{Sym}(JE^{\vee}_{\text{string}}) = \bigoplus_{k\geq 0} \mathrm{Sym}^k(JE^{\vee}_{\text{string}}),$$

where $JE^{\vee}_{\text{string}}$ denotes the appropriate dual vector bundle. (Some care is required here because JE_{string} is a pro-finite rank vector bundle.) To unpack this assertion a little, note that a smooth section λ of $JE^{\vee}_{\text{string}}$ can be evaluated on the ∞-jet of a field to obtain a smooth function on the Riemann surface; it thus determines a linear functional of fields with values in functions on the surface. Similarly, a polynomial in such λ determines a nonlinear functional on fields with values in smooth functions. In other words, it is a Lagrangian. If we multiply it against a density, then we obtain a Lagrangian density and hence a local functional.

Note that sections of $\mathrm{Sym}(JE^{\vee}_{\text{string}})$ naturally form a graded-commutative algebra, since polynomials can be multiplied. We denote it by $\mathcal{O}(JE_{\text{string}})$. The shifted pairing on fields determines a shifted Poisson bracket on $\mathcal{O}(JE_{\text{string}})$, which we will denote $\{-,-\}$, since the construction is parallel to the BV bracket.

We will restrict our attention from hereon to $\Sigma = \mathbb{C}$, on which $\mathrm{d}^2 z$ determines a natural volume form. The classical action functional S thus determines a Lagrangian (simply divide by this volume form) that we will abusively denote S as well. The operator $\{S,-\}$, known as the BRST operator in physics, is square-zero by construction. Hence, we obtain a commutative dg algebra

$$(\mathcal{O}(JE_{\text{string}}), \{S,-\})$$

in left D-modules, where a flat connection is inherited by the dual bundle and hence by the symmetric powers.

Elements of $(\mathcal{O}(JE_{\text{string}}), \{S,-\})$ are not observables of the classical theory, since they are just Lagrangians. There is, however, a natural way to produce observables from Lagrangians. Essentially, a Lagrangian can be multiplied by a de Rham current; for instance, evaluating a delta function with a Lagrangian produces a pointlike observable (or local operator in the terminology of physics).

We choose to encode this idea in the following way. Consider $\Omega^*(\mathbb{C}, \mathcal{O}(JE_{\text{string}}))$, the de Rham complex of our D-module equipped with the total differential $\nabla^{jet} + \{S,-\}$. It consists of smooth de Rham forms with values in Lagrangians. These determine observables supported on closed submanifolds, as follows. For a closed submanifold $C \subset \mathbb{C}$, fix a tubular neighborhood N_C. Integration along C then determines a map

$$\int_C : \Omega^*(N_C, \mathcal{O}(JE_{\text{string}})) \to \text{Obs}^{\text{cl}}_{\text{string}}(N_C) \tag{14}$$

of cochain complexes.

Many of the most familiar observables arise in this fashion. For instance, take C to be the point at the origin and consider

$$F(\gamma, \beta, b, c) = \frac{1}{n!}(\partial_z^n \gamma)\big|_0,$$

which returns the coefficient of z^n in the Taylor expansion of γ around the origin. (Note that evaluation at 0 denotes integration at the origin.) It is easy here to factor F into a Lagrangian term and a de Rham form term: the Lagrangian is the linear functional that returns the function $\frac{1}{n!}\partial_z^n\gamma$, and the form is the constant function 1.

As a closely related example, take C to be the unit circle and consider

$$F(\gamma, \beta, b, c) = \frac{1}{2\pi i}\int_C \gamma(z)\frac{\mathrm{d}\, z}{z^{n+1}}.$$

For "on-shell" γ fields—i.e., when $\gamma \in C^\infty(\mathbb{C})$ is holomorphic)—this observable F returns the coefficient of z^n in the Taylor expansion around the origin. It is easy here to factor F into a Lagrangian term and a de Rham form term: the Lagrangian is the linear functional that simply returns the function γ, and the form is $(1/2\pi i)\,\mathrm{d}\, z/z^{n+1}$, which is well-defined on any small annulus around C.

These concrete examples exhibit a compelling virtue of this process for producing observables: it encompasses the observables typically discussed in physics, particularly in conformal field theories. In our situation it is straightforward to show that the entire vertex algebra $\text{Gr}\,\mathcal{V}_{\text{string}}$ is realized by the image of the map (14). (Such an argument is given in Part III of [**GGW**] for the free $\beta\gamma$ system.)

After this lengthy build-up of notions and notations, we now finally turn to describing descent. Our interpretation is that it is a process for promoting pointlike operators to more general observables, available for any translation-invariant theory. We will describe it for this theory and then apply it to give an explanation for the Gerstenhaber structure on $H^*\mathcal{V}_{\text{string}}$, identified by Lian-Zuckerman.

Definition 5.10. A *pointlike operator* is an element of $\text{Sym}(JE^\vee_{\text{string}})_0$, the fiber at the origin $0 \in \mathbb{C}$ of the bundle $\text{Sym}(JE^\vee_{\text{string}})$.

Equivalently, it is an element of the algebra $\text{Sym}((J_0E_{\text{string}})^*)$ of polynomial functions on the fiber at 0 of JE_{string}. Since our theory is translation-invariant, any point in the plane would serve as well as the origin. Note that this definition is equivalent to our earlier, heuristic notion.

Construction 5.11 (Method of "descent"). Any pointlike operator O *descends* to an element

$$\widetilde{O} = \widetilde{O}^0 + \widetilde{O}^1 + \widetilde{O}^2$$

in $\Omega^*(\mathbb{C}, \mathcal{O}(JE_{\text{string}}))$. We construct it as follows. First, because our theory is translation-invariant, there is a natural trivialization of the bundle $\text{Sym}(JE^\vee_{\text{string}})$, and hence there is a canonical element $\widetilde{O}^0 \in C^\infty(\mathbb{C}, \mathcal{O}(JE_{\text{string}}))$, given by a constant section whose value at the origin is O. In formulas, we write

$$\widetilde{O}^0(z) = \tau_z O,$$

where $\tau_z : \mathbb{C} \to \mathbb{C}$ is the translation sending a point w to $w+z$. (This operator acts on fields by pullback, and so on the jets of fields as well.) Using the homotopies $\eta, \overline{\eta}$, we define the 1-form part as

$$\widetilde{O}^1 = \mathrm{d}\, z\, \widetilde{(\eta O)}^0 + \mathrm{d}\, \overline{z}\, \widetilde{(\overline{\eta} O)}^0 ,$$

where ηO denotes the image of O under the map η. Similarly, the 2-form part is

$$\widetilde{O}^2 = \mathrm{d}\, z\ \mathrm{d}\, \overline{z}\, \widetilde{(\eta\overline{\eta} O)}^0 .$$

By construction, if O is closed for the classical differential $\{S, -\}$, then

$$(\mathrm{d}_{dR} + \{S, -\})\widetilde{O} = 0,$$

so the total form $\widetilde{O}$ is closed as well.

REMARK 5.12. The terminology "descent" is due to Witten [**Wit88**]. Our construction above is a method of solving what he refers to as the topological descent equations. Mathematically, we start with a constant, BRST-invariant Lagrangian and then construct a cocycle in the double complex given by the de Rham complex of the flat vector bundle of Lagrangians, by a zigzag or staircase process well-known in homological algebra. In the horizontal direction, the Lagrangians are equipped with the BRST operator $\{S, -\}$. In the vertical direction there is the differential induced from the flat connection, and we use an explicit homotopy retract of the flat connection to "descend" in the direction opposite the vertical differential.

Combining the construction with the map (14), we find that a pointlike operator O and a closed submanifold C determine an observable

$$\int_C \widetilde{O} \in \mathrm{Obs}^{\mathrm{cl}}_{\mathrm{string}}(N_C).$$

Note that if O has cohomological degree k and C is of dimension l, then $\int_C \widetilde{O}$ has degree $k-l$. We remark that every element of $\mathrm{Gr}\, \mathcal{V}_{\mathrm{string}}$ can be realized by applying descent to the pointlike operators and then evaluating at the origin.

Extending this whole package to quantum observables is nontrivial. The map (14) makes sense at the level of graded vector spaces, but it is not easy to equip $\mathcal{O}(JE_{\mathrm{string}})$ with a BV Laplacian in such a way that the map (14) intertwines with the differential on the quantum observables. For linear observables, however, no such issues arise (*cf.* Part III of [**GGW**]), and those are sufficient to identify the Gerstenhaber bracket, the problem to which we now turn.

5.4.4. *Formula for the Gerstenhaber bracket.* A Gerstenhaber algebra is a graded commutative algebra with a Lie bracket of cohomological degree -1 that is a graded biderivation for the product. In this section we show how to explicitly write down the product and bracket on the local observables (i.e., the observables on any disk) and compare our answer to the work of Lian-Zuckerman [**LZ93**].

As explained just before Proposition 5.6, the Gerstenhaber operad *Gerst* is the operad $H_*(E_2)$ arising by taking homology of the E_2 operad. Recall that $E_2(2)$ parametrizes the space of binary operations as the configuration space of disjoint two disks in the unit disk in $\mathbb{R}^2$. This space deformation retracts onto S^1. Hence

$$Gerst(2) = H_*(E_2(2)) \cong H_*(S^1).$$

(Note that we view homology of spaces as concentrated in nonpositive degrees, since it is viewed as the linear dual to cohomology.) The degree zero operation—corresponding to a commutative product—matches with a zero-dimensional cycle of S^1, and the degree -1 operation—corresponding to the shifted Poisson bracket—matches with a one-dimensional cycle of S^1.

Thus, to obtain the commutative product on $H^* \operatorname{Obs}^{\mathrm{q}}$, we need only pick an embedding of two disjoint disks inside a larger disk, which is precisely such a zero-cycle in $E_2(2)$. Then the factorization product

$$\operatorname{Obs}^{\mathrm{q}}(D) \otimes \operatorname{Obs}^{\mathrm{q}}(D') \to \operatorname{Obs}^{\mathrm{q}}(D'').$$

induces the commutative product

$$\cdot : H^* \operatorname{Obs}^{\mathrm{q}} \otimes H^* \operatorname{Obs}^{\mathrm{q}} \to H^* \operatorname{Obs}^{\mathrm{q}}.$$

Since this configuration space $E_2(2)$ is connected, we could use any other choice of embeddings and get the same answer at the level of cohomology. In particular, we could have put D' on the opposite side of D, which is why the product must be commutative. (A topologist would call this the Eckmann-Hilton argument, as it is the same argument one uses to show that the homotopy group $\pi_2(X)$ is always abelian.)

To construct the shifted Poisson (i.e., Gerstenhaber) bracket, we need to pick a one-cycle in the configuration space $E_2(2)$. To describe the associated binary operation, we use descent along this one-cycle in conjunction with the fact—s remarked just after Construction 5.11—that the underlying vector space of the vertex algebra $\mathcal{V}_{\text{string}}$ is generated by pointlike operators.

Let O and O' be two pointlike operators supported at $0 \in \mathbb{C}$. Fix a disk D centered at zero. We can view O as an element in $\operatorname{Obs}^{\mathrm{q}}(D)$.

Now fix a loop C that wraps around D but is disjoint. Let N_C be a tubular neighborhood of C that does not intersect D. Via descent we have the observable $\int_C \widetilde{O}'$ in $\operatorname{Obs}^{\mathrm{q}}(N_C)$. Pick a bigger disk D' containing D and N_C. Then $\int_C \widetilde{O}'$ also determines an observable in $\operatorname{Obs}^{\mathrm{q}}(D')$.

We suppose that these observables are cocycles, which lets us identify them with elements $[O]$ and $[O']$ of the vertex algebra.

Consider now the factorization product

$$\mu : \operatorname{Obs}^{\mathrm{q}}(D) \otimes \operatorname{Obs}^{\mathrm{q}}(N_C) \to \operatorname{Obs}^{\mathrm{q}}(D')$$

corresponding to $D \sqcup N_C \hookrightarrow D'$. We define a bracket by

$$\{[O], [O']\}_{\text{Ger}} := \mu\left(O, \int_C \widetilde{O}'\right).$$

Note that if $\deg(O) = k$ and $\deg(O') = k'$, then $\deg(\{[O], [O']\}_{\text{Ger}}) = k + k' - 1$, so we obtain a bracket of the correct degree to define a Gerstenhaber structure.

REMARK 5.13. This construction manifestly involves picking a 1-cycle, here C, to exhibit the bracket, and it should be clear geometrically how we could relate to any other choice C'. If C and C' do not intersect, they bound an annulus and hence determine cohomologous observables. (One may have to shrink D in the construction, but that is no issue by local constancy.) If they do intersect, one can choose a C'' that does not intersect either, and then one has a pair of cohomologous terms. As the terms are cohomologous, they induce the same brackets at the level of cohomology.

We now connect our constructions with well-known approaches.

PROPOSITION 5.14. *The bracket $\{-,-\}_{\text{Ger}}$ together with the product $\cdot$ determine the structure of a Gerstenhaber algebra on $H^*\mathcal{V}_{\text{string}}$, the cohomology of the dg vertex algebra $\mathcal{V}_{\text{string}}$. This Gerstenhaber structure is isomorphic to the one found by Lian-Zuckerman* [**LZ93**].

PROOF. The vertex algebra construction of [**CG17**] extracts $\mathcal{V}_{\text{string}}$ as the direct sum of the weight spaces of $\text{Obs}^{\text{q}}_{\text{string}}(D)$, where D is a disk centered at the origin and we take weight space for the rotation action of S^1 on $\mathbb{C}$. The bracket and product restrict to this subspace of $\text{Obs}^{\text{q}}(D)$, manifestly playing nicely with this eigenspace decomposition. Hence they descend to the cohomology of $\mathcal{V}_{\text{string}}$.

Let V_{LZ} be the Gerstenhaber algebra considered by Lian-Zuckerman. As vector spaces, both $H^*\mathcal{V}_{\text{string}}$ and V_{LZ} are isomorphic to the state space of the $\beta\gamma$ vertex algebra.

According to the construction of a vertex algebra from a holomorphic factorization algebra in Chapter 6 of [**CG17**], the factorization product of two disks is what defines the operator product map $Y(-,z) : V \otimes V \to V((z))$ of a vertex algebra. It is this operator product that Lian-Zuckerman use to define the commutative product. Thus, as commutative algebras, the algebras coincide.

The brackets coincide by noting that the derivation η trivializing $\mathrm{d}/\mathrm{d}z$ agrees with Lian-Zuckerman's trivialization. □

6. The holomorphic string on closed Riemann surfaces

Thus far we have discussed the local behavior of the holomorphic string, such as its quantization on a disk and the concomitant vertex algebra. Now we turn to its global behavior, particularly the observables on a closed Riemann surface, and the relationship with certain natural holomorphic vector bundles on the moduli space of Riemann surfaces. This local-to-global transition is where Costello's BV/factorization package really shines. On the one hand, the theory of factorization algebras provides a conceptual characterization of the local-to-global relationship, much like the understanding of sheaf cohomology as the derived functor of global sections. On the other hand, the examples from BV quantization provide computable, convenient models for the global sections, much as the de Rham or Dolbeault complexes do for the cohomology of sheaves that arise naturally in differential or complex geometry.

As we will explain, the answers we recover for the holomorphic string can be related quite cleanly to natural determinant lines on the moduli of Riemann surfaces, hence providing a bridge from the Feynman diagrammatic anomaly computations to the index-theoretic computations.

6.1. The free case. Before jumping to the holomorphic string, we will work out the global observables in the simpler case of the $bc\beta\gamma$ system, introduced in Remark 2.1. The global *classical* observables on a Riemann surface Σ are given by the symmetric algebra on the continuous linear dual to the fields,

$$\text{Sym}\left(\Omega^{0,*}(\Sigma,V)^\vee \oplus \Omega^{1,*}(\Sigma,V^\vee)^\vee \oplus \Omega^{0,*}(\Sigma,T[1])^\vee \oplus \Omega^{1,*}(\Sigma,T^*_\Sigma[-2])^\vee\right),$$

with the differential $\bar{\partial}$ extended as a derivation. Hence the cohomology is

$$\text{Sym}\left(H^*(\Sigma,V)^\vee \oplus H^*(\Sigma,\omega\otimes V^\vee)^\vee \oplus H^*(\Sigma,T[1])^\vee \oplus H^*(\Sigma,\omega^{\otimes 2}[-2])^\vee\right),$$

where ω denotes the canonical bundle. Although this expression might look complicated, it can be readily unpacked in the setting of algebraic geometry, particularly when Σ is closed. In that case, this graded commutative algebra is a symmetric algebra on a finite-dimensional graded vector space, which encodes the derived tangent space of the moduli of Riemann surfaces at Σ and of holomorphic functions to V.

As this theory is free, it admits a canonical BV quantization. Denote by $\mathrm{Obs}^{\mathrm{q}}_{free}$ be the corresponding factorization algebra. One can compute its global sections on Σ by using a spectral sequence whose first page is the global classical observables. The result of Theorem 8.1.4.1 of [**CG17**] states that the cohomology of this free theory along a closed Riemann surface with values in *any* line bundle is one-dimensional concentrated in a certain cohomological degree. In our case, it the calculation implies that we get a shifted determinant of the cohomology of the fields:

$$H^*\left(\mathrm{Obs}^{\mathrm{q}}_{free}(\Sigma)\right) \cong \det\left(H^*(\Sigma;\mathcal{O}_\Sigma)\right)^{\otimes \dim(V)} \otimes \det\left(H^*(\Sigma;T^{1,0}_\Sigma)\right)^{-1}[d(\Sigma)]$$

where

$$\begin{aligned} d(\Sigma) = \dim(V)&\left(\dim H^0(\Sigma;\mathcal{O}_\Sigma) + \dim H^1(\Sigma;\mathcal{O}_\Sigma)\right) \\ &+ \dim(H^0(\Sigma;T^{1,0}_\Sigma)) - \dim(H^1(\Sigma;T^{1,0}_\Sigma)). \end{aligned}$$

REMARK 6.1. The shift $d(\Sigma)$ here likely looks funny. In this case at least, the meaning can be unpacked pretty straightforwardly. The BV complex for an ordinary finite-dimensional vector space is equivalent to the de Rham complex shifted down by the dimension of the vector space, so that the top forms are in degree 0. (Abstracting this situation is one way to "invent" the BV formalism.) For the σ-model, the global solutions to the equations of motion are $H^0(\Sigma,\mathcal{O})\otimes V$ for the γ fields and $H^0(\Sigma,\omega)\otimes V^\vee$ for the β fields. For Σ closed, these are finite-dimensional, and thus we get the shift

$$\dim(V)\left(\dim H^0(\Sigma;\mathcal{O}_\Sigma) + \dim H^1(\Sigma;\mathcal{O}_\Sigma)\right).$$

For the ghost system (the bc fields), the BV complex recovers the Euler characteristic

$$\dim(H^0(\Sigma;T^{1,0}_\Sigma)) - \dim(H^1(\Sigma;T^{1,0}_\Sigma))$$

as it encodes the de Rham complex on the formal quotient stack $B\mathfrak{g} = */\mathfrak{g}$ for the Lie algebra of symmetries $\mathfrak{g}$.

The computation here works for any Riemann surface Σ and, indeed, for any family of Riemann surfaces. Hence it implies that the global observables of the free $bc\beta\gamma$ system determine a determinant line bundle on the moduli $\mathcal{M}$ of Riemann surfaces.

We can work out the first Chern class of this determinant line bundle using the Grothendieck-Riemann-Roch (GRR) theorem as follows. Consider the universal Riemann surface $\pi\colon C \to \mathcal{M}$ over the moduli space, and consider the bundles $\mathcal{O}_C \otimes V$ and $\mathcal{T}_\pi = \mathcal{T}_{C/\mathcal{M}}$, which one can view the universal γ fields and c fields, respectively. The first Chern class of the derived pushforward $R\pi_*(\mathcal{O}_C\otimes V)$ is given by the first Chern class of $\det(H^*(\mathcal{O}_C\otimes V)) \cong \det(\mathcal{O}_C)^{\otimes \dim V}$, since the first Chern class of a vector bundle is the first Chern class of its determinant bundle. The Grothendieck-Riemann-Roch theorem states that for a complex of coherent sheafs

$\mathcal{F} = \mathcal{F}^*$ on C, the Chern character $\mathrm{ch}(R\pi_*\mathcal{F})$ of its derived pushforward $R\pi_*\mathcal{F}$ is given by

$$\pi_*(\mathrm{ch}(\mathcal{F})\,\mathrm{Td}(T_\pi)) = \pi_*\left(\left(\sum_i (-1)^i \,\mathrm{ch}(\mathcal{F}^i)\right)\mathrm{Td}(T_\pi)\right)$$

$$= \pi_*\left(\left(\sum_i (-1)^i(\mathrm{rk}(\mathcal{F}^i) + c_1(\mathcal{F}^i) + \frac{1}{2}(c_1(\mathcal{F}^i)^2) + \cdots\right)(1 + \frac{1}{2}c_1(T_\pi) + \frac{1}{12}c_1(T_\pi)^2)\right)$$

where T_π denotes the relative tangent bundle along π, which is here just the tangent line bundle of a Riemann surface. The first Chern class is the component of cohomological degree 2. For instance, when $\mathcal{F} = \mathcal{F}^0$ is concentrated in degree zero, the above simplifies to:

$$\frac{1}{12}\mathrm{rk}(\mathcal{F})c_1(T_\pi)^2 + \frac{1}{2}c_1(\mathcal{F})c_1(T_\pi) + \frac{1}{2}c_1(\mathcal{F})^2.$$

When $\mathcal{F} = T_\pi^{\otimes n}[1]$, the expression for the first Chern class is $-\frac{1+6n+6n^2}{12}c_1(T_\pi)^2$. When $\mathcal{F} = \mathcal{O} \otimes V$ we simply get $\dim(V)$.

Hence, when $\mathcal{F} = T[1] \oplus \mathcal{O} \otimes V$, for the determinant line of global observables $H^*\left(\mathrm{Obs}^{\mathrm{q}}_{free}(C)\right)$ as a bundle over C we obtain

$$c_1\left(H^*\left(\mathrm{Obs}^{\mathrm{q}}_{free}(\Sigma)\right)\right) = \frac{1}{12}(\dim(V) - 13)c_1(T_\pi)^2.$$

It is worthwhile to point out that the above argument based on GRR for identifying the first Chern class of this determinant line bundle resonates with our computation of the anomaly of the bosonic string on the disk. Indeed, this is a manifestation of "Virasoro uniformization." Also, notice that the above calculation assumed that there was no deformation, so that we were working with a free theory. However, deforming the action from free $bc\beta\gamma$ system to holomorphic bosonic string doesn't affect the line bundles, since those are continuous parameters and Chern classes are discrete.

6.2. The anomaly and moduli of quantizations on an arbitrary Riemann surface. We have already seen that the holomorphic string *on a disk* admits a BV quantization if and only if the target is a complex vector space of dimension 13. Here we will explain why this anomaly calculation is actually enough to show the existence of a quantization on an *arbitrary* Riemann surface. An argument using the Grothendieck-Riemann-Roch theorem was given in the above section. In this section we give a proof using only the perspective of BV quantization. One can view this as giving a proof of the Grothendieck-Riemann-Roch theorem using Feynman diagrams (and will be the topic of future work).

Our diagrammatic arguments showed that only wheels with c legs appear in the anomaly, and these arguments did not depend on the choice of Σ. Hence the anomaly will be purely a functional on the c fields. So we restrict ourselves to the piece of the deformation complex only involving such fields.

When Σ is a disk, a corollary of the calculations in Section 3 is that this deformation complex for the c-fields is quasi-isomorphic to $\mathrm{C}^*_{\mathrm{Lie,red}}(W_1)[2]$. A classical calculation of Gelfand and Fuks shows that the reduced Lie algebra cohomology

$H^*_{red}(W_1)$ of formal vector fields is one-dimensional concentrated in degree 3. Thus, the cohomology of the deformation complex is

$$H^*_{\mathrm{red}}(W_1)[2] = \mathbb{C}[-1].$$

The generator of this cohomology can be taken to be $\lambda_{-1}\lambda_0\lambda_1$, where λ_i is dual to the formal vector field $L_i = z^{i+1}\partial_z$. Some readers might recognize this cocycle as a manifestation of the usual cocycle for the Virasoro Lie algebra.

As shown in Proposition 5.3 of [**Wil17**], this situation generalizes from disks to arbitrary Riemann surfaces: on any Riemann surface Σ, the deformation complex is equivalent to the derived global sections of $\mathbb{C}_\Sigma[-1]$, the constant sheaf on Σ with a cohomological shift. In particular, the cohomology of the deformation complex is equal to the shift of the de Rham cohomology $H^*(\Sigma)[-1]$. Hence, the first cohomology group of the deformations—where the anomaly lives—is spanned by the constant functions. When Σ is connected, we thus know the anomaly, up to scale. Locally, this anomaly cocycle agrees with the usual expression for the generator of $H^3(W_1)$, but since this description depends on the choice of a coordinate, the global version is somewhat subtle. See Section 5 of [**Wil17**] for an extensive discussion.

The sheaf-theoretic nature of the deformation complex works to our advantage here. As the construction of BV quantization is manifestly *local-to-global* on spacetime anomalies inherit this property: the anomaly computed on an open set $U \subset \Sigma$ is equal to the anomaly of the theory on Σ restricted to U. In our case, this global anomaly is a 1-cocycle for the derived global sections of the shifted constant sheaf, and hence is determined by a constant function on Σ. Thus, it suffices to compute the anomaly on an arbitrary open, such as on a flat disk. But this is precisely the context in which we computed the anomaly in Section 4, so we know the anomaly is simply the dimension of the target vector space. Thus, a quantization of the holomorphic string exists on any Riemann surface provided $\dim_{\mathbb{C}}(V) = 13$.

Now we ask how many such quantizations are possible, i.e., what is the moduli of theories. By the calculation in Section 3, we know that, up to BV equivalence, the possible one-loop terms in the quantized action functional are parametrized by

$$H^0(\Sigma) \otimes \Omega^1(V) \oplus H^1(\Sigma) \otimes \Omega^2_{cl}(V).$$

(That is, these vector spaces are the first cohomology group of the relevant deformation complex.) This space of deformations corresponds to continuous parameters we can vary in the action functional. As the isomorphism classes of line bundles form a discrete set, varying these continuous parameters will not change the class of the line bundle of global observables. In conclusion, no matter what one-loop quantization we choose, the cohomology of the global observables will be the same.

6.3. Global observables for the holomorphic string. Now, let us consider the global observables of the holomorphic string $\mathrm{Obs}^{\mathrm{q}}(\Sigma)$. Consider the filtration on the quantum observables induced by the polynomial degree of the functional. There is a spectral sequence abutting to the cohomology of the global observables $H^* \mathrm{Obs}^{\mathrm{q}}(\Sigma)$ with E_1 page given by the cohomology of the global observables of the free $bc\beta\gamma$ system which we have already computed:

$$\begin{aligned} E_2 &\cong \det\left(H^*(\Sigma; T_\Sigma[1])\right) \otimes \det\left(H^*(\Sigma; \mathcal{O}_\Sigma)^{\oplus 13}\right) \\ &\cong \det\left(H^1(\Sigma; T_\Sigma)\right) \otimes \det\left(H^0(\Sigma; T_\Sigma)\right)^{-1} \otimes \det\left(H^0(\Sigma; K_\Sigma)\right)^{-13} \end{aligned}$$

where we have used the fact that $H^0(\Sigma; \mathcal{O}) \cong \mathbb{C}$ for any Σ. Since this page is concentrated in a single line, we see that the spectral sequence degenerates at this page.

Let Σ_g be a surface of genus g. Then for $g = 1$ the above simplifies to

$$\det\left(H^1(\Sigma_1; T_{\Sigma_1})\right) \otimes \det\left(H^0(\Sigma_1; K)\right)^{-14}.$$

If $g \geq 2$ one has

$$\det\left(H^1(\Sigma_1; T_{\Sigma_1})\right) \otimes \det\left(H^0(\Sigma_1; K)\right)^{-13}.$$

Thus the above expressions give the global observables for the holomorphic string for genus $g = 1$ and $g \geq 2$, respectively. Compare these formulas to Witten's analysis of the bosonic string in [**Wit15**].

7. Looking ahead: Curved targets

In this section we briefly advertise our future work, which is to provide a complete analysis of the bosonic string with a complex manifold as the target. Our approach is a modification of our treatment of the curved $\beta\gamma$ system given in [**GGW**]. The main idea there was to consider the $\beta\gamma$ system with target a formal disk $\widehat{D}^n$. Then, in the style of Gelfand and Kazhdan's treatment of formal geometry, we show how working equivariantly for formal automorphisms allows us to globalize this theory to a complex manifold. In general, we find an obstruction to doing this, which is measured by the second component of the Chern character of the tangent bundle of the complex manifold. The appearance of the characteristic class is expected from the theory of chiral differential operators. In fact, we show that the factorization algebra of observables descends to the sheaf of chiral differential operators on the target manifold.

We will give a similar argument for the bosonic string. The key difference to the $\beta\gamma$ system is that even in the case of a flat target, the bosonic string is an interacting theory. Nevertheless, the theory of BV quantization that is equivariant for formal automorphisms can still be applied and we arrive at the following result.

THEOREM 7.1 ([**GW**]). *Consider the holomorphic bosonic string with target a complex manifold X. There exists a one-loop exact quantization if and only if*

(1) $\dim_{\mathbb{C}} X = 13$,
(2) $\mathrm{ch}_2(T_X) = 0$, *and*
(3) $c_1(T_X) = 0$.

Moreover, if the conditions above hold, the space of all quantizations is a torsor for the abelian group

$$H^1(X, \Omega^2_{cl}) \oplus H^0(X, \Omega^1).$$

There are two further directions we intend to address in our future work:

(1) We have seen that the local observables for the case of a flat target return the semi-infinite BRST cohomology of the $\beta\gamma$ vertex algebra. We expect that the local observables in the case of a curved target produce a sheafy refinement of semi-infinite cohomology. This should produce a sensitive invariant of the target manifold and gives a variant of quantum sheaf cohomology.

(2) The partition function of the curved $\beta\gamma$ system on elliptic curves is known to produce the Witten genus of the target manifold [**Cos**]. For flat space, the partition function of the string is given by the Mumford form [**BlM86**]. We propose that the partition function for the curved string produces an invariant of the target manifold analogous to the Witten genus.

Appendix A. Calculation of anomaly

In this section we compute the functional $F[L]$ and $G[L]$ mentioned in the proof of Proposition 4.6, hence completing the calculation of the anomaly.

We have reduced the calculation to the weight of two wheel diagrams: A) with internal edges labeled by the bc heat kernel and propagator, respectively. B) with internal edges labeled by the $\beta\gamma$ heat kernel and propagator, respectively. The weight of A gives the functional we called $F[L]$, and the weight of B gives the functional we called $\dim_{\mathbb{C}}(V)G[L]$.

We will utilize the following version of Wick expansion to evaluate the integrals below.

LEMMA A.1. *Let Φ be a smooth compactly supported function on $\mathbb{C}$ and suppose $\tau > 0$. Then*

$$\int_{\xi\in\mathbb{C}} \Phi(\xi)e^{-\tau|\xi|^2/4}\,\mathrm{d}^2\,\xi = 4\pi\cdot\tau^{-1}\left(\exp\left(\tau^{-1}\frac{\partial}{\partial\xi}\frac{\partial}{\partial\overline{\xi}}\Phi\right)_{\xi=0}\right).$$

We start with the weight of diagram A. Use coordinates z, w to denote the coordinates at each of the vertices. Denote the inputs of the weight by the compactly supported vector fields $f(z)\partial_z$ and $g(w)\,\mathrm{d}\overline{w}\partial_w$. (Note that the diagram is only nonzero if the total degree of the elements is $+1$.) If $c(z)\partial_z$ is another vector field, the action by $f(z)\partial_z$ is given by

$$[f(z)\partial_z, c(z)\partial_z] = f(z)\partial_z c(z)\partial_z - c(z)\partial_z f(z)\partial_z.$$

Thus, the weight of diagram A can be written as the $\ell\to 0$ limit of

$$\begin{aligned} &\int_{z,w} f(z)\partial_z P_\ell^L(z,w)g(w)\partial_w K_\ell(z,w)\\ -\;&\int_{z,w} \partial_z f(z)P_\ell^L(z,w)g(w)\partial_w K_\ell(z,w)\\ -\;&\int_{z,w} f(z)\partial_z P_\ell^L(z,w)\partial_w g(w)K_\ell(z,w)\\ +\;&\int_{z,w} \partial_z f(z)P_\ell^L(z,w)\partial_w g(w)K_\ell(z,w). \end{aligned} \tag{15}$$

We label the integrals in each line above as I,II, III, IV, respectively.

Using the form of the propagator in (8) we see that line I is given by

$$\mathrm{I} = \frac{1}{(4\pi)^2}\int_{(z,w)\in\mathbb{C}\times\mathbb{C}}\int_{t=\ell}^{L} f(z)g(w)\frac{1}{\epsilon^2}\frac{1}{t^3}\frac{(\overline{z}-\overline{w})^3}{8}\exp\left(-\frac{1}{4}\left(\frac{1}{\ell}+\frac{1}{t}\right)|z-w|^2\right)$$

(we are omitting volume factors for simplicity). To evaluate this integral we change variables and apply the Wick expansion, Lemma A.1, to one of the variables of integration. Indeed, introduce $\xi = z - w$, and notice that the integral simplifies to

$$\mathrm{I} = \frac{1}{(4\pi)^2}\int_{w\in\mathbb{C}}\int_{\xi\in\mathbb{C}}\int_{t=\ell}^{L} f(\xi+w)g(w)\frac{1}{\epsilon^2}\frac{1}{t^3}\frac{\overline{\xi}^3}{8}\exp\exp\left(-\frac{1}{4}\left(\frac{1}{\ell}+\frac{1}{t}\right)|\xi|^2\right).$$

Applying Lemma A.1 to the ξ-integral we see that this simplifies to

$$\mathrm{I} = \frac{1}{4\pi}\int_{w\in\mathbb{C}} \partial_w^3 f(w)g(w)\int_{t=\ell}^{L}\frac{\ell^2 t}{(\ell+t)^4} + O(\ell)$$

where the terms $O(\ell)$ are of order ℓ so are zero in the limit $\ell \to 0$. On the other hand, we can evaluate the remaining t-integral and see that in the limit $\ell \to 0$, Line I becomes

$$\lim_{\ell \to 0} \mathrm{I} = \frac{1}{4\pi} \frac{1}{12} \int_{w \in \mathbb{C}} \partial_w^3 f(w) g(w) \, \mathrm{d}^2 w.$$

We evaluate II, III, and IV in a similar fashion.

After changing coordinates and performing the Wick type integral, we obtain

$$\mathrm{II} = \frac{1}{4\pi} \int_{w \in \mathbb{C}} \partial_w^3 f(w) g(w) \, \mathrm{d}^2 w \int_{t=\ell}^{L} \frac{\ell t}{(\ell + t)^3} \, \mathrm{d}t + O(\ell).$$

Evaluating the remaining t integral and taking $\ell \to 0$, this becomes

$$\lim_{\ell \to 0} \mathrm{II} = \frac{1}{4\pi} \frac{3}{8} \int_{w \in \mathbb{C}} \partial_w^3 f(w) g(w) \, \mathrm{d}^2 w.$$

Integral III is given by

$$\frac{1}{4\pi} \int_{w \in \mathbb{C}} \partial_w^3 f(w) g(w) \, \mathrm{d}^2 w \int_{t=\ell}^{L} \frac{\epsilon^2}{(\epsilon + t)^3} \, \mathrm{d}t + O(\ell).$$

In the limit $\ell \to 0$ we obtain

$$\lim_{\ell \to 0} \mathrm{III} = \frac{1}{4\pi} \frac{1}{8} \int_{w \in \mathbb{C}} \partial^3 f(w) g(w) \, \mathrm{d}^2 w.$$

Finally, integral IV is

$$\frac{1}{4\pi} \int_{w \in \mathbb{C}} \partial_w^3 f(w) g(w) \, \mathrm{d}^2 w \int_{t=\ell}^{L} \frac{\epsilon}{(\epsilon + t)^2} \, \mathrm{d}t + O(\ell).$$

In the limit $\ell \to 0$, we obtain

$$\lim_{\ell \to 0} \mathrm{IV} = \frac{1}{4\pi} \frac{1}{2} \int_{w \in \mathbb{C}} \partial_w^3 f(w) g(w) \, \mathrm{d}^2 w.$$

In total, the functional $F[L]$ applied to $(f(z)\partial_z, g(w) \, \mathrm{d}\overline{w} \partial_w)$ is given by

$$F[L](f(z)\partial_z, g(w) \, \mathrm{d}\overline{w} \partial_w) = -\frac{1}{4\pi} \frac{13}{12} \int_{w \in \mathbb{C}} \partial_w^3 f(w) g(w) \, \mathrm{d}^2 w.$$

Note that this functional is independent of L.

Diagram B is similar to A, except the internal edges are labeled by the $\beta\gamma$ propagator. Applied to the input vector fields $(f(z)\partial_z, g(w) \, \mathrm{d}\overline{w} \partial_w)$ the weight is given by the dimension of V times the integral we computed in I. Thus

$$G[L](f(z)\partial_z, g(w) \, \mathrm{d}\overline{w} \partial_w) = \frac{1}{4\pi} \frac{1}{12} \int_{w \in \mathbb{C}} \partial_w^3 f(w) g(w) \, \mathrm{d}^2 w.$$

The proposition follows.

Acknowledgments

We learned this approach to perturbative field theory as students of Kevin Costello. He guided us towards this theory of the holomorphic string, and he pointed out key results and features visible through this BV/renormalization formalism. OG spent some time on the holomorphic string in graduate school, partly in collaboration with Yuan Shen, whom he thanks for illuminating discussions and computations. The authors also wish to thank Si Li for his typically incisive comments and insights on CFT, which clarified some of the trickier technical aspects.

We thank the referee for their careful reading of an earlier version of this manuscript, and for suggestions that improved the paper. Finally, we must express deep gratitude toward the Max Planck Institute for Mathematics, which supported OG throughout his work on this project and which contributed substantially to the efficacy of this collaboration by allowing BW to visit repeatedly over recent years.

References

[BK86] A. A. Belavin and V. G. Knizhnik, *Algebraic geometry and the geometry of quantum strings*, Phys. Lett. B **168** (1986), no. 3, 201–206, DOI 10.1016/0370-2693(86)90963-9. MR830618

[BlM86] A. A. Beĭlinson and Yu. I. Manin, *The Mumford form and the Polyakov measure in string theory*, Comm. Math. Phys. **107** (1986), no. 3, 359–376. MR866195

[Boc87] Marco Bochicchio, *Gauge fixing for the field theory of the bosonic string*, Phys. Lett. B **193** (1987), no. 1, 31–36, DOI 10.1016/0370-2693(87)90451-5. MR899933

[CFMP85] C. G. Callan, D. Friedan, E. J. Martinec, and M. J. Perry, *Strings in background fields*, Nuclear Phys. B **262** (1985), no. 4, 593–609, DOI 10.1016/0550-3213(85)90506-1. MR819433

[CG] Kevin Costello and Owen Gwilliam, *Factorization algebras in quantum field theory, Volume 2.*, available at `http://people.mpim-bonn.mpg.de/gwilliam`.

[CG17] Kevin Costello and Owen Gwilliam, *Factorization algebras in quantum field theory. Vol. 1*, New Mathematical Monographs, vol. 31, Cambridge University Press, Cambridge, 2017. MR3586504

[Cos] Kevin Costello, *A geometric construction of the Witten genus, II*, available at `http://arxiv.org/abs/1112.0816`.

[Cos11] Kevin Costello, *Renormalization and effective field theory*, Mathematical Surveys and Monographs, vol. 170, American Mathematical Society, Providence, RI, 2011. MR2778558

[DEF+99] Pierre Deligne, Pavel Etingof, Daniel S. Freed, Lisa C. Jeffrey, David Kazhdan, John W. Morgan, David R. Morrison, and Edward Witten (eds.), *Quantum fields and strings: a course for mathematicians. Vol. 1, 2*, American Mathematical Society, Providence, RI; Institute for Advanced Study (IAS), Princeton, NJ, 1999. Material from the Special Year on Quantum Field Theory held at the Institute for Advanced Study, Princeton, NJ, 1996–1997. MR1701618

[DP88] Eric D'Hoker and D. H. Phong, *The geometry of string perturbation theory*, Rev. Modern Phys. **60** (1988), no. 4, 917–1065, DOI 10.1103/RevModPhys.60.917. MR969998

[DVV91] Robbert Dijkgraaf, Herman Verlinde, and Erik Verlinde, *Notes on topological string theory and 2D quantum gravity*, String theory and quantum gravity (Trieste, 1990), World Sci. Publ., River Edge, NJ, 1991, pp. 91–156. MR1164017

[FBZ01] Edward Frenkel and David Ben-Zvi, *Vertex algebras and algebraic curves*, Mathematical Surveys and Monographs, vol. 88, American Mathematical Society, Providence, RI, 2001. MR1849359

[FHL93] Igor B. Frenkel, Yi-Zhi Huang, and James Lepowsky, *On axiomatic approaches to vertex operator algebras and modules*, Mem. Amer. Math. Soc. **104** (1993), no. 494, viii+64, DOI 10.1090/memo/0494. MR1142494

[Fuk86] D. B. Fuks, *Cohomology of infinite-dimensional Lie algebras*, Contemporary Soviet Mathematics, Consultants Bureau, New York, 1986. Translated from the Russian by A. B. Sosinskiĭ. MR874337

[Get94] E. Getzler, *Batalin-Vilkovisky algebras and two-dimensional topological field theories*, Comm. Math. Phys. **159** (1994), no. 2, 265–285. MR1256989

[GG14] Owen Gwilliam and Ryan Grady, *One-dimensional Chern-Simons theory and the $\hat{A}$ genus*, Algebr. Geom. Topol. **14** (2014), no. 4, 2299–2377, DOI 10.2140/agt.2014.14.2299. MR3331615

[GG18] Ryan Grady and Owen Gwilliam, *Lie algebroids as L_∞ spaces*, Journal of the Institute of Mathematics of Jussieu (2018), 1–49.

[GGW] Owen Gwilliam, Vassily Gorbounov, and Brian Williams, *Chiral differential operators via Batalin-Vilkovisky quantization*, available at `https://arxiv.org/abs/1610.09657`.

[GH18] Owen Gwilliam and Rune Haugseng, *Linear Batalin–Vilkovisky quantization as a functor of ∞-categories*, Selecta Math. (N.S.) **24** (2018), no. 2, 1247–1313, DOI 10.1007/s00029-018-0396-0. MR3782422

[GJF] Owen Gwilliam and Theo Johnson-Freyd, *How to derive Feynman diagrams for finite-dimensional integrals directly from the bv formalism*, available at `https://arxiv.org/abs/1202.1554`.

[GLL17] Ryan E. Grady, Qin Li, and Si Li, *Batalin-Vilkovisky quantization and the algebraic index*, Adv. Math. **317** (2017), 575–639, DOI 10.1016/j.aim.2017.07.007. MR3682678

[GSW12a] Michael B. Green, John H. Schwarz, and Edward Witten, *Superstring theory. Vol. 1. Introduction*, 25th anniversary edition, Cambridge University Press, Cambridge, 2012. MR3155203

[GSW12b] Michael B. Green, John. H. Schwarz, and Edward Witten, *Superstring theory. Vol. 2. Loop amplitudes, anomalies and phenomenology*, anniversary ed., Cambridge University Press, Cambridge, 2012. MR3155202

[GW] Owen Gwilliam and Brian Williams, *The holomorphic bosonic string, II: the case of a curved target*, to appear.

[KLS91] M. Knecht, S. Lazzarini, and R. Stora, *On holomorphic factorization for free conformal fields*, Phys. Lett. B **262** (1991), no. 1, 25–31, DOI 10.1016/0370-2693(91)90637-6. MR1114271

[Li] Si Li, *Vertex algebras and quantum master equation*, available at `https://arxiv.org/abs/1612.01292`.

[LL16] Qin Li and Si Li, *On the B-twisted topological sigma model and Calabi-Yau geometry*, J. Differential Geom. **102** (2016), no. 3, 409–484. MR3466804

[LMZ06] Andrei S. Losev, Andrei Marshakov, and Anton M. Zeitlin, *On first-order formalism in string theory*, Phys. Lett. B **633** (2006), no. 2-3, 375–381, DOI 10.1016/j.physletb.2005.12.010. MR2198165

[Lur] Jacob Lurie, *Higher Algebra*, available at `http://www.math.harvard.edu/~lurie/papers/higheralgebra.pdf`.

[LZ93] Bong H. Lian and Gregg J. Zuckerman, *New perspectives on the BRST-algebraic structure of string theory*, Comm. Math. Phys. **154** (1993), no. 3, 613–646. MR1224094

[LZ96] Bong. H. Lian and Gregg J. Zuckernam, *Algebraic and geometric structures in string backgrounds*, Strings '95 (Los Angeles, CA, 1995), World Sci. Publ., River Edge, NJ, 1996, pp. 323–335. MR1660720

[Pol98] Joseph Polchinski, *String theory. Vol. I*, Cambridge Monographs on Mathematical Physics, Cambridge University Press, Cambridge, 1998. An introduction to the bosonic string. MR1648555

[Sch] Claudia Scheimbauer, *Factorization homology as a fully extended topological field theory*, on `https://people.maths.ox.ac.uk/scheimbauer/ScheimbauerThesis.pdf`.

[Sch75] J. Scherk, *An introduction to the theory of dual models and strings*, Rev. Modern Phys. **47** (1975), 123–164, DOI 10.1103/RevModPhys.47.123. MR0459409

[Wil17] Brian Williams, *The Virasoro vertex algebra and factorization algebras on Riemann surfaces*, Lett. Math. Phys. **107** (2017), no. 12, 2189–2237, DOI 10.1007/s11005-017-0982-7. MR3719639

[Wit88] Edward Witten, *Topological quantum field theory*, Comm. Math. Phys. **117** (1988), no. 3, 353–386. MR953828

[Wit15] Edward Witten, *Notes on holomorphic string and superstring theory measures of low genus*, Analysis, complex geometry, and mathematical physics: in honor of Duong H. Phong, Contemp. Math., vol. 644, Amer. Math. Soc., Providence, RI, 2015, pp. 307–359, DOI 10.1090/conm/644/12786. MR3372474

[WZ92] Edward Witten and Barton Zwiebach, *Algebraic structures and differential geometry in two-dimensional string theory*, Nuclear Phys. B **377** (1992), no. 1-2, 55–112, DOI 10.1016/0550-3213(92)90018-7. MR1168630

Max Planck Institute for Mathematics, 53111 Bonn, Germany
Email address: `gwilliam@mpim-bonn.mpg.de`

Department of Mathematics, Northwestern University, Evanston, Illinois 60208
Email address: `brianwilliams2012@u.northwestern.edu`

Selected Published Titles in This Series

718 **David Ayala, Daniel S. Freed, and Ryan E. Grady, Editors,** Topology and Quantum Theory in Interaction, 2018

716 **Alex Martsinkovsky, Kiyoshi Igusa, and Gordana Todorov, Editors,** Surveys in Representation Theory of Algebras, 2018

715 **Sergio R. López-Permouth, Jae Keol Park, S. Tariq Rizvi, and Cosmin S. Roman, Editors,** Advances in Rings and Modules, 2018

714 **Jens Gerlach Christensen, Susanna Dann, and Matthew Dawson, Editors,** Representation Theory and Harmonic Analysis on Symmetric Spaces, 2018

713 **Naihuan Jing and Kailash C. Misra, Editors,** Representations of Lie Algebras, Quantum Groups and Related Topics, 2018

712 **Nero Budur, Tommaso de Fernex, Roi Docampo, and Kevin Tucker, Editors,** Local and Global Methods in Algebraic Geometry, 2018

711 **Thomas Creutzig and Andrew R. Linshaw, Editors,** Vertex Algebras and Geometry, 2018

710 **Raphaël Danchin, Reinhard Farwig, Jiří Neustupa, and Patrick Penel, Editors,** Mathematical Analysis in Fluid Mechanics, 2018

709 **Fernando Galaz-García, Juan Carlos Pardo Millán, and Pedro Solórzano, Editors,** Contributions of Mexican Mathematicians Abroad in Pure and Applied Mathematics, 2018

708 **Christian Ausoni, Kathryn Hess, Brenda Johnson, Ieke Moerdijk, and Jérôme Scherer, Editors,** An Alpine Bouquet of Algebraic Topology, 2018

707 **Nitya Kitchloo, Mona Merling, Jack Morava, Emily Riehl, and W. Stephen Wilson, Editors,** New Directions in Homotopy Theory, 2018

706 **Yeonhyang Kim, Sivaram K. Narayan, Gabriel Picioroaga, and Eric S. Weber, Editors,** Frames and Harmonic Analysis, 2018

705 **Graham J. Leuschke, Frauke Bleher, Ralf Schiffler, and Dan Zacharia, Editors,** Representations of Algebras, 2018

704 **Alain Escassut, Cristina Perez-Garcia, and Khodr Shamseddine, Editors,** Advances in Ultrametric Analysis, 2018

703 **Andreas Malmendier and Tony Shaska, Editors,** Higher Genus Curves in Mathematical Physics and Arithmetic Geometry, 2018

702 **Mark Grant, Gregory Lupton, and Lucile Vandembroucq, Editors,** Topological Complexity and Related Topics, 2018

701 **Joan-Carles Lario and V. Kumar Murty, Editors,** Number Theory Related to Modular Curves, 2018

700 **Alexandre Girouard, Dmitry Jakobson, Michael Levitin, Nilima Nigam, Iosif Polterovich, and Frédéric Rochon, Editors,** Geometric and Computational Spectral Theory, 2017

699 **Mark L. Agranovsky, Matania Ben-Artzi, Catherine Bénéteau, Lavi Karp, Dmitry Khavinson, Simeon Reich, David Shoikhet, Gilbert Weinstein, and Lawrence Zalcman, Editors,** Complex Analysis and Dynamical Systems VII, 2017

698 **Alexander M. Blokh, Leonid A. Bunimovich, Paul H. Jung, Lex G. Oversteegen, and Yakov G. Sinai, Editors,** Dynamical Systems, Ergodic Theory, and Probability: in Memory of Kolya Chernov, 2017

697 **Fabrizio Broglia, Françoise Delon, Max Dickmann, Danielle Gondard-Cozette, and Victoria Ann Powers, Editors,** Ordered Algebraic Structures and Related Topics, 2017

696 **Ara S. Basmajian, Yair N. Minsky, and Alan W. Reid, Editors,** In the Tradition of Ahlfors–Bers, VII, 2017

For a complete list of titles in this series, visit the AMS Bookstore at **www.ams.org/bookstore/conmseries/**.